HANDBOOK OF
MASS
MEASUREMENT

HANDBOOK OF

MASS MEASUREMENT

FRANK E. JONES
RANDALL M. SCHOONOVER

CRC Press
Taylor & Francis Group
Boca Raton London New York

CRC Press is an imprint of the
Taylor & Francis Group, an **informa** business

A CHAPMAN & HALL BOOK

Front cover drawing is used with the consent of the Egyptian National Institute for Standards, Gina, Egypt. Back cover art from *II Codice Atlantico di Leonardo da Vinci nella Biblioteca Ambrosiana di Milano, Editore Milano Hoepli 1894–1904*. With permission from the Museo Nazionale della Scienza e della Tecnologia Leonardo da Vinci Milano.

First published 2004 by Chapman & hall / CRC Press

Published 2019 by CRC Press
Taylor & Francis Group
6000 Broken Sound Parkway NW, Suite 300
Boca Raton, FL 33487-2742

©2004 by Taylor & Francis Group, LLC
CRC Press is an imprint of Taylor & Francis Group, an Informa business

First issued in paperback 2019

No claim to original U.S. Government works

ISBN-13: 978-0-367-45499-9 (pbk)
ISBN-13: 978-0-8493-2531-1 (hbk)

Visit the Taylor & Francis Web site at
http://www.taylorandfrancis.com

and the CRC Press Web site at
http://www.crcpress.com

Library of Congress Cataloging-in-Publication Data

Jones, Frank E.
    Handbook of mass measurement / Frank E. Jones, Randall M. Schoonover
        p.   cm.
    Includes bibliographical references and index.
    ISBN 0-8493-2531-5 (alk. paper)
    1. Mass (Physics)—Measurement. 2. Mensuration. I. Schoonover, Randall M. II. Title.

QC106 .J66 2002
531'.14'0287—dc21
                                                                        2002017486
                                                                        CIP

Preface

"A false balance is abomination to the Lord: but a just weight is his delight."

— Proverbs 11.1

The purpose of this handbook is to provide in one location detailed, up-to-date information on various facets of mass measurement that will be useful to those involved in mass metrology at the highest level (at national standards laboratories, for example), in science and engineering, in industry and commerce, in legal metrology, and in more routine mass measurements or weighings. We have pursued clarity and hope that we have in some measure succeeded.

Literature related to mass measurement, historical and current, has been cited and summarized in specific areas. Much of the material in this handbook is our own work, in many cases previously unpublished.

We take this opportunity to recognize the considerable contributions to mass measurement of the late Horace A. Bowman, including the development of the National Bureau of Standards (NBS) 2 balance with an estimate of standard deviation of 1 part per billion (ppb) and the development of the silicon density standard with estimate of standard deviation of 2 parts per million (ppm), adopted worldwide. In addition, he was mentor to each of us and positively affected our careers.

Chapter 1 introduces mass and mass standards. Historical background material in Section 1.2 is an excerpt from NBS monograph, "Mass and Mass Values," by Paul E. Pontius, then chief of the U.S. NBS section responsible for mass measurements.

Chapter 2 presents recalibration of the U.S. National Prototype Kilogram and the Third Periodic Verification of National Prototypes of the Kilogram.

Chapter 3 discusses contamination of platinum-iridium mass standards and stainless steel mass standards. The literature is reviewed and summarized. Carbonaceous contamination, mercury contamination, water adsorption, and changes in ambient environmental conditions are studied, as are various methods of analysis.

Cleaning of platinum-iridium mass standards and stainless steel mass standards are discussed in Chapter 4, including the BIPM (Bureau International des Poids et Mesures) Solvent Cleaning and Steam Washing procedure. Results of various cleaning methods are presented.

In Chapter 5, the determination of mass differences from balance observations is treated in detail.

In Chapter 6, a glossary of statistical terms that appear throughout the book is provided.

The U.S. National Institute of Standards and Technology (NIST) guidelines for evaluating and expressing the uncertainty of measurement results are presented in Chapter 7. The Type A and Type B evaluations of standard uncertainty are illustrated.

In Chapter 8, weighing designs are discussed in detail. Actual data are used for making calculations.

Calibration of the screen and the built-in weights of direct-reading analytical balances is described in Chapter 9.

Chapter 10 takes a detailed look at the electronic balance. The two dominant types of electronic balance in use are the hybrid balance and the electromagnetic force balance. Features and idiosyncrasies of the balance are discussed.

In Chapter 11, buoyancy corrections and the application of buoyancy corrections to mass determination are discussed in detail. For illustration, the application of buoyancy corrections to weighings of titanium dioxide powder in a weighing bottle on a balance is demonstrated.

The development of the air density equation for use in calculation of values of air density to be used in making buoyancy corrections is presented in detail in Chapter 12. The development of the air density equation by Jones is used as background material. Then, the BIPM 1981 and the BIPM 1981/1991 equations are presented and discussed. Direct determination of air density, experimental determination of air density in weighing on a 1-kg balance in air and in vacuum, a practical approach to air density determination, and a test of the air density equation at differing altitude are summarized from original papers and discussed.

Chapter 13 discusses the continuation of programs undertaken by NIST to improve hydrostatic weighing and to develop a density scale based on the density of a solid object. Central to this development is the classic paper, "Procedure for High Precision Density Determinations by Hydrostatic Weighing," by Bowman and Schoonover. Among the subjects discussed in Chapter 13 are the principles of use of the submersible balance, determination of the density of mass standards, an efficient method for measuring the density or volume of similar objects, and the measurement of liquid density.

The calculation of the density of water is the subject of Chapter 14. Redeterminations of the density of water and corresponding equations developed by three groups of researchers were corrected for changes in density of water with air saturation, compressibility, and isotopic concentration.

In Chapter 15, the conventional value of weighing in air, its concept, intent, benefits, and limitations are discussed. Examples of computation are included.

Comparison of error propagations for mass and the conventional mass is presented in detail in Chapter 16. OIML Recommendation R111 is used for the comparison.

Parameters that can cause error in mass determinations are examined in detail in Chapter 17. Subjects covered are mass artifacts, mass standards, mass comparison, the fundamental mass relationship, weighing designs, uncertainties in the determination of the mass of an object, buoyancy, thermal equilibrium, atmospheric effects, cleaning of mass standards, magnetic effects, and the instability of the International Prototype Kilogram.

In Chapter 18, the problem of assigning mass values to piston weights of about 590 g nominal mass with the goal of accomplishing an uncertainty in mass corresponding to an error in the maximum pressure generated by the piston-gauge rotating assembly of 1 ppm is discussed. The mass was determined with a total uncertainty of 0.1 ppm.

The response of apparent mass to thermal gradients and free convective currents is studied in Chapter 19, based on the known experimental fact that if an artifact is not at thermal equilibrium with the balance chamber the apparent mass of the artifact deviates from the value at thermal equilibrium.

In Chapter 20, magnetic errors in mass metrology, that is, unsuspected vertical forces that are magnetic in origin, are discussed.

The "gravitational configuration effect," which arises because for weights of nominally equal mass the distance of the center of gravity above the base of each weight depends on the size and shape of the weight, is examined in Chapter 21.

In Chapter 22, the "between-time" component of error in mass measurements is examined. The between-time component manifests itself between groups of measurements made at different times, on different days, for example.

Chapter 23 illustrates the key elements for the most rigorous mass measurements.

In Chapter 24, control charts are developed and used to demonstrate attainment of statistical control of a mass calibration process.

Tolerance testing of mass standards is discussed in Chapter 25. Procedures to be followed for determining whether or not mass standards are within the tolerances specified for a particular class of weights are reviewed.

Surveillance testing of weights is discussed in Chapter 26. Surveillance looks for signs that one or more members of a weight set may have changed since the latest calibration.

Chapter 27 describes a project to disseminate the mass unit to surrogate laboratories using the NIST portable mass calibration package. A surrogate laboratories project began with the premise that a NIST-certified calibration could be performed by the user in the user's laboratory. The very informal, low-budget project was undertaken to expose the technical difficulties that lay in the way.

In Chapter 28, the concept that the mass of an object can be adequately determined (for most applications) by direct weighing on an electronic balance *without* the use of external mass standards is examined.

A piggyback balance experiment, an illustration of Archimedes' principle and Newton's third law, is described in Chapter 29.

In Chapter 30, the application of the electronic balance in high-precision pycnometry is discussed and illustrated.

The Appendices are Buoyancy Corrections in Weighing (a course); Examination for Buoyancy Corrections in Weighing Course; Answers for Examination for Buoyancy in Weighing Course; OIML R111 Maximum Permissible Errors; OIML R111 Minimum and Maximum Limits for Density of Weights; Density and Coefficient of Linear Expansion of Pure Metals, Commercial Metals, and Alloys; and Linearity Test.

The Authors

Frank E. Jones is currently an independent consultant. He received a bachelor's degree in physics from Waynesburg College, Pennsylvania, and a master's degree in physics from the University of Maryland, where he has also pursued doctoral studies in meteorology. He served as a physicist at the National Bureau of Standards (now National Institute of Standards and Technology, NIST) in many areas, including pressure measurements, flow measurements, standardizing for chemical warfare agents, chemical engineering, processing of nuclear materials, nuclear safeguards, evaporation of water, humidity sensing, evapotranspiration, cloud physics, helicopter lift margin, moisture in materials, gas viscosity, air density, density of water, refractivity of air, earthquake research, mass, length, time, volume, and sound.

He began work as an independent consultant upon retirement from NIST in 1987. He is author of more than 90 technical publications, four books, and holds two patents. The diverse titles of his previous books are *Evaporation of Water*, *Toxic Organic Vapors in the Workplace*, and *Techniques and Topics in Flow Measurement*. A senior member of the Instrument Society of America and of the Institute for Nuclear Materials Management, he has been associated with other technical societies from time to time as they relate to his interests.

Randall M. Schoonover was an employee of the National Bureau of Standards (currently National Institute of Standards and Technology) for more than 30 years and was closely associated with mass and density metrology. Since his retirement in 1995 he has continued to work as a consultant and to publish scientific work. He attended many schools and has a diploma for electronics from Devry. During his career he authored and coauthored more than 50 scientific papers. His most notable work was the development, along with his colleague Horace A. Bowman, of the silicon density standard as part of the determination of Avogadro's constant; the silicon density standard is now in use throughout the world. He has several inventions and patents to his credit, among them are the immersed electronic density balance and a unique high-precision load cell mass comparator.

We are pleased to dedicate this handbook to our wives

Virginia B. Jones and Caryl A. Schoonover.

Contents

1

Mass and
Mass Standards

1.1 Introduction

1.1.1 Definition of Mass

The following quotation of Condon and Odishaw[1] is presented here as a succinct definition of mass: "The property of a body by which it requires force to change its state of motion is called inertia, and *mass* is the numerical measure of this property."

1.1.2 The Mass Unit

According to Maxwell,[2] "every physical quantity [mass in the present case] can be expressed as the product of a pure number and a unit, where the unit is a selected reference quantity in terms of which all quantities of the same kind can be expressed." The fundamental unit of mass is the international *kilogram*. At present the kilogram is realized as an artifact, i.e., an object. Originally, the artifact was designed to have the mass of 1 cubic decimeter of pure water at the temperature of maximum density of water, 4°C. Subsequent determination of the density of pure water with the air removed at 4°C under standard atmospheric pressure (101,325 pascals) yielded the present value of 1.000028 cubic decimeters for the volume of 1 kilogram of water.

1.1.3 Mass Artifacts, Mass Standards

The present embodiment of the kilogram is based on the French platinum kilogram of the Archives constructed in 1792. Several platinmum-iridium (Pt-Ir) cylinders of height equal to diameter and nominal mass of 1 kg were manufactured in England. These cylinders were polished and adjusted and compared with the kilogram of the Archives. The cylinder with mass closest to that of the kilogram of the Archives was sent to the International Bureau of Weights and Measures (Bureau International des Poids et Mesures, BIPM) in Paris and chosen as the International Prototype Kilogram (IPK) in 1883. It was ratified as the IPK by the first General Conference of Weights and Measures (CPGM) in 1899. Other prototype kilograms were constructed and distributed as national prototypes. The United States received prototypes Nos. 4 and 20. All other mass standards in the United States are referred to these. As a matter of practice, the unit of mass as maintained by the developed nations is interchangeable among them.

Figure 1.1 is a photograph of a building at BIPM, kindly provided by BIPM. Figure 1.2 is U.S. prototype kilogram K20, Figure 1.3 is a collection of brass weights, Figure 1.4 is a stainless steel weight set, and Figure 1.5 is a collection of large stainless steel weights that, when assembled, become a deadweight force machine.

References

1. Condon, E. U. and Odishaw, H., *Handbook of Physics*, McGraw-Hill, New York, 1958, 2.
2. *The Harper Encyclopedia of Science*, Harper & Row, Evanston Sigma, New York, 1967, 223.

FIGURE 1.1 Building at Bureau International des Poids et Mesures (BIPM) in Paris, France. (Photograph courtesy of BIPM.)

1.2 The Roles of Mass Metrology in Civilization*

Paul E. Pontius

1.2.1 The Role of Mass Measurement in Commerce

1.2.1.1 Prior to the Metric System of Measurement Units

The existence of deliberate alloys of copper with lead for small ornaments and alloys of copper with varying amounts of tin for a wide variety of bronzes implies an ability to make accurate measurements with a weighing device ca. 3000 B.C. and perhaps earlier.[1] That trade routes existed between Babylonia and India, and perhaps the Persian Gulf and Red Sea countries, at about the same time implies a development of commercial enterprise beyond barter.[2] Economic records were the earliest documents and these in turn influenced both the development of the written language and the development of numbering systems.[3,4] The transition between the tradition of an illiterate craftsman working with metals and a universally accepted commercial practice is largely conjecture.

The impartial judgment of the weighing operation was well known ca. 2000 B.C., as evidenced by the adoption of the balance as a symbol of social justice,[5] a practice that continues today. Then, as now, the weighing operation will dispense equal value in the form of equal quantities of the same commodity. It was, and still is, easy to demonstrate that the comparison, or weighing out, has been accomplished within the practical limit of plus or minus a small weight or a few suitably small objects such as grains of wheat or barley. In the beginning, there would have been no requirement that a standard quantity of one commodity should have any relation to the standard quantity of another commodity. The small weight

---

* This material of historical interest is extracted, with minor alterations, from NBS Monograph 133, Mass and Mass Values, 1974, by Paul E. Pontius, who was at that time Head of the NBS Mass Group.

FIGURE 1.2 U.S. kilogram No. 20.

FIGURE 1.3 Brass weight set.

or object used to verify the exactness of comparison could have been accepted by custom. Wealthy families, early rulers, or governments may have fostered the development of ordered weight sets to account for and protect their wealth. Measurement practices associated with collecting taxes in kind would likely be adopted in all other transactions.

FIGURE 1.4 Stainless steel weight set.

FIGURE 1.5 Large stainless steel weights that when assembled become a deadweight force machine.

Ordered sets of weights were in use ca. 2000 B.C.[6] In these sets, each weight is related to the next larger weight by some fixed ratio. To develop such a set was a substantial undertaking. Individual weights were adjusted by trial and error until both the one-to-one and summation equalities were satisfied within the precision of the comparison process. Ratios between weights varied with preference to numbers that had many factors.[7,8] For example, if 12 *B* were to be equivalent to *A*, then in addition to intercomparing the 12 *B* weights with *A*, the *B* weights could be intercompared one by one, two by two, three by three, four by four and six by six. Once established, it was not difficult to verify that the ratios were proper, nor was it difficult to duplicate the set.

Precious metals were used for exchange from the earliest times.[9] "To weigh" meant payment in metal and "to measure" meant payment in grain.[10] Simple barter had become in essence sales. Goods of one sort being exchanged for goods of another sort were separately valued to a common standard, and these values brought to a common total.[11] Overseas trade involved capitalization, letters of credit, consignment, and payment of accounts on demand.[12] There is evidence that a mina weight ca. 2100 B.C. was propagated by duplication over a period of 1500 years (to ca. 600 B.C.).[13]

Maspero[14] gives the following description of an Egyptian market transaction:

Exchanging commodities for metal necessitated two or three operations not required in ordinary barter. The rings or thin bent strips of metal which formed the "tabnu" and its multiples did not always contain the regulation amount of gold or silver, and were often of light weight. They had to be weighed at every fresh transaction in order to estimate their true value, and the interested parties never missed this excellent opportunity for a heated discussion: after having declared for a quarter of an hour that the scales were out of order, that the weighing had been carelessly performed, and that it should be done over again, they at last came to terms, exhausted with wrangling, and then went their way fairly satisfied with one another. It sometimes happened that a clever and unscrupulous dealer would alloy the rings, and mix with the precious metal as much of a baser sort as would be possible without danger of detection. The honest merchant who thought he was receiving in payment for some article, say eight tabnu of fine gold, and who had handed to him eight tabnu of some alloy resembling gold, but containing one-third of silver, lost in a single transaction, without suspecting it, almost one-third of his goods. The fear of such counterfeits was instrumental in restraining the use of tabnu for a long time among the people, and restricted the buying and selling in the markets to exchange in natural products or manufactured objects.

The impact of coinage guaranteed by the government (ca. 500 B.C.) was profound and is still with us today.[15,16] One normally thinks that measurements associated with the exchange of goods in commerce are ordering worth. This is only partly true from the viewpoint of the ultimate consumer. The establishment of a monetary system permitted a third party to enter the transaction without the difficulty of physically handling the material to be traded. Assigning a money value to a unit measure of a commodity permitted the establishment of a much broader market, which was not generally concerned with each local transaction but which, nonetheless, established in part the money value for each commodity in the local market. The customer, then as now, must pay the asked price, the measurement process merely determining how much the total transaction will be.

Commerce thrives on the variation of commodity values with time and location.[17] This variation, coupled with confusion and perhaps a willful lack of communication on matters concerning money value and measurement units, is a happy situation for the enterprising entrepreneur. As far as the normal customer is concerned, the only element he has in common with the seller is the measurement process and perhaps some preferential treatment associated with social status, profession, or some other factor totally unrelated to the value of the commodity. Emphasis on the exactness of the measurement can mask more important factors such as the quality of the product offered for sale.

Uniform weights and measures, and common coinage were introduced throughout the Roman Empire.[18–20] Yet, perhaps with the exception of doing business with the government, it was not until the early part of the 18th century that the first real efforts toward a mandatory usage of uniform measures was started. Many leaders through the ages have made profound statements relating to the need for uniform measures. Little, however, was done except in the control of the quality of the coinage. No one ruler had been powerful enough to change the customary measures and practices of his land. This was changed in France with the establishment of the metric system of measurement units.

1.2.1.2 The Kilogram and the Pound[1]

It is not generally emphasized that the prime motivation for establishing the metric system of measurements was the utter chaos of the French marketplace.[2] It was not that the conditions in the French marketplace were any different than in any other marketplace, but it was these conditions coupled with two other factors that eventually brought about the reform. These factors were the French Revolution whose great objective was the elimination of all traces of the feudal system and royalty, and the influence of the natural philosophers of the time who realized the international importance of such a forward step in creating a common scientific language. Other powerful influences objected vigorously to the mandatory standards plan. After the new standards had been completed they were not readily accepted. Severe penalties were necessary to enforce their usage in the common measurements of the time. On the other hand, the metric system of measurements almost immediately became the measurement language of all science.

As with all previous artifacts that eventually reached the status of measurement standards, the choice for the basis of the metric standards was arbitrary. With the idea of constancy and reproducibility in mind, the choice for the length unit finally came down to either a ten-millionth part of the length of a quadrant of the Earth's meridian, or the length of a pendulum with a specified period. The nonconcurrence of most of the important foreign powers who had been invited to participate in establishing the measurement system left the French to proceed alone.

From the measurements of a segment of a meridian between points near Barcelona and Dunkirk, it was determined by computation that the meridianal distance between the pole and the equator was 5,130,740 toises, from which the ten-millionth part, or the meter, was 3 pieds 11.296 lignes. A unit for mass was defined in terms of length and the density of water. The concept of mass was relatively new to science, and completely new in the history of weighing, which had heretofore been concerned with quantities of material rather than the properties of matter. With the meter established in customary units, using hydrostatic weighings of carefully measured cylinders, it was determined that a mass of one kilogram was 18827.15 grains with respect to the weights of the Pile of Charlemagne. With these relationships defined in terms of customary units of measurement, it was then possible to proceed with the construction and adjustment of new standards for the metric units.

The first task was the construction of provisional metric standards. The construction of the kilogram and the meter of the Archives followed, the kilogram of the Archives no doubt being adjusted[3] with the same weights used to adjust the provisional kilogram. The kilogram of the Archives, as it was later discovered, had been adjusted prior to a precise determination of its displacement volume. This important measurement was not made after adjustment because of the fear that the water in a hydrostatic weighing

---

[1]This section is essentially an abstract of two papers. The Moreau paper[53] is an excellent general paper on the development of the metric standards and the work of the International Bureau of Weights and Measures. The Miller paper[54] is a comprehensive work describing the reconstruction of the Imperial Standard Pound. Reference to specific passages are made in this section.

[2]At that time there was no shortcoming in the ability to make measurements as evidenced by the use of existing equipment and measurement techniques to establish the new standards. A comprehensive study of density, hydrometry, and hydrostatic weighing had been published in the 12th century.[55] Instructions for adjusting weights for use in assay work published in 1580 are just outlines, implying that the techniques of weighing and the precision of the equipment are common knowledge among assayers.[56]

[3]Adjusting a weight is adding or removing material from a weight to establish a one-to-one relationship with an accepted standard. In the case of one-piece weights, such as the prototype kilogram, the weight to be adjusted is usually initially heavier than the standard. Material is carefully removed until the one-to-one relationship is established, or until the difference is some small part of the on-scale range of the instrument being used.

would leach out some of the inclusions that were typical of the platinum of the time. While the technical developments were going on, the Treaty of the Meter was consummated, and the General Conference of Weights and Measures was established to review and finally accept the work.

Techniques were developed prior to the construction of the prototype standards that resulted in more homogeneous material (introduction of the oil-fired furnace and the use of cold working). From a small group of kilograms made from the new material and adjusted in the same manner as the kilogram of the Archives, the one that was most nearly identical to the kilogram of the Archives, as deduced from the data resulting from direct comparisons, was chosen to be the prototype standard defined to embody a mass of exactly one kilogram. (This standard is now generally called the international prototype kilogram, designated by $\mathfrak{R}$, to differentiate it from other prototype kilograms, which are designated by number or letter-number combinations and used as transfer standards.) The task of manufacturing, adjusting, and establishing the mass values of the prototype standards for distribution to the nations that were participating in the metric convention was long and tedious. The survey to determine the length of the arc of the meridian had been started in June 1792. The General Conference[4] formally sanctioned the prototype meter and kilogram and the standards for distribution in September of 1889.

A second major effort in the construction of standards for measurement was going on within this same period. In 1834 all of England's standards of volumetric measure and weight were either totally destroyed or damaged by fire in the House of Parliament to such an extent that they were no longer suitable for use as standards. The Imperial standard troy pound was never recovered from the ruins. A commission, appointed to consider the steps to be taken for the restoration of the standards, concluded that while the law provided for reconstructing the standard of length on the basis of the length of a pendulum of specified period and for the reconstruction of the standard of weight on the basis of the weight of water, neither method would maintain the continuity of the unit.

In the case of length, there were difficulties in carrying out the specified experiment. In the case of weight, differences based on the best determinations of the weight of water by French, Austrian, Swedish, and Russian scientists amounted to a difference on the order of one-thousandth of the whole weight, whereas the weighing operation could be performed with a precision smaller than one-millionth of the whole weight. Therefore, it was recommended that the reconstruction could best be accomplished by comparison with other weights and length measures that had previously been carefully compared with the destroyed standards. It was further recommended that the new standard should be the avoirdupois pound in common usage rather than the destroyed troy pound. In 1843, a committee was appointed to superintend the construction of the new standards.

This work resulted in the construction of a platinum avoirdupois pound standard and four copies, the copies to be deposited in such a manner that it would be unlikely that all of them would be lost or damaged simultaneously. It was decreed that "the Commissioners of Her Majesty's Treasury may cause the same to be restored by reference to or adoption of any of the copies so deposited."[21] Careful work determined the relationship between the avoirdupois pound and the kilogram. While it was not until 1959 that the English-speaking nations adopted an exact relation between the pound and the kilogram, this work provided the basis for coexistence of the two sets of measurement units.[22] The relationship adopted differed only slightly from that established as a part of the reconstruction program. (It was in this work that it was discovered that the displacement volume of the kilogram[23] of the Archives had not been precisely determined before final adjustment.)

The entire reconstruction was based on the existence of weights RS and SP of known displacement volume, which had been compared with *U*. The average air temperature and barometric pressure for several hundred comparisons (used in the above definition) established a standard air density ρ_o. Knowing the displacement volume of the weight, *T*, used to construct the new standard, from comparisons with

---

[4]The General Conference of Weights and Measures (CGPM), assisted by the International Committee of Weights and Measures (CIPM) and the Consultative Committee for Unit. (CCU), makes decisions and promulgates resolutions, recommendations, and declarations for the International Bureau of Weights and Measures (BIPM). Ref. 57 reproduces in chronological order the decisions promulgated since 1889.

RS and SP in air of known density, one can compute the weight that T would appear to have if it were possible to compare it with U in air of density without knowing the density of U. In like manner, W above is a fictitious weight of 7000 grains of the same density as U, the lost Imperial standard; thus, the displacement volumes of weights must be known in order to compute values relative to the commercial pound, W.

This work included the construction and distribution of brass avoirdupois pound standards to approximately 30 countries, including the countries of the British Empire. Recognizing the practical difficulties that would arise because of the platinum defining standard and the brass standards for normal use, the platinum standard was defined to be one pound "in a vacuum"[5] and a commercial standard pound was defined as follows[24]:

> The commercial standard lb is a brass weight which in air (temperature 18.7°C, barometric pressure 755.64 mm) ... appears to weigh as much as W. ... For in air having the above mentioned temperature and pressure, the apparent weight of such a lb would be 7000/5760 of that of the lost standard.

The density of each of the new standards, both platinum and brass, was carefully determined. The assigned values, as computed from the comparison data, were expressed in the form of corrections, or deviations from a nominal value of 1, both on the basis as if compared with PS "in a vacuum," and as if compared with W in air of the defined density. For example, the correction for PS in a vacuum was expressed as 0.00000 since under this condition it is defined as 1 pound; however, because of its small displacement volume, if compared with W in air of specific gravity log delta = 7.07832 − 10 (air density approximately 1.1977 mg/cm³), it would appear to be 0.63407 grain heavy; thus on this basis the assigned correction was +0.63407 grain. This action firmly established two bases for stating values, one used to verify values assigned to standards with reference to the defining standard, and one to maintain the continuity of established commercial practices.

1.2.1.3 In the Early United States

In 1828, the Congress of the United States enacted legislation to the effect that the troy pound obtained from England in 1827 be the standard to be used in establishing the conformity of the coinage of the United States.[25] Apparently it was declared by Captain Kater, who had made the comparison with the Imperial pound standard that was later destroyed, to be an "exact" copy.[26] It is assumed that it was given the assigned value of 1 troy pound, the uncertainty of the comparison, or the announced correction, if any, being considered negligible. In 1830, the Senate directed the Secretary of the Treasury to study the weights and measures used at the principal Customhouses.[27] As a result of this study, the Treasury Department set out on its own to bring about uniformity in the standards of the Customhouses.

As a part of this work, Hassler constructed, along with other standards, a 7000 grain avoirdupois pound based on the troy pound of the mint. It was reported later[28,29] that Hassler's pound agreed very well with the copy of the standard pound furnished to the United States by England, as mentioned earlier. Eventually, this program was expanded by resolution[30] of Congress to include equipping the states with weights, measures, and balances. In 1866, the Congress enacted[31] that "no contract or dealing, or pleading in any court shall be deemed invalid or liable to objection because the weights or measures expressed or referred to therein are weights or measures of the metric system." In due course the states were also furnished metric standards.

Gross changes in the form of the economy of the United States have occurred. America has been profoundly influenced by the nearness of the people to the soil and the leadership that an agrarian society develops.[32] As late as 1830, approximately 70% of the working population of the United States was involved in agriculture and other forms of food production, and in producing raw materials. Only about 20%

---

[5]Weighings are not actually made in a vacuum. By properly accounting for the buoyant forces acting on the objects being compared, the data can be adjusted to obtain the result expected if the weighing had been made in a vacuum. One can also include in the weighing a small weight that is nearly equivalent to the difference in buoyant forces acting on the objects being compared.

was involved in manufacturing.[6] In such an environment weights and measures had a meaning in the value structure somewhat similar to that of ancient times. Now, something on the order of 30% is all that are involved in the area that includes producing food, raw materials, the manufacturing of both durable and nondurable goods, and construction. Thus, the number of items in which weights and measures have any relation to the value structure is very few, the major cost to the consumer being associated with value added rather than quantity.

The normal consumer can only choose from those products offered, selecting on the basis of asking price. The products offered, because of the high cost associated with establishing a large-scale production, are only those that have a high probability of being desirable to the buying public. While measurement may be necessary to establish the price to the customer, there is no meaningful relationship between the weights and measures and the unit price one must pay to acquire the item. One does not weigh automobiles or television sets. Where measurements are a part of the transaction, they are, in essence, merely counting operations similar in nature to counting out a dozen where items are priced by the dozen. Under these circumstances, the virtues of precise measurement and the exactness of the standard do not guarantee equity in the marketplace.

1.2.1.4 Summary

In retrospect at this point, it seems clear that both the construction of the kilogram and the reconstruction of the pound were essentially scientific efforts directed toward assuring the longevity of the respective mass units. Both efforts required precise definitions and detail work far beyond that usually associated with the previous history of weighing. Having established platinum standards, the assignment of values to weights of other materials (mostly brass) required as much as, if not more, attention to procedural detail.

The above two efforts, establishment and maintenance of the unit and calibration, together with normal usage has, in effect, polarized activities into separate groups — one group that works with defining mass standards and one group that works with practical everyday weighings — and in the middle a group that ostensibly translates the scientific into the practical. The degree to which such a hierarchy can be effective is related to the extent to which a specific end use can be characterized. If a measurement process requirement can be completely specified, one can devise a plan that will reduce a complex measurement to a simple operational routine. Such an engineered system, however, is not always adequate and may be completely misleading in other areas of usage.

The intellectual elegance of the metric system was lost almost from the start. A careful redetermination of the density of water created a situation in which, according to the original definition, the value assigned to the prototype kilogram would be in error by about 28 parts in a million. To change the value of the prototype and all of its copies was unthinkable; therefore, a new "volume" unit was proposed to replace the cubic centimeter. By conference action in 1901 (3d CGPM, 1901), the unit of volume, for high-accuracy determinations, was defined as the volume occupied by a mass of 1 kg of pure water at its maximum density and at standard pressure, this volume being called the liter [at present, 1 milliliter is equal to 1 cm^3]. While it is doubtful that the discrepancy was at all significant in common measurement, the liter has been accepted almost universally. This caused no end of problems concerning both volume and density measurement. The circle has been complete, for in 1961 (CIPM, 1961) the cubic decimeter was declared the unit for precise volume measurement, relegating the liter to the realm of customary units that still prevail.

Quite apart from the use of weights in commerce, various technologies over the centuries used weights as a convenient way to generate forces. The use of suspended or stacked weights to measure the draw of a bow, the ability of a structure to support a given load, and to characterize the strength of various materials has been prevalent throughout history and continues today. This led to an ambiguity in both

---

[6]The percentages have been estimated from various census reports. Because of the different classifications used over the years, they are only approximate. They are, however, valid indicators of a shift from an agrarian to an urban society in a very short time span.

the names assigned to the units and to the comparison operations. In 1901 (3d CGPM, 1901), the General Conference considered it necessary to take action to put an end to the "ambiguity which in current practice still subsists on the meaning of the word weight, used sometimes for mass and sometimes for mechanical force."

The Conference declared: "The kilogram is the unit of mass, it is equal to the mass of the international prototype kilogram. The word *weight* denotes a quantity of the same nature as force, the weight of a body is the product of its mass and the acceleration due to gravity, in particular, the *standard weight* of a body is the product of its mass and the standard acceleration due to gravity."

This did not end the confusion.[33,34] Such a statement made no sense at all to those who were concerned with commercial weighing. To officially sanction such a definition of weight is to refuse to recognize that at some time the use of a standard acceleration of gravity in lieu of the appropriate local acceleration of gravity would introduce significant systematic errors in many measurements.[35]

The situation has been rectified by including the newton as an accepted unit for force in the supplementary units of the International System of Units, known as the SI system (11th CGPM, 1960). By this action, the meaning of the words weight and weighing could revert to more general meanings, for example: weight — an object which embodies a mass or mass related property of interest; weighing — to make a quantitative comparison.[7] While this action may in time discourage practices such as introducing the term "massing"[36] as meaning to make a mass measurement, universal acceptance may never be achieved because of the natural tendency of the literature to propagate what has gone on before.

1.2.2   The Role of Measurement in Technology

It is difficult to trace the details of the various crafts. The Sumerians, for example, thought that all knowledge came from the gods; therefore, it was sacred and could not be communicated. The priest passed on instructions orally being careful to limit instructions to the exact steps to be followed.[37] For the craftsman, his knowledge was his livelihood. Traditions were passed from father to son. Families became noted for their particular crafts. Later, where products and trades were concerned, to divulge details was to invite economic disaster from competition. The impressive state of development reached, however, can be observed in the artifacts produced and the longevity of some of the techniques. An example of the latter is the 11 "touchstone" tests for purity of gold and silver alloys that made possible the issuance of coinage. Agricola described in 1556 essentially the same tests, indicating a longevity in excess of 2000 years.[38]

In terms of the development of the crafts and the dissemination of the products, the Roman Empire was remarkable. While somewhat short on invention, the Romans perfected masonry, tiling, road building, surveying, molded pottery, blown glass, watermill, and a host of others.[39] The use of glass, for example, in a wide variety of applications including commercial packaging reached a scale unmatched before the 19th century.[40] That these could not be accomplished without measurement clearly emphasizes the fact that, where function is the main concern, all measurements are relative. Things work because relative geometry, proportion, or properties of materials are correct, not because of any particular choice of measurement units. Mortar, for example, lasts through the ages because the ingredients have the right properties and are combined in the right proportions. Machinery works because each part has the right characteristics and the relative dimensions are correct. Each craft had to develop its own methods for determining and describing the parameters that were critical to its particular trade or profession.

Early crafts encompassed the entire operation from raw material to finished products. As the demand for finished products increased, the time the craftsmen could afford to spend in making ready raw materials lessened. In some instances, the materials in a product came from several distant sources. These

---

[7]A facsimile of the first edition of Webster's Dictionary[58] gives the following definitions: mass — a lump; weight — a mass by which bodies are weighed; weigh — to try the weight, consider, examine, judge … etc.

situations led to the development of early industries concerned basically with raw materials such as charcoal and metallic ores, and with quarrying, lumbering, and weaving. This action was the first breach in the tight security of the craft system. Craft guilds appeared during the Medieval Age, and the resulting "codes" were probably more directed toward protection from competition than convincing the possible clients of the perfection of the product. For example, in 1454 the penalty for divulging the secrets of Venetian glass was death.[41] Craft mysteries persisted until the Industrial Revolution ca. 1750.[42] The inventions of the 18th and 19th centuries brought about changes that are considered to be the Industrial Revolution. These changes can be summarized as follows: (1) a shift from animal and wind power to coal and steam, (2) the effects of this shift on the iron and textile industries,[43] and (3) the change from working for a livelihood to working for a profit.[44]

The forerunners of industry as we know it today stem from the military. The first large-scale demand for standardized goods was the provision of uniforms for large standing armies.[45] The use of interchangeable parts in the assembly of muskets and rifles was demonstrated by LeBlanc in France, and Whitney in the United States.[46] Through the years, the dividing line between raw material supply and preprocessing, such as the production of pig iron, steel, and cloth, and product manufacturing has become more prominent, with the preprocessed materials becoming more like other commercial commodities. Most items that are procured today, either by the individual or by the government, are the results of the combined efforts of many throughout the world. Industrial subdivision, or compartmentalization with its large economic benefits, has created a special role for measurement. Material or preprocessed material suppliers enlarge their market by resolving small differences in requirements among their customers. In time, the terminology of the supplier must be accepted by all who use the material; hence, measurements become wed to marketing requirements rather than functional requirements.

Subdivision of a task requires detailed delineation of what is to be done by each subunit. This can take the form of organization charts, specifications, detailed drawings, samples, and the like. Many ways are used depending upon the nature of the item and its function in the overall task. If someone else is to provide the service, some limits must be established for judging that the offered product will perform as intended in the overall endeavor. Determining the dividing line between success and failure is not always easy. These limits, once established and regardless of whether they were established by lengthy experiments, good engineering judgment, or by sheer guess, become fixed restraints on the next element of the subdivision. The effect is a dilution of the ability to make function-related judgments. In complex situations, no one person knows the full scope of the task; therefore no one can instigate changes of any sort without fear of jeopardizing the entire venture.

It is a tendency for tolerances to be tightened by each organizational element through which the task must pass. In the procurement-production stage, the product must comply (within the tolerance) to the specification or drawing. Compliance is defined by a set of procedures, usually measurements, which supposedly will assure the buyer of the suitability of the product for its intended use. The net result is that the most precise measurement processes are frequently used to differentiate between scrap and acceptable parts in order to consummate a particular contract, the sorting limits in many cases having little relation to the function the parts must perform. Troubles are merely transferred to the gauge if the measurements are differences between the part in question and a pseudo standard or gauge. Difficult problems occur when a specification attempts to describe a complex part completely by dimensions or specification verbiage.

The mechanism for verifying specification compliance is created for the most part by those who do not fully understand either the measurement or the function. Many procedures rely on ritualistic documentation with little attention given to the characteristics of the measurement processes that are used. In many instances the status of the source of the documentation becomes more important than problems relating to the environment in which the required measurements may be valid and the environment in which the measurements of the product are to be made. It is not unusual to find that a prerequisite for doing business is the possession of such documents and precise measurement facilities, which often do not relate to the completion of the task at hand.

However, in those cases where measurement data are really critical, the most important measurement is that on the production floor. The part or assembly will either operate properly or not regardless of the supporting hierarchy. The most precise measurements could, if necessary, be moved directly to the production floor to achieve the desired function.

Today, there is little doubt that the solutions of the most difficult and challenging measurement problems are being carried out in an environment of strict industrial security. This is similar to development in the days of the guilds. However, now external communications are necessary. The present economic facts of life make it necessary to know what is going on in related science and industry so that each new task is not a "re-invention of the wheel." A recent report suggests that innovations important to one industry may come from a completely nonrelated industry.[47] On the other hand, to divulge certain information at the developmental level is almost certain to result in an economic setback, perhaps even a catastrophe in the raw materials market, the product market, or in the capital market, sometimes in all three.

1.2.3 The Role of Measurement in Science

In sharp contrast to both previous areas of discussion, the advancement of science depends completely upon a free and open exchange of information.[48] Thus, having agreed to accept an arbitrary set of measurement units, it is imperative that the continuity of the units be maintained. By constructing a minimal set of units and constants from which all measurement quantities of interest can be derived, ambiguities are removed. By defining a means to realize each unit, in principle one can construct the units one needs without introducing ambiguity into the measurement system. What happens in practice is, of course, another story.

Most defining experiments are complex and tedious and not always related to the problems of measuring things or describing phenomena. Having established a definition of the unit of time based on an atomic phenomenon, and having constructed the hardware to realize the unit, the ease by which the unit can be disseminated by broadcast makes it highly unlikely that more than a few would seriously consider duplicating the effort. Mass, on the other hand, is and will no doubt for some time be embodied in a prototype standard to be disseminated by methods that are in essence many thousands of years old.

By international agreement, the SI-defined measurement units together with a substantial group of auxiliary units have replaced and augmented the original three — length, mass, and volume — of the metric system. Having accepted the structure of the SI, the definition, or redefinition, of the measurement units, insofar as possible, must maintain the continuity of the original arbitrary units. Further, the uncertainty of the unit as realized must be compatible with the exploratory experiments in which the unit may be used.

One requirement for a phenomenon to be considered in redefining a unit is that, under the contemplated definition, the newly defined unit would be more stable than the unit under the current definition.[49] Having verified that this would be the case, the next task is to determine the unit in terms of the new phenomenon to a degree such that the uncertainty of the unit as expressed by the new phenomenon is within the uncertainty limits associated with the unit as expressed by the old phenomenon. The important point is this action relates only to the definition of the unit, and may not be extendible in any form to the manner in which the unit is used to make other kinds of measurement. Because all units are candidates for redefinition, and because one is now able to evaluate the performance characteristics of a wide variety of measurement processes,[50] a new definition for the "best" measurement process must be established.

In the distant past, a weight was attested, or certified, to be an exact copy of another by the reputation or position of the person making the comparison, and by the stamp of the person on the weight. Having obtained such a verification, one was free to use the marked weight as he wished. The report of calibration from a currently existing measurement facility is in essence no different. Throughout history, the status of the standard with which the unknown was compared and the status of the facility doing the comparison established the quality of the work. Since all methods of comparison were essentially the same, to refute all criticism one might decide to pay more and wait longer in order to utilize the highest status facility of the land. Little attention was given to the consistency of the measurements at the operating level because there was no way to manipulate the masses of data required to evaluate a single measurement process, let alone a whole series of interconnected processes. One was paying for a judgment.

It has been well known from the beginning of precise measurement that repeated measurements often produce different numbers. The man who put his mark on the weight was in effect saying that it is close enough to some standard to be considered as an exact replica. The report of calibration says "call it this number," the number sometimes being accompanied by an uncertainty that is ridiculously small with reference to any practical usage, or when stated as a deviation from some nominal value, the deviation or the number being so small that the user may consider the item as exactly the nominal value.

It is now possible to look in detail at the performance characteristics of a measurement process[51] and at the consistency of measurement at any point in the entire system.[52] Further, the cost of relating a measurement to the manner in which the unit is defined may be prohibitive if indeed it is at all possible. Under these circumstances, the definition of the best process must start from the end use rather than the defining standard. Having first established that a particular measurement is necessary to the success of the venture at hand, the best process is that which produces these results in the most economical manner, based on verification by demonstration. This applies equally to the most complex scientific study or the simplest measurement. As a point of departure, it is necessary to make it clear to all the basis on which certain mass values are stated.

References

1. Childe, G. V., The prehistory of science: archaeological documents, Part 1, in *The Evolution of Science, Readings from the History of Mankind*, Metraux, G. S. and Crouzet, F., Eds., New American Library/Mentor Books, New York, 1963, 66–67.
2. Gandz, S. and Neugebauer, O. E., Ancient Science, in *Toward Modern Science*, Vol. I, Palter, R. M., Ed., Farrar, Straus & Cudahy/Noonday Press, New York, 1961, 8–9, 20–21.
3. Neugebauer, O. E., *Exact Sciences in Antiquity*, Harper & Brothers/Harper Torchbooks, New York, 1962, 18–19.
4. Woolley, Sir Leonard, *The Beginnings of Civilization*, Vol. I, Part 2, *History of Mankind, Cultural and Scientific Development*, New American Library/Mentor Books, New York, 362–364.
5. Breasted, J. H., *The Dawn of Conscience*, Charles Scribners's Sons, New York, 1933, 188–189.
6. Chadwick, J., Life in Mycenaean Greece, *Sci. Am.*, 227(4), 37–44, 1972.
7. Durant, W., *Our Oriental Heritage*, Part 1, *The Story of Civilization*, Simon and Schuster, New York, 1954, 79.
8. Maspero, G., *The Dawn of Civilizatioin, Egypt and Chaldaea*, Macmillan, New York, 1922, 772–774.
9. Forbes, R. J., Metals and early science, in *Toward Modern Science*, Vol. 1, Paltern, R. M., Ed., Farrar, Straus & Cudahy, New York, 1961, 30–31.
10. Woolley, Sir Leonard, in reference 4, pp. 748–751.
11. Woolley, Sir Leonard, in reference 4, pp. 330–333.
12. Woolley, Sir Leonard, in reference 4, pp. 340–341.
13. Berriman, A., *Historical Metrology*, E. P. Dutton, New York, 1953, 58–59.
14. Maspero, G., in reference 8, pp. 323-326.
15. Durant, W., in reference 7, p. 289.
16. Childe, G. V., *What Happened in History*, Pelican Books, New York, 1942, 192–193.
17. Maspero, G., in reference 8, pp. 752–753.
18. Kisch, B., *Scales and Weights*, Yale University Press, New Haven, CT, 1965, 11.
19. Durant, W., *Caesar and Christ*, Part III, *The Story of Civilization*, Simon and Schuster, New York, 1935, 78–79.
20. Pareti, L., *The Ancient World*, Vol. II, *History of Mankind*, Harper & Row, New York, 1965, 138.
21. Miller, W. H., On the construction of the new imperial standard pound, and its copies of platinum; and on the comparison of the imperial standard pound with the kilogram des Archives, *Philos. Trans. R. Soc. London*, 146(3), 943–945, 1856.

22. Astin, A. V., Refinement values for the yard and the pound, *Fed. Regis.,* July 1, 1959.

23. Miller, W. H., in reference 21, p. 875.

24. Miller, W. H., in reference 21, p. 862.

25. Tittmann, O. H., The National Prototypes of the Standard Metre and Kilogramme, Appendix No. 18 — Report for 1890, Coast and Geodetic Survey (U.S.), 1891, 736–737.

26. Tittmann, O. H., in reference 25, pp. 738–739.

27. Fischer, L. A., History of the Standard Weights and Measures of the United States, National Bureau of Standards (U.S.), Miscellaneous Publication No. 64, 1925, 5, 7.

28. Tittman, O. H., in reference 25, p. 739.

29. Miller, W. H., in reference 21, pp. 735, 945.

30. Fischer, L. A., in reference 27, pp. 10, 12.

31. Smith, R. W., The Federal Basis for Weights and Measures, Nat. Bur. Stand. (U.S.), Circ. 593, pp. 12–13 (1958).

32. Wright, L. B., *The Cultural Life of the American Colonies 1607–1763,* Harper & Row/Harper Torchbooks, New York, 1962, 1–3.

33. Huntington, E. V., Agreed upon units in mechanics, *Bull. Soc. Promotion Eng. Ed.,* 11(4), 171, 1920.

34. Huntington, E. V., Bibliographical note on the use of the word mass in current textbooks, *Am. Math. Mon.,* 25(1), 1918.

35. Tate, D. R., Gravity Measurements and the Standards Laboratory, National Bureau of Standards (U.S.), Technical Note 491, 1969, 10 pp.

36. Maracuccio, P., Ed., *Science and Children,* private communication, December 1970.

37. Contenau, G., Mediterranean antiquity, in *A History of Technology & Invention,* M. Daumas, Ed., Vol. I, Part II, Crown Publishers, New York, 1969, chap. 6, 116.

38. Agricola, G., *De Re Metallica* (translated from the first Latin edition of 1556 by H. C. Hoover and L. H. Hoover), Dover Publications, New York, 1950, 252–260.

39. Duval, P.-M., The Roman contribution to technology, Ch. 9, in reference 37, pp. 256–257.

40. *Encyclopaedia Brittanica,* Vol. 10, *Game to Gun Metal,* Glass in Rome, William Benton, Chicago, 1959, 410.

41. Durant, W., The Renaissance, *The Story of Civilization,* Part V, Simon and Schuster, New York, 1953, 313.

42. Ubbelohde, A. R. J. P., *Change from Craft Mystery to Science, A History of Technology,* Vol. IV, in Childe, Gordon V., *Early Forms of Society,* C. Singer, E. J. Holmyard, and A. R. Hall, Eds., Oxford University Press, New York, 1954, 663–669.

43. *Encyclopaedia Brittanica,* Vol. 12, *Hydrozoa to Jeremy,* Epistle of Industrial Revolution, William Benton, Chicago, 307–310.

44. Mumford, L., *The Paleotechnic Phase, Technics and Civilization,* Harcourt, Brace and Co., New York, 1946, Ch. IV, 151–159.

45. Mumford, L., Agents of Mechanization, in reference 44, p. 92.

46. Usher, A. P., *A History of Mechanical Inventions,* Beacon Press, Boston, 1959, 378–380.

47. Technological Innovation: Its Environment and Management, Department of Commerce (U.S.), Panel Report, 7, January 1967.

48. Astin, A. V., Standards of measurement, *Sci. Am.,* 218(6), 5, June 1968.

49. Huntoon, R. D., Status of the national standards for physical measurement, *Science,* 50, 169, October 1965.

50. Eisenhart, C., Realistic evaluation of the precision and accuracy of instrument calibration systems, *J. Res. Natl. Bur. Stand.* (U.S.), 67C (Eng. and Instr.), 2, 161–187, April–June 1962.

51. Pontius, P. E. and Cameron, J. M., Realistic Uncertainties and the Mass Measurement Process, Natl. Bur. Stand. (U.S.), Monograph 103, 17 pp., 1967.

52. Pontius, P. E., Measurement Philosophy of the Pilot Program for Mass Calibration, Natl. Bur. Stand. (U.S.), Tech. Note 288, 39 pp., 1968.

53. Moreau, H., The genesis of the metric system and the work of the International Bureau of Weights and Measures, *J. Chem. Educ.*, 30, 3, January 1953.
54. Miller, W. H., in reference 21, p. 753.
55. Winter, H. J. J., Muslim mechanics and mechanical appliances, *Endeavor*, XV(57), 25–28, 1956.
56. Sisco, A. G. and Smith, C. S., *Lazarus Ercker's Treatise on Ores and Assaying*, University of Chicago Press, Chicago, 1951, 89–91, 210.
57. Page, C. W. and Vigoureux, P., Eds., The International System of Units (SI), Natl. Bur. Stand. (U.S.), Spec. Publ. 330, 42 pp., January 1971.
58. Webster, N., *A Compendious Dictionary of the English Language,* A facsimile of the first (1806) edition, Crown Publishers, Inc./Bounty Books, New York, 1970.

1.3 Report by John Quincy Adams

Extract from the Report on Weights and Measures by the Secretary of State, made to the Senate on February 22, 1821:

Weights and measures may be ranked among the necessaries of life to every individual of human society.

They enter into the economical arrangements and daily concerns of every family.

They are necessary to every occupation of human industry; to the distribution and security of every species of property; to every transaction of trade and commerces; to the labor of the husbandman; to the ingenuity of the artificer; to the studies of the philospher; to the researches of the antiquarian; to the navigation of the mariner, and the marches of the soldier; to all the exchanges of peace, and all the operations of war.

The knowledge of them, as in established use, is among the first elements of education, and is often learned by those who learn nothing else, not even to read and write.

This knowledge is riveted in the memory by the habitual application of it to employments of men throughout life.

2

Recalibration of
Mass Standards

2.1 Recalibration of the U.S. National Prototype Kilogram[*]

2.1.1 Introduction

In 1984, the U.S. National Prototype Kilogram, K20, and its check standard, K4, were recalibrated at the Bureau International des Poids et Mesures (BIPM). Two additional kilograms, designated CH-1 and D2, made of different alloys of stainless steel, were also included in the calibrations.

The mass of K20 was stated to be 1 kg − 0.039 mg in an 1889 BIPM certification; the mass of K4 was stated to be 1 kg − 0.075 mg in an 1889 BIPM certification. K20 was recalibrated at BIPM in 1948 and certified to have a mass of 1 kg − 0.019 mg. K4 had never before been recalibrated.

The nominal masses of the stainless steel kilograms were 1 kg + 13.49 mg for D2 and 1 kg − 0.36 mg for CH-1.

The four 1-kg artifacts were hand-carried from the National Bureau of Standards, NBS (now National Institute of Standards and Technology, NIST), Gaithersburg, MD to BIPM on commercial airlines. The carrying case for K20 was an enclosure in which the kilogram was held firmly on the top and bottom and clamped gently at three places along the side. Clamped areas, conforming to the contour of the adjacent kilogram surfaces, were protected by low-abrasive tissue paper backed by chamois skin, which had previously been degreased through successive soakings in benzene and ethanol. The outer case of the container was metal, the seal of which was not airtight.

In the carrying case for K4, of simpler design, the artifact was wrapped in tissue, then wrapped in chamois skin, and finally placed in a snug-fitting brass container. The container seal was not airtight.

The stainless steel kilograms were wrapped in tissue paper and were then padded with successive layers of cotton batting and soft polyethylene foam. The outer container was a stiff cardboard tube. The kilogram was held fast within the tube by the padding.

2.1.2 Experimental

The balances used in the 1984 comparisons were NBS-2 (at BIPM), a single-pan balance designed and built at NBS (now NIST) and then permanently transferred to BIPM in 1970; and V-1 (at NIST), the primary kilogram comparator of NBS (NIST), manufactured by the Voland Corporation of Hawthorne, NY. Both balances, similar in design, were based on design principles established by Bowman and colleagues[2,3] during the 1960s. The estimate of the standard deviation of the measurement of the difference of two mass artifacts being compared is approximately 1 μg on NBS-2 and approximately 4 μg on V-1.

---

[*]Chapter is based on Ref. 1.

2.1.3 1984 BIPM Measurements

The four NBS standards were compared to two platinum-iridium standards of BIPM, first in the state in which they arrived at BIPM. Then they were compared after cleaning with benzene. Platinum-iridium prototypes K4 and K20 were, in addition, washed under a steam jet of doubly distilled water.

In the course of each weighing, the density of moist air was calculated using the "formula for the determination of the density of moist air (1981)."[4] The parameters in the formula, temperature, pressure, relative humidity, and carbon dioxide concentration in the balance chamber were measured using a platinum resistance thermometer, an electromanometer, a hygrometer transducer, and an infrared absorption analyzer, respectively.

The mass values found at BIPM for the four artifacts are as follows:

	Before Cleaning	After Cleaning
K20	1 kg – 0.001 mg	1 kg – 0.022 mg
K4	1 kg – 0.075 mg	1 kg – 0.106 mg
CH-1	1 kg – 0.377 mg	1 kg – 0.384 mg
D2	1 kg + 13.453 mg	1 kg + 13.447 mg

The estimate of the standard deviation of each of the before cleaning results was 1.2 µg. The estimate of the standard deviation of each of the after cleaning results was 1.3 µg.

2.1.4 1984 NBS Measurements

After return to NBS, K20 and K4 were compared with two platinum-iridium check standards, KA and K650, in some preliminary measurements. After the measurements on K20 and K4 before cleaning, the two artifacts were cleaned with benzene and then they were washed in a vapor jet of doubly distilled water. After cleaning, the artifacts were again compared with KA and K650.

The results showed that K20 was unchanged by the cleaning, whereas K4 lost approximately 4 µg. KA and K650 were not cleaned for these measurements.

Using the six weights, a set of 18 symmetrized observations was then made.

CH-1 and D2 were then cleaned by vapor degreasing and observations 13 through 18 were then repeated, after which the new results were compared with the original observations. If it were that the masses of K20 and K4 were invariant during these weighings of CH-1 and D2, the results may be interpreted as CH-1 having lost 16.5 µg and D2 having lost 19.3 µg as a result of the vapor degreasing cleaning.

The 1984 NBS mass values for K20, K4, CH-1, and D2 after cleaning are listed below:

K20	1 kg – 0.022 mg
K4	1 kg – 0.103 mg
CH-1	1 kg – 0.3887 mg
D2	1 kg + 13.4516 mg

The estimates of the standard deviation for CH-1 and D2 were 4.8 µg.

Davis[1] suggested that, based on the results of the measurements:

1. "It appears that long-term measurements of platinum- iridium artifacts based on K20 can be stable to 10 micrograms provided that the artifact is vigorously cleaned before use, according to the BIPM method."

2. "Mass values can be supplied to stainless steel weights with an uncertainty of about 30 micrograms. This includes all known sources of uncertainty as well as an additional 'between times' component."

2.1.5   Recommendations

Davis[1] recommended, to improve the ability to make reproducible mass measurements and to improve the prospects for understanding the effects of influencing factors, that:

1. It was desirable for NBS to use stainless steel working standards for routine calibrations.
2. A balance (preferably automated) must be made available, which has a standard deviation of 1 µg or better.
3. The balance chamber should be hermetically sealed.
4. A cleaner environment for storing and using the weights should be considered.

2.2   Third Periodic Verification of National Prototypes of the Kilogram

2.2.1   Introduction

The Third Periodic Verification of National Prototype Kilograms was conducted beginning in the summer of 1989 and ending in the autumn of 1992.[5] Initially, the International Prototype Kilogram, $\mathfrak{R}$, was compared with the six official copies [K1; Nos. 7, 8(41), 32, 43, and 47], and the copies used by BIPM (Nos. 25, 9, 31, and 67).

2.2.2   Preliminary Comparisons

The International Prototype Kilogram, its official copies, and prototype No. 25 were compared to prototypes Nos. 9 and 31, before and after two successive cleanings and washings. These treatments resulted in changes in mass. The observed changes confirmed previous measurements, made since 1973, on platinum-iridium standards sent to BIPM to compare with the BIPM working standards. As a function of time since previous cleaning and washing, the changes in mass with time could be fitted to a straight line with slope of −1 µg/year. The line did not pass through the origin, indicating that the effect of surface pollution with time may be more rapid just after cleaning and washing.

Consequently, the change in mass of the international prototype immediately after cleaning and washing was followed. The study also included official copy No. 7 and two standards (Nos. 67 and 73) manufactured by diamond machining; changes in mass were measured relative to working standards Nos. 9 and 31.

The change in mass of the international prototype during the first 120 days was linear with a value of +0.0368 µg/day, a value that was adopted for all the prototypes during the third verification.

2.2.3   Comparisons with the International Prototype

All the weighings of the third periodic verification were made using the NBS-2 balance, which accommodated six standards on its weight exchanger.

The following was the comparison plan:

1. $\mathfrak{R}$ with five prototypes
2. $\mathfrak{R}$ with the remaining five prototypes
3. Three prototypes, including No. 31 from group 1; and three prototypes, including No. 9 from group 2
4. Prototype No. 31 with the remaining prototypes of group 1 and prototype No. 9 with the remaining two prototypes from group 2

The mass of the international prototype was taken as exactly 1 kg. The results of these comparisons led the CIPM (Comité International des Poids et Mesures) to take two decisions:[6]

1. The mass of the international prototype of the kilogram for the purposes of the 1889 definition is that just after cleaning and washing by the method used at the BIPM.[7] Its subsequent mass is determined, under certain conditions, by taking into account the linear coefficient given above. This interpretation of the definition is adopted for the third periodic verification but the BIPM makes clear that the interpretation does not in any way constitute a new definition of the kilogram.

2. The Working Group on Mass Standards of the Consultative Committee for Mass should meet to give its opinion on whether the national prototypes should be cleaned and washed. [In November 1989, the Working Group replied in the affirmative.]

2.2.4 Verification of the National Prototypes

The plan adopted for the third verification of national prototypes of the kilogram is outlined in Ref. 5.

2.2.5 Conclusions Drawn from the Third Verification

The change in mass of prototypes the fabrication of which dates from 1886 (Nos. 1 to 40) seems steady and confirms the values obtained in the second periodic verification (1946–1953). Prototypes Nos. 2, 16, and 39 were accidentally damaged and so could not be taken into account; the behavior of No. 23 since 1948 was peculiar.

The national prototypes in this batch had been apparently well stored and carefully used. The mass had grown, on the average, by 0.25 µg per year.

The mass of prototypes participating in the third verification with numbers between 44 and 55 showed a change of about 0.9 µg since the second verification.

The changes were said to include changes due to wear from use.

Prototype No. 34 was sealed within its travel container. Its change in mass of +0.027 µg between 1950 and 1992 could be considered to be significant and unequivocal.

The cleaning of some of the prototypes was partial; prototype No. 6 had the treatments:

17 June 1985   "wipe and washing-organic solvents"
13 September 1986   "jet steam cleaning"

The two cleanings of this prototype at BIPM in October 1991 resulted in a loss of 0.032 µg.

Prototypes Nos. 4 and 20 belonging to the United States had in 1983 first been rubbed with benzene, and then washed as at BIPM. The treatment at BIPM caused their mass to decrease by 0.031 and 0.021 mg, respectively.

Each member of the Convention du Metre that possessed a platinum-iridium prototype was able to send it to BIPM for the third periodic verification.

The mass of each prototype was determined with respect to the international prototype with a combined uncertainty of 2.3 µg.

The final results calculated for the prototypes in the third periodic verification are shown in Table 2.1.

BIPM has shown that there is a possibility that $\Re$ has changed by 50 µg per 100 years.

TABLE 2.1 Results of the Third Periodic Verification of National Prototype Kilograms

$\mathfrak{R}$	1 kg	$\mathfrak{R}$	1 kg
Official Copies		BIPM Prototypes	
K1	1 kg + 0.135 mg	No. 25	1 kg + 0.158 mg
No. 8(41)	1 kg + 0.321 mg	No. 31	1 kg + 0.131 mg
No. 43	1 kg + 0.330 mg	No. 9	1 kg + 0.312 mg
No. 7	1 kg − 0.481 mg		
No. 32	1 kg + 0.139 mg		
No. 47	1 kg + 0.403 mg		

Other Prototypes

$\mathfrak{R}$		1 kg
No. 2	Rumania	1 kg − 1.127 mg
No. 3	Spain	1 kg + 0.077 mg
No. 5	Italy	1 kg + 0.064 mg
No. 6	Japan	1 kg + 0.176 mg
No. 12	Russian Federation	1 kg + 0.100 mg
No. 16	Hungary	1 kg + 0.012 mg
No. 18	Royaume-Uni	1 kg + 0.053 mg
No. 20	United States	1 kg − 0.012 mg
No. 21	Mexico	1 kg + 0.068 mg
No. 23	Finland	1 kg + 0.193 mg
No. 24	Spain	1 kg − 0.146 mg
No. 34	Academy of Sciences of Paris	1 kg − 0.051 mg
No. 35	France	1 kg + 0.189 mg
No. 36	Norway	1 kg + 0.206 mg
No. 37	Belgium	1 kg + 0.258 mg
No. 38	Switzerland	1 kg + 0.242 mg
No. 39	Republic of Korea	1 kg − 0.783 mg
No. 40	Sweden	1 kg − 0.035 mg
No. 44	Australia	1 kg + 0.287 mg
No. 46	Indonesia	1 kg + 0.321 mg
No. 48	Denmark	1 kg + 0.112 mg
No. 49	Austria	1 kg − 0.271 mg
No. 50	Canada	1 kg − 0.111 mg
No. 51	Poland	1 kg + 0.227 mg
No. 53	Low Countries	1 kg + 0.121 mg
No. 54	Turkey	1 kg + 0.203 mg
No. 55	Federal Republic of Germany	1 kg + 0.252 mg
No. 56	South Africa	1 kg + 0.240 mg
No. 57	India	1 kg − 0.036 mg
No. 58	Egypt	1 kg − 0.120 mg
No. 60	Peoples Republic of China	1 kg + 0.295 mg
No. 62	Italy (IMGC)	1 kg − 0.907 mg
No. 64	Peoples Republic of China	1 kg + 0.251 mg
No. 65	Slovak Republic	1 kg + 0.208 mg
No. 66	Brazil	1 kg + 0.135 mg
No. 68	Peoples Democratic Republic of Korea	1 kg + 0.365 mg
No. 69	Portugal	1 kg + 0.207 mg
No. 70	Federal Republic of Germany	1 kg − 0.236 mg
No. 71	Israel	1 kg + 0.372 mg
No. 72	Republic of Korea	1 kg + 0.446 mg
No. 74	Canada	1 kg + 0.446 mg
No. 75	Hong Kong	1 kg + 0.132 mg

References

1. Davis, R.S., Recalibration of the U.S. National Prototype Kilogram, *J. Res. Natl. Bur. Stand.* (U.S.), 90, 263, 1985.
2. Bowman, H.A. and Macurdy, L.B., Gimbal device to minimize the effects of off-center loading on balance pans, *J. Res. Natl. Bur. Stand.* (U.S.), 64, 277, 1960.
3. Bowman, H.A. and Almer, H.E., Minimization of the arrestment error in one-pan two-knife systems, *J. Res. Natl. Bur. Stand.* (U.S.), 64, 227, 1963.
4. Giacomo, P., Formula for the determination of the density of moist air (1981), *Metrologia*, 18, 33, 1982.
5. Girard, G., Third Periodic Verification of National Prototypes of the Kilogram, May, 1993.
6. *Proc.-verb. Com. Int. Poids et Mesures*, 57, 15, 1989.
7. Girard, G., Le nettoyage-lavage des prototypes du kilogramme au BIPM, BIPM, Sèvres, France, 1990.

3

Contamination of Mass Standards

3.1 Platinum-Iridium Mass Standards

3.1.1 Growth of Carbonaceous Contamination on Platinum-Iridium Alloy Surfaces, and Cleaning by Ultraviolet–Ozone Treatment

3.1.1.1 Introduction

At the surface of platinum-iridium mass artifacts, three main types of contamination are thought to occur[1]:

1. Sorption of water vapor
2. Carbonaceous contamination
3. Mercury contamination

Because of exposure of the surface to atmospheric oxygen, there is also a certain mass of metal oxide and metal hydroxide.

The likely physical mechanism of the "buildup" of carbonaceous contamination on platinum-iridium mass artifacts was examined, and historical weighing data were analyzed by Cumpson and Seah[1] using this model mechanism to deduce the rate of increase of the contamination with time.

An approach to the cleaning of platinum-iridium mass artifacts, involving exposure to ultraviolet (UV) light and ozone in air at room temperature and pressure, was investigated.

Optimum UV intensities, ozone concentration, and cleaning times were determined. Recommendations for experiments to validate the cleaning procedure on reference kilograms were discussed.

This UV–ozone approach was investigated as a possible alternative to the Bureau International des Poids et Mesures (BIPM) method[2] for cleaning Pt–10%Ir reference kilograms. The BIPM method involves manual rubbing by chamois-leather soaked in methanol and ether followed by washing in recondensed water droplets from a jet of steam (*nettoyage/lavage*), see Chapter 4.

3.1.1.2 Ultraviolet–Ozone Cleaning

Two routes are possible to improve the stability of the mass of platinum-iridium mass artifacts:[1]

1. "Improve the storage environment of the Pt-Ir prototype so as to remove the carbonaceous contamination from the atmosphere; or,
2. Use a simple, repeatable cleaning procedure which can be used routinely prior to weighings in order to remove contamination and to return the prototype to its nominal [defining] mass."

Route 1 was considered to be extremely difficult, and complete elimination of carbonaceous contamination from the air was considered to be almost impossible.

24

Therefore, consideration was given to the development of a "new 'push-button' cleaning method which fulfils the requirements of (1) removing carbonaceous contamination, (2) being sufficiently standardized, simple and repeatable to be employed by all national standards laboratories."

The cleaning method considered was a noncontact cleaning method based on UV light and ozone. The method is a procedure for removing carbonaceous contamination from surfaces by exposure to short-wave UV light together with ozone at parts-per-million concentration, in air at room temperature and pressure. It is a direct photochemical oxidation process.

This method was adapted from a method used in the microelectronics industry for cleaning wafers. The advantages of this, UV/O_3, method are as follows:

(a) UV/O_3 cleaning can be performed in ordinary air at room temperature and pressure.

(b) UV/O_3 cleaning is less aggressive with respect to the bulk material; hydrocarbon contamination is removed without removing any inorganic material from the system being cleaned.

(c) UV/O_3 cleaning is a straightforward "push button" procedure requiring no expertise and no high-purity, wet-chemical reagents with their associated lengthy, documented procedures.

(d) Since UV light intensity and ozone concentration are easily measured and reproduced, the cleaning method itself is very reproducible.

In investigating the method, the aim was to minimize the effort involved in weighing trials to confirm the applicability of the method.

UV/O_3 cleaning very effectively removed carbonaceous contamination but left a single atomic layer of mixed metal oxide/metal carbide at the surface.

[Excerpts above taken with permission from *Metrologia*.]

3.1.1.3 Optimum Cleaning Conditions

Optimum operating conditions for a UV/O_3 cleaning apparatus are as follows:

1. 5 ppm O_3
2. 50 watts (W)/m² UV at about 250 nm
3. Cleaning exposure for 2 h

If a Pt-Ir prototype has been cleaned no more than 10 years previously, this cleaning treatment followed by an appropriate stabilization period will reset the mass of hydrocarbon contamination to <3 µg.

This mass of contamination compares well with the uncertainty of ±2.3 µg (at 1 standard deviation) achieved for all the national prototypes involved in the third verification.[3]

Therefore, UV/O_3 cleaning has the potential to reduce the uncertainty due to long-term contamination of mass standards to the same magnitude of the other weighing uncertainties for Pt-Ir prototypes.

3.1.1.4 Conclusions

1. Growth of carbonaceous contamination on Pt-Ir prototype masses leads to a mass increase, easily measurable using state-of-the-art balances, which limits the precision with which mass standards can be disseminated.

2. The kilogram prototypes of various nations appear to gain in mass very significantly due to carbonaceous contamination. This mass gain is diffusion limited and will not "saturate" within any reasonable time.

3. The precise rates of contamination vary, possibly because of the different types of carbonaceous species in the atmosphere in different locations.

4. The BIPM *nettoyage/lavage* process includes a manual rubbing, which is not easy to standardize and reproduce.

5. Carbonaceous contamination on Pt-Ir surfaces is very effectively removed by the UV/O_3 cleaning method.

6. Optimum cleaning conditions are described above.

3.1.1.5 Recommendations

1. In particular, by storage in an enclosure with a sub-micron filter, exposure of Pt-Ir prototypes to sources of carbonaceous contamination should be minimized.
2. UV/O_3 cleaning of Pt-Ir prototype kilograms should be considered for adoption by weighing laboratories.
3. Weighing trials are required
   a. To prove the UV/O_3 method to be repeatable and reliable,
   b. To determine the length of the stabilization period required after UV/O_3 cleaning, before weighings using the reference mass can recommence.
4. UV/O_3 cleanings should be repeated at least every 10 years.

3.1.2 Progress of Contamination and Cleaning Effects

3.1.2.1 Introduction

Surface analytic techniques, including X-ray photoelectron spectroscopy (XPS), were used by Ikeda et al.[4] on specimens cut from prototype material of Pt–10%Ir alloy to examine the stability of prototypes of the kilogram after cleaning. Steam-jet cleaning (SJC) and ultrasonic cleaning with solvents (UCS) were compared in terms of cleaning effects and progress of contamination.

The specimens were cut from a single lot of prototype material supplied by BIPM. The analysis of contamination continued for a period of 6 months.

3.1.2.2 Problems with Steam-Jet Cleaning

The main problems with SJC were considered to be as follows:

(a) oxidation of surface metals after ion sputtering and the reduction, or dissolving in water, of oxidized metals in the reducing environment of SJC;

(b) effectiveness of SJC in removing contaminants compared with that of sputtering and the progress of contamination;

(c) effectiveness of SJC in removing contaminants and the rate of contamination relative to that following ultrasonic cleaning with solvent(s);

(d) chemisorption of water immediately after cleaning. [Excerpts taken with permission from *Metrologia*.]

For the analysis of items (a), (b), and (c), XPS was chosen; for item (d), thermal desorption spectroscopy (TDS) was chosen.

3.1.2.3 Steam-Jet Cleaning Procedure

A simplified SJC cleaner was made from a distiller, consisting of a flask with a nozzle made of fused quartz and an electric furnace.

Under constant steam generation, each surface of the specimen was held 5 mm in front of the nozzle for 5 min with a temperature of approximately 90°C. After cleaning, specimens were dried naturally.

3.1.2.4 Ultrasonic Cleaning with Solvents Procedure

Ultrasonic cleaning was performed on the specimens for 5 min in beakers of 50-cm³ capacity each half filled with a different solvent (acetone or ethanol) and placed in a bath half filled with distilled water.

After cleaning in organic solvents, the specimens were given a final ultrasonic treatment in distilled water. Cleaned specimens were dried naturally in the room or in the draft from an electric fan.

3.1.2.5 Results

Causes of mass change in prototypes of the kilogram were examined, mainly by XPS, with the following results:

1. In ambient air, oxidation of platinum was not observed.
2. Oxidation of iridium to the extent of one monolayer was observed in an accelerated heat test.
3. Adsorbed substances, which included carbon, were identified as hydrocarbons from the ambient air.
4. SJC reduced the amount of carbon on the surfaces of as-received specimens to about two thirds, and to about one half by UCS.
5. Cleaning by UCS was found to be superior to SJC in reducing carbon deposits and the adsorption of water; the cleaning power of acetone when used in UCS was great enough to eliminate the need for SJC.
6. Hydrocarbon contamination increased with elapsed time after cleaning; the mass gain in the first month was several times that found in the succeeding 5 months.
7. For SJC, the mass of contaminants remaining after cleaning and the mass gain due to contaminating hydrocarbons after a 6-month exposure to air were estimated to be 37 and 4.3 µg; for UCS, the corresponding figures were 24 and 16 µg.
8. The experimental results indicated that improvement in conservation of prototypes is far more important than cleaning in maintaining the stability of mass standards.

3.1.3 Effects of Changes in Ambient Humidity, Temperature, and Pressure on "Apparent Mass" of Platinum-Iridium Prototype Mass Standards

3.1.3.1 Introduction

In an attempt to quantify the surface-related influence of changes in ambient relative humidity, temperature, and pressure on the "apparent mass" of platinum-iridium prototype mass standards, an experimental study was carried out at BIPM in France.[5]

Two 1-kg mass standards of diamond-machined platinum-iridium were used in the study. One of the standards was a right-circular cylinder; the other was made up of four disks. The total surface area of the four disks was close to 150 cm², which was twice the surface area of the cylinder.

The two standrds were compared on a flexure strip balance.

3.1.3.2 Experimental Procedures and Results

3.1.3.2.1 *Surface Effects in Ambient Conditions*

The experimental study was carried out at BIPM.

While keeping the pressure constant at 100 kPa and the temperature constant at 22°C, the relative humidity of the air in the balance case was varied in the range 37 to 58%.

Then, while keeping both the relative humidity and the temperature constant, the ambient pressure was varied in the range 99 to 103 kPa.

Finally, keeping both the relative humidity and pressure constant, the temperature was varied in the range 19 to 23°C.

From these three sets of measurements, the changes in mass difference between the two standards Δm_h, Δm_p, Δm_T as functions of relative humidity, h, pressure, p, and temperature, T, respectively, were deduced.

For a difference of surface area between the standards of 75 cm², the following were obtained:

$$\left(\Delta m_h / \Delta h\right) = \left(1.8 \pm 0.6\right) \mu g$$

$$\left(\Delta m_p / \Delta p\right) = \left(-0.18 \pm 0.04\right) ng / Pa$$

$$\left(\Delta m_T / \Delta T\right) = \left(0.3 \pm 0.1\right) \mu g / °C$$

The authors, Quinn and Picard,[5] concluded that for diamond-machined 1-kg platinum-iridium mass standards the effects of changes in ambient conditions of humidity, pressure, and temperature were small and could easily be kept below 0.1 µg.

There remained the variation in mass with time. This variation was not correlated with changes in ambient conditions and the origin of the variation was at that time unknown.

The magnitude of the variation with time was similar to that previously observed and reported for Pt-Ir 1-kg standards in the months following cleaning and washing. The two mass standards in this study had been cleaned a few months before the measurements were begun.

3.1.3.2.2 *Reproducibility of Mass between Ambient Conditions and Vacuum*

Starting in June 1992, five successive measurements of the mass difference between the same two mass standards were made alternately in "vacuum" (pressure of about 1 Pa) and at atmospheric pressure. During the measurements at atmospheric pressure, no particular care was taken to reproduce the pressure, temperature, and relative humidity because the previous study had shown relative insensitivity of the mass to these parameters.

The period of time of measurements in vacuum and at atmospheric pressure was usually a few days; however, after the third measurement in vacuum the atmospheric pressure was maintained for 14 weeks.

Reproducibility of mass difference of better than 0.5 µg was indicated by these preliminary results.

The changes in mass difference with temperature and pressure were similar to those in a previous study; however, the coefficient obtained for relative humidity was about ten times greater than that in the previous measurements made entirely at atmospheric pressure.

Possible explanations for the greater humidity coefficient and the apparent increase in mass difference between the two mass standards upon going from atmospheric pressure to vacuum were being investigated.

The data were corrected to standard conditions using the coefficients for pressure, temperature, relative humidity, and time. The scatter remaining in the data had a standard deviation of about 30 ng.

3.1.4 Evidence of Variations in Mass of Reference Kilograms Due to Mercury Contamination

3.1.4.1 Introduction

Samples of Pt–10%Ir were studied by XPS by Cumpson and Seah[6] to assess environmental contamination of reference kilogram masses at the National Physical Laboratory (NPL) of the United Kingdom. Samples were sputtered clean by 7-keV argon ions and distributed in four places where reference kilograms were kept.

3.1.4.2 Results

All four specimens showed carbon and oxygen contamination, and mercury contamination with levels that varied from venue to venue. Mercury had the highest vapor pressure of the metals used in the laboratory, and mercury reacts with platinum. In the most-contaminating venue, the case enclosing the U.K. primary balance, the effective mass of a reference kilogram increased by 0.26 µg/day. The mercury was not removed by washing or by scrubbing with a chamois leather soaked in ethanol and ether.

Over a period of a month, up to 50% of a monolayer of mercury might be taken up from mercury vapor in the air, leading to a mass increase of 14 µg on a reference kilogram. Although reasonably stable, this mass increase might be increased by exposure to a fresh environment of higher mercury content.

It did not appear that the cleaning procedures used at BIPM[2] would remove this mass increase. For stainless steels exposed to the same environments, no mercury contamination was observed.

It was recommended[6] that:

All laboratories in which Pt or Pt-Ir reference kilograms are maintained or weighed are monitored by XPS for probable mercury contamination and that procedures are developed either to remove mercury from weighing environments or to define a stabilization exposure for use in final stage of manufacture of Pt-Ir masses. [Excerpts taken with permission from *Metrologia*.]

3.1.5 Mechanism and Long-Term Effects of Mercury Contamination

3.1.5.1 Introduction

In a second study by Cumpson and Seah[7] of mercury contamination on platinum-iridium mass standards from environmental contamination, the mass uptake per unit area of Pt–10%Ir exposed to mercury vapor was measured as a function of time using the mass response of a quartz crystal microbalance with electrodes of Pt–10%Ir. Mass increases equivalent to less than 0.1 µg on a prototype kilogram could be detected with accuracy.

3.1.5.2　Results and Conclusions

1. Quartz crystal microbalance measurements showed that atmospheric mercury is adsorbed onto, and then absorbed into, surfaces of Pt–10%Ir alloys.
2. Significant, irreversible mass increases for reference masses of Pt–10%Ir can occur in typical laboratory environments containing 1 to 5 µg/m³ of mercury.
3. The two distinct phases in the sorption of mercury by the Pt–10%Ir surface are (a) an initial rapid chemisorption of a monolayer of mercury, followed by (b) a slow mass increase proportional to the square root of time, caused by mercury diffusing into the layer of surface damage introduced by polishing or diamond machining.
4. The initial rapid adsorption of mercury on reference masses probably takes place within a few months of manufacture. If this monolayer is retained for the entire life of the prototype, it might not affect the accuracy of mass comparisons.
5. The long-term mass increase, proportional to the square root of time, is likely to continue for many years and was unlikely to have terminated for any of the prototypes then in service.
6. The observed mass gains of the prototypes might be due partly to mercury and partly to phys-isorbed carbonaceous contamination, which might also be expected to display a "root *t*" time dependence.

3.1.5.3　Recommendations

1. Pt–10%Ir reference masses should be kept in an environment as free as possible from mercury; this requires mercury levels well below current health and safety limits.
2. Storage areas should be monitored for mercury contamination.
3. A nondestructive chemically specific technique is required to measure the subsurface mercury on Pt–10%Ir reference kilograms in service.
4. For new prototypes, it might be possible to insert a processing step after diamond machining but before final cleaning/washing, to help prevent mercury sorption by removing the damaged layer.

3.1.6　Water Adsorption Layers on Metal Surfaces

3.1.6.1　Introduction

In particular with changes in the humidity of the air, the stability of the mass of very high precision weights is influenced by the H_2O adsorption layer.[8]

A water adsorption layer forms on the surface of solids. The thickness of the adsorption layer, in the nanometer range, depends on the humidity of the ambient air, as well as on the material, the surface condition, and surface impurities.[8]

Adsorption of water on various metal surfaces was investigated by Kochsiek[8] at the PTB (Physikalisch-Technische Bundesanstalt) in Germany to determine mass variations due to adsorption of water at more or less constant temperatures (in the range 20 to 22°C) and atmospheric pressures (in the range 99,000 to 101,800 Pa).[8]

Figure 3.1 illustrates the increase in mass with relative humidity for various metal surfaces.

3.1.6.2　Experimental Procedures

Investigations were carried out on plates (of thickness greater than or equal to 0.2 mm) of the following materials:

Platinum-iridium 90/10
Austenitic steel X 5 Cr Ni 18 9
Brass 63
Aluminum 99

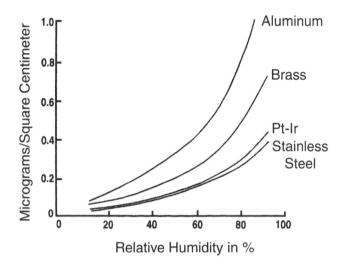

FIGURE 3.1 Increase in mass with relative humidity for various metal surfaces.

Coated Surface	Layer Thickness, μm
Base material: brass surface	
a. Nickel-plated	10
b. Electrogilded	3
c. Gilded by vapor deposition	0.3
d. Nickel-plated with subsequent chromium plating	10 μm Ni + 5 μm Cr
e. Chromium-plated	10
Base material: aluminum surface	
a. Anodized aluminum	Between 5 and 20

The variation of mass at varying water adsorption was determined by weighing. The mass of the individual samples was determined in at least three weighing operations, after the measuring equipment reached a suitable air-conditioned state and the samples were cleaned and arranged.

The zero of the balance was checked before each weighing by remotely unloading the balance and correcting for any deviation.

For each series of measurements, air relative humidities of 12.4, 33.6, 54.9, 75.5, and 93.2% (established by saturated salt solutions) were adjusted one after the other and finally the relative humidity adjusted first was reset.

The measuring cycle took about 300 h; the tests were carried out over a period of 4 years.

The samples were cleaned by:

1. Removing dust particles with a brush; adhering particles were removed by rubbing the sample with soft leather
2. Cleaning in a methyl alcohol ultrasonic bath
3. Cleaning with solvents such as methyl alcohol

3.1.6.3 Results

1. There could be an uncertainty of up to 10% when the ratio of the mass of adsorbed water to the mass of the sample was 5×10^{-6} and the relative standard deviation was 2×10^{-7} (4 μg at 20 g).
2. In general, the change of the adsorbed water layer on the sample occurs within a few hours following a sudden change of the humidity of air, for example, from 12 to 93% relative humidity, or for the change in the reverse direction, depending upon the possible existence of oxide layers.

3. Large scattering and variations, which could not be explained, were observed for Pt/Ir 90/10. The processes were reversible except for metals that oxidize even under normal environmental conditions.
4. For various materials, adsorptive behavior was similar when other influencing parameters including environmental conditions were kept constant.
5. For steel × 5 Cr Ni 18 9:
   a. The roughness of the surface was not a sufficient measure of the area of the active surface.
   b. The influence of tarnish layers was small.
   c. A small amount of impurities on the surface substantially increased the adsorption on the surface.
6. The adsorption of water and the resulting change in mass are critically influenced by the type of surface cleaning.
7. For anodized surfaces, water adsorption increased considerably with coating thickness.
   a. For anodized surfaces, there was a decrease in water adsorption after 1 or 2 years.
8. The results permitted changes in mass due to the adsorption of gases other than water vapor and the adsorption of aerosols to be neglected.

The author concluded that "for high-precision weighing, e.g., comparison with mass standards, it is possible to make an overall estimate of the effect of adsorption under diverse measuring conditions and for different materials."

[Excerpts taken with permission from *Metrologia*.]

3.2   Stainless Steel Mass Standards

3.2.1   Precision Determination of Adsorption Layers on Stainless Steel Mass Standards — Introduction

Long-term stability of the mass of high-precision mass standards, such as Pt-Ir prototype kilograms, depends essentially on surface adsorption effects.[9,10]

Investigations have been made by Schwartz[9,10] "to determine the adsorption layers on 1-kilogram stainless steel standards as directly and precisely as possible in terms of the influencing factors: relative humidity, material (stainless steel composition), surface cleanliness, roughness, and ambient temperature."[9,10] [Excerpt taken with permissiom from *Metrologia*.]

For this purpose, ellipsometry, an optical method of surface analysis was combined with mass comparison. Ellipsometry directly determines the absolute layer thickness on the surface examined. The weighing method can ascertain changes of the adsorption layers, only if at least two specimens with different geometrical surfaces are compared.

Ellipsometry can be applied under normal ambient conditions and in vacuum, enabling adsorption effects to be studied as a function of air relative humidity as the most important parameter of influence.

3.2.2   Adsorption Measurements in Air

3.2.2.1   Experimental Setup

Two independent computer-controlled measuring instruments, a 1-kg mass comparator and an ellipsometer, were arranged together in a vacuum-tight chamber.

Humidity was adjusted to defined relative humidities between 3 and 77% in the chamber. Alternatively, the chamber could be evacuated to defined pressure between 0.005 and 100,000 Pa.

3.2.2.2   Mass Comparator

The 1-kg mass comparator had a resolution of 0.1 μg and a standard deviation of about 2 μg. The comparator was equipped with an automatic mass exchange mechanism. The operation of the balance was computer controlled as was the data logging, including pressure, temperature, and humidity.

3.2.2.3 Ellipsometer

Except for the light source, the ellipsometer was set up in the immediate vicinity of the mass comparator inside the vacuum chamber.

One series of measurements consisted of scanning of 24 measurement points of the specimen surface. The time of measurement per point was approximately 5 min and the standard deviation of a single measurement corresponded to a variation of less than 0.003 nm or 1% of a water monolayer.

3.2.2.4 Measurement of Air Parameters and Humidity Control

To determine both air density for air buoyancy corrections for the weighings and adsorption isotherms, accurate control and measurement of temperature, pressure, and relative humidity in the vacuum chamber was necesary.

The uncertainty of the temperature measurement was 0.01°C. The relative humidity during the mass comparisons in the weighing chamber was measured using a capacitive humidity sensor calibrated against a dew-point device. An oscillating quartz barometer was used for pressure measurement under normal ambient conditions, with an uncertainty of 5 Pa.

3.2.2.5 Mass Standards and Sorption Artifacts

Commercial 1-kg weights of International Organization of Legal Metrology (OIML) classes E_1 and E_2 were used for the ellipsometric measurements. Two pairs of special 1-kg artifacts were prepared for the mass comparisons. These artifacts were composed of 16 and 8 disks solidly shrunk on carrying rods. The material of the weights and the artifacts were two austenitic steels.

The geometrical surfaces were approximately six or three times those of a 1-kg weight. The surface structures of the mass standards and the artifacts were almost identical. The results of the sorption investigations could be directly transferred to precision mass standards of stainless steel.

The standard uncertainty for the adsorbed mass per area measured by mass comparison was approximately 0.005 µg/cm². Prior to the sorption investigations, all specimens were cleaned and dried by the same procedure:

1. Wiping with a linen cloth soaked with ethanol and diethylether
2. Ultrasonic cleaning in ethanol for 15 min
3. Drying in a vacuum oven at 50°C and 50 Pa for 4 h

3.2.2.6 Summary and Conclusions

The standard uncertainties of the measurements were ≤0.005 µg/cm². The sensitivity of the automatic ellipsometer against variations of adsorption layers was ≤0.001 µg/cm².

Long-term drifts of the adsorption layers still played an important role even months after the specimens had been cleaned.

The long-term drifts were considered to be probably due to chemisorption of water in conjunction with slight growth of the oxide layer on the stainless steel surface.

The sorption behavior of carefully polished stainless steel surfaces was typical of hydrophilic surfaces.

The sorption behavior of precision stainless steel mass standards was mainly influenced by the degree of surface cleanliness. Uncleaned mass standards with an absolute mass of the adsorption layer per surface area of ≥0.7 µg/cm² had sorption-induced mass variations greater by a factor of up to 2.6 relative to clean surfaces.

Ellipsometry "proved to be a valuable technique for the direct, absolute and precise determination of adsorption layers on high-level mass standards ... and appears to be predestined for the precise investigation of the sorption and long-term behavior of prototype kilograms of platinum-iridium."[9]

3.2.3 Sorption Measurements in Vacuum

3.2.3.1 Introduction

Of particular metrological interest are mass determinations in vacuum; however, they require exact knowledge of the sorption-induced mass changes in the transition from normal pressure to vacuum.

Investigations have been undertaken to determine the adsorption layers on 1-kg stainless steel mass standards in vacuum as directly and precisely as possible in terms of relative humidity, surface cleanliness, and stainless steel composition.[10]

Two independent measuring techniques,[9] mass comparison and ellipsometry, were used to make measurements.

3.2.3.2   Results for Cleaned Specimen

Ellipsometric measurements were made directly on commercial 1-kg stainless steel mass standards of OIML accuracy classes E_1 and E_2.

Mass comparisons were made for two pairs of special 1-kg stainless steel artifacts with the same material properties and surface finish.

In two successive cycles, atmospheric pressure was gradually varied by steps of approximately one power of 10. Relative humidity was measured or estimated throughout. The specimens had been cleaned about 7 months before measurements were made.

Meaurements made using two independent measurements, ellipsometry and mass comparison, were strongly correlated, showing clearly that changes in adsorption layers could be identified as a function of pressure and relative humidity.

3.2.3.3   Sorption Isotherms for Cleaned Polished Surfaces

After correction for irreversible changes in high vacuum, a reversible sorption isotherm was derived from which it could be concluded that adsorption layers on stainless steel mass standards change *reversibly* during repeated evacuation and ventilation if sources of contamination are carefully eliminated.

Distinct hysteresis between desorption and adsorption curves was exclusively attributed to the influence of different residual humidities in the vacuum chamber.

3.2.3.4   Factors Influencing Adsorption Isotherms

The influence of steel composition and cleanliness on sorption isotherms was investigated. Sorption measurements in air had already shown that sorption behavior is influenced only insignificantly by other factors, such as ambient temperature and surface roughness.

3.2.3.5   Influence of Steel Composition

A 1-kg mass standard (E_1) and two 1-kg sorption artifacts made of austenitic material, X 2 NiCrMoCu 25 20, were used to study the influence of steel composition.

Taking into consideration the standard uncertainties of measurements (≤ 0.005 $\mu g/cm^2$), the difference between the coefficients for reversible adsorption isotherms for the two materials seemed not to be significant.

3.2.3.6   Influence of Surface Cleanliness

The influence of relatively large surface coverings (≥ 0.8 $\mu g/cm^2$) on sorption behavior was investigated using uncleaned mass standards and artifacts.

It was concluded that stainless steel mass standards with these coverages showed sorption-induced mass variations greater than those of a cleaned surface by a factor of about 2.5. This result was analogous to measurements showing variation of adsorption with relative humidity at normal pressure.[10]

3.2.3.7   Summary and Conclusions

By using mass comparison and ellipsometry, adsorption layers on 1-kg stainless steel mass standards and artifacts were determined directly and precisely as a function of pressure in the range of 0.005 to 0.1 Pa.

For clean polished surfaces, reversible sorption isotherms with a coefficient of 0.024 ± 0.005 $\mu g/cm^2$ due to the transition from normal pressure (at a relative humidity of 3%) to the pressure of 0.1 Pa were found.

For uncleaned surfaces, this coefficient rose by a factor of about 2.5.

Independent of surface cleanliness, at pressures greater than or equal to 0.005 Pa and less than 0.1 Pa, irreversible adsorption with a constant adsorption rate was observed. This effect was attributed to continuous condensation of oil particles originating from the turbomolecular pump used.

The investigations showed that ellipsometry (a method of analysis that can be applied both under normal ambient conditions and in vacuum) is a valuable technique giving a noncontact, absolute, and precise determination of adsorption layers on well-polished mass standards.

3.2.4 Effect of Environment and Cleaning Methods on Surfaces of Stainless Steel and Allied Materials

3.2.4.1 Introduction

The surfaces of the alloys Immaculate 5, En58AM, and Nimonic 105 have been studied by Seah et al.[11] by XPS after cleaning using simple washing methods and after contamination in the laboratory environment for periods of up to 156 days, using filtered and unfiltered laboratory air.

All surfaces were covered by a thin oxide layer under a thin layer of carbonaceous contamination. The carbonaceous layer was considered to be of atmospheric origin.

These three materials were considered to be possibilities for mass standards.

Immaculate 5 (also known as AISI Type 310) is a high-nickel, high-chromium steel used for quality mass standards.

En58AM (also known as AISI Type 303) is a machinable grade of 18/8 stainless steel widely used for mass standards.

Nimonic 105 is a high-quality nickel-based alloy that was developed for its resistance to oxidation, for use in jet engines, for example.

3.2.4.2 Results and Conclusions

1. The results of the study, showing the same general behavior for the three materials, were relevant to the choice of possible materials for mass standards.
2. After being washed initially in boiling water, the masses were exposed to the environment leading to an oxide film of fairly stable thickness.
3. A carbonaceous layer of contamination grew on top of the oxide film, increasing with time according to a parabolic or logarithmic law.
4. The carbonaceous layer contained carboxyl groups with adsorbed water layers on top.
5. Maintaining the masses in a static environment with an air leak through a filter largely removed the growth of the carbonaceous contamination.
6. Leaving the masses exposed to moving laboratory air led to significant contamination in a period of 1 month.
7. Boiling water and washing in Micro™ (an emulsion of anionic and non-ionic surfactants, stabilizing agents, alkalis, sequestering agents, and builders, in water) effectively removed the carbonaceous contamination to the level of a 1-day exposure; boiling water was less successful.
8. Micro seemed to be the best of the washing materials tried, but may also dissolve cations.

It was recommended that

1. After manufacture and before use, stainless steel mass standards should be cleaned in boiling water for 5 min.
2. The standards should be kept in closed glass containers with a filtered vent.
3. If masses become contaminated, Micro can be used as an effective cleaner, followed by an ultrapure water rinse. The action of Micro should be confirmed by weighing experiments.
4. If a mass loss due to leaching of cations is indicated by such experiments, then simpler, pure surfactant solutions should be used.

3.2.5   Studies of Influence of Cleaning on Stability of XSH Alacrite Mass Standards

3.2.5.1   Introduction

The Institut National de Metrologie (INM) of France produced secondary mass standards of XSH alacrite and adopted a mechanical cleaning method using ethanol and isopropanol. The composition (mass fraction) of XSH alacrite is 0.20 chromium, 0.15 tungsten, 0.10 nickel, 0.001 carbon, and the remainder is cobalt. Its density is 9150 kg/m^3, higher than that of stainless steel. The material was originally developed for use in aircraft.

The INM spent about 2 years investigating the stability of its alacrite secondary standards and, in particular, studied the effect of cleaning.

The use of the gravimetric (weighing) method to study the effect of cleaning using alcohols on two standard kilograms of XSH alacrite was reported in 1994/1995 by Pinot.[12]

The alacrite kilograms, designated 07 and 09, were made from the same bar, machined, adjusted, and polished in the same way and at virtually the same time. The density of the two kilograms was identical, 9148.39 kg/m^3.

The national secondary kilogram, 07, did not leave INM until 1987 when it was calibrated at BIPM. The national tertiary standard, 09, was sent to a number of laboratories between 1987 and 1992 as part of a comparison campaign.

3.2.5.2   Investigation of Stability

In an investigation of the characteristics of the alacrite, especially stability, two mass standards were sent to BIPM and alacrite specimens were sent to Instituto di Metrologia "G. Colonetti" (IMGC).

The alacrite standards were cylindrical in shape, with height equal to diameter (52 mm), of surface area 127 cm^2. They were polished at INM, and their density was determined at INM by hydrostatic weighing.

Each kilogram, resting on treated chamois leather covered by filter paper, was stored in a stainless steel container. Filtered air could enter through the tops of the containers, which were deposited inside a safe installed in the air-conditioned clean room of the mass comparator.

The method of cleaning involved wiping the surface of the alacrite standards fairly briskly with optical paper impregnated with 99.45% pure ethyl alcohol and then with 99.7% pure isopropyl alcohol. The standards were then dried naturally by exposure to air. After alcohol cleaning, they were wiped using dry optical paper to remove whitish marks on the surface visible to the unaided eye.

3.2.5.3   INM Mass Comparator

To relate the alacrite reference standards to the French national platinum-iridium standard, a mass comparator was constructed at INM. The comparator consisted of a single-pan constant-load balance, the beam of which rested on sapphire knife edges and planes.

The difference in apparent mass between two or more mass standards was determined from differences in servo currents required to hold the beam in a horizontal position. For each series of comparisons, a separate measurement was made of the sensitivity of the comparator.

All weighings were carried out in air, so a correction for the buoyancy of the air was obtained by an indirect method that involved calculation of air density (using the BIPM equation[13]) from measurements of temperature, pressure, relative humidity, and carbon dioxide content.

Two platinum-iridium reference standards were used in this study. Kilogram JM15, made in 1975, was found to be quite stable, with an increase in mass of 9 μg in 8 years. The mass of Kilogram 13, made in 1884, decreased by 19 μg in 8 years after two cleanings using the BIPM method.[6]

Kilogram 07 was also calibated against the French national standard kilogram, but Kilogram JM15 was used as a reference for all the INM mass values in these studies. There was no access to the value of mass of an object within about 10 h of cleaning.

3.2.5.4 Results and Conclusions

The results for the XSH alacrite kilograms that had not been cleaned with alcohol for several years led to the following conclusions:

1. During the weeks following the cleaning process, the increase in mass was a logarithmic function of time.
2. The surface mass loss immediately after alcohol cleaning was 0.3 μg/cm^2 for Kilogram 09 and 0.6 μg/cm^2 for Kilogram 07.
3. In the hours following alcohol cleaning, the recontamination rate was at least of the order of 0.1 μg/cm^2/day.
4. In the weeks or months following the first cleaning, a second alcohol cleaning resulted in a smaller decontamination of approximately 0.16 μg/cm^2.
5. Several months after alcohol cleaning, the mass of a standard finally tended toward a stable value, which might differ significantly from one cleaning to another.

The two kilograms behaved differently with regard to alcohol cleaning.
An overall conclusion was that:

It is perfectly clear that the mechanical cleaning of alacrite standards using ethanol and isopropanol resulted in considerable mass instability for several months, leading to a final mass value that is stable but not reproducible.

More than six months after alcohol cleaning, the alacrite kilograms show good stability, with their mass changing at less than 2 micrograms a year.

It is obvious that this type of cleaning should not be used systematically but only occasionally.[12] [Excerpt taken with permission of *Metrologia*.]

3.2.5.5 BIPM Cleaning/Washing Method

Following the work in which alcohol was used for cleaning, the cleaning/washing method recommended by the BIPM on platinum-iridium standards[2] was applied to alacrite standards.

The BIPM method involves wiping the mass standard with a treated chamois leather impregnated with a mixture of equal volumes of ethanol and ethyl ether, and then washing with a jet of steam from doubly distilled water. Droplets condensing on the surface of the standard were removed with a jet of pressurized nitrogen. The temperature could have reached as high as 70°C during the operation.

A gravimetric study was made of the effect of the BIPM method of cleaning and washing XSH alacrite kilogram standards, and XPS analysis was made of surface contaminations as a function of the cleanings.

The initial results of these studies showed that the BIPM method had a greater cleaning power than that using alcohols and that there is no short-term instability. However, it was found that during the first year following this type of cleaning, the relative mass of the standard increased by about 5×10^{-9} and the reproducibility of the method could not be better than 4×10^{-9}.

References

1. Cumpson, P. J. and Seah, M. P., Stability of reference masses. IV: Growth of carbonaceous contamination on platinum-iridium alloy surface, and cleaning by UV/ozone treatment, *Metrologia*, 33, 507, 1996.
2. Girard, G., The washing and cleaning of kilogram prototypes at the BIPM, Bureau International des Poids et Mesures, Sèvres, France, 1990.
3. Girard, G., The third verification of national prototypes of the kilogram (1988–1992), *Metrologia*, 31, 317, 1994.
4. Ikeda, S. et al., Surface analytical study of cleaning effects and the progress of contamination on prototypes of the kilogram, *Metrologia*, 30, 133, 1993.

5. Quinn, T. J. and Picard, A., Surface effects on Pt-Ir mass standards, BIPM Document CCM/93-6, 1993.

6. Cumpson, P. J. and Seah, M. P., Stability of reference masses. I: Evidence for possible variations in mass of reference kilograms arising from mercury contamination, *Metrologia*, 31, 21, 1994.

7. Cumpson, P. J. and Seah, M. P., Stability of reference masses. III: Mechanism and long-term effects of mercury contamination on platinum-iridium mass standards, *Metrologia*, 31, 375, 1994/95.

8. Kochsiek, M., Measurement of water adsorption layers on metal surfaces, *Metrologia*, 18, 153, 1982.

9. Schwartz, R., Precision determination of adsorption layers on stainless steel mass standards by mass comparison and ellipsometry. Part I: Adsorption isotherms in air, *Metrologia*, 31, 117, 1994.

10. Schwartz, R., Precision determination of adsorption layers on stainless steel mass standards by mass comparison and ellipsometry. Part II: Sorption phenomena in vacuum, *Metrologia*, 31, 129, 1994.

11. Seah, M. P. et al., Stability of reference masses. II: The effect of environment and cleaning methods on the surfaces of stainless steel and allied materials, *Metrologia*, 31, 93, 1994.

12. Pinot, P., Stability of mass standards made of XSH alacrite: gravimetric study of the influence of cleaning, *Metrologia*, 31, 357, 1994/95.

13. Giacomo, P., Equation for the density of moist air (1981), *Metrologia*, 18, 33, 1982.

4

Cleaning of
Mass Standards

4.1 Introduction

G. Girard of BIPM has reviewed cleaning of platinum-iridium prototypes at Bureau International des Poids et Mesure (BIPM).[1] In this section we shall draw extensively from Girard's paper, with permission from the Bureau International des Poids et Mesures.

The international platinum-iridium prototype kilogram was first used in 1888. It was then used in 1939 and again for the second periodic verification of national kilogram prototypes in 1946. It was used most recently for the third periodic verification in 1988–1989. For the first verification between 1899 and 1910, the international prototype had not been involved and only a small number of other prototypes had been involved.

The first 40 platinum-iridium prototypes were fabricated from a Johnson–Matthey alloy and were compared among themselves in numerous combinations between 1882 and 1889. Then they were compared separately with the International Prototype Kilogram, designated $\mathfrak{K}$.

The prototypes were washed in ethanol vapor and steam before they were used in definitive studies. They were then dried in the presence of anhydrous potassium hydroxide under a bell jar.

After 1889, preliminary washing was not used when prototypes were compared. Rather, either simple dusting or, in some cases, wiping with solvent took the place of the washing procedure.

In 1939, the prototypes were cleaned by rubbing all surfaces with a chamois leather that had first been soaked in ethanol and then in redistilled petrol.

Subsequently, the question of cleaning mass standards was reviewed and a study of the original washing procedures was carried out. As a result, a procedure combining solvent cleaning followed by steam washing was developed.[2] The procedure was referred to as a "cleaning and washing" (*nettoyage-lavage*). Solvent cleaning is done with a mixture of equal parts ethanol and ether.

In 1946, in a second verification of national prototypes, the international prototype was compared with its official copies and with the BIPM working standards after they had all been cleaned and washed. In the following years, a cleaning and washing procedure has been used on all prototypes returned to BIPM for verification.

4.2 Solvent Cleaning and Steam Washing (*Nettoyage-Lavage*)[1]

The following is the treatment of solvent cleaning and steam washing procedure now used at BIPM.

4.2.1 Solvent Cleaning

For cleaning, chamois leather, which has been soaked in a mixture of equal parts of ethanol and ether for 48 h, is used. The absorbed solvent is wrung out of the chamois leather after the soaking. This preliminary soaking removes impurities that might otherwise be deposited on the mass standards; a second and a third soaking are then required to clean the chamois leather sufficiently.

To clean the standards, clean chamois leather that has been saturated with the ethanol–ether mixture is used to rub each standard over its entire surface "fairly hard" by hand (with an estimated applied pressure of the order of 10 kPa). (Between 1946 and 1990, a mixture of benzene and ether was used.)

4.2.2   Steam Washing

Steam washing follows the solvent cleaning. Doubly distilled water is heated to boiling in a round-bottomed Pyrex™ flask of 1-l capacity filled three quarters full. As the water boils, steam passes through a tube, which terminates in a small orifice (about 2 mm in diameter) directed toward the mass standard.

The mass standard is placed on a disk of platinum-iridium in a shallow bowl at the top of a tripod. The upper part of the tripod can both turn about a vertical axis and be displaced vertically by several centimeters.

The steam jet (with the water in the flask boiling away at about 0.5 l/h) is first pointed toward the upper surface of the standard, which can be rotated about the vertical axis. The steam jet is successively directed to all parts of the upper surface top. After a few minutes, the prototype is rotated and displaced vertically and the jet is swept about the cylindrical surface. A distance of about 5 mm is maintained between the tapered glass tube and the surface of the prototype throughout these operations.

The steam washing of the cylindrical surface lasts about 15 to 20 min. Water that has condensed on the surface of the prototype and has not run off is absorbed by putting an edge of high-purity filter paper in contact with each drop, and the water is allowed to flow into the filter paper by capillary action. Alternatively, a jet of clean gas can blow away the water.

The standard is then inverted (resting on the base that has just been cleaned), the steam washing is continued at the upper surface, and the steam washing ends with a second washing of the cylindrical surface. The above procedure for steam washing a prototype takes about 50 min. The disk upon which the prototype rested was previously cleaned with solvent and steam-washed by the same method used for a mass standard.

The prototype is subsequently stored beneath a bell jar on its support. No chemical desiccant is used in the bell jar.

4.2.3   Effect of Solvent Cleaning and Steam Washing

For at least the past 25 years, all prototypes going to BIPM for verification have been compared with the working standards of BIPM, then solvent-cleaned and steam-washed, and then finally compared again with the working standards. The mass of a collection of platinum kilograms as a function of time is shown in Figure 4.1.[3] It was not until 1973 when the NBS-2 balance was used to make comparison that BIPM was able to determine within a few micrograms (and more precisely recently) the effect of the treatment.

The various prototypes are indicated by number on the figure. A straight line (with a slope of –1 µg/year) is drawn through the points labeled ●. Points labeled +, representing the international prototype and its official copies, seem also to follow the line. The points labeled ○ designate prototypes of relatively poor surface. To isolate the effect of the treatment, other factors must be considered: conditions under which the prototype is stored; the state of its surface; and frequency of use of the prototype.

The fact that the straight line (when extended) does not pass through the origin is consistent with the hypothesis that surface contamination is more rapid just after solvent cleaning and steam washing.

The results of a 1989 study carried out on the international prototype and prototypes Nos. 7, 67, and 73 indicated that the mass of prototypes increased by 1 µg/month during the first 3 or 4 months after solvent cleaning and steam washing. Consequently, national prototypes should, after their return from BIPM to their respective laboratories, be solvent-cleaned and steam-washed again before they are compared to other standards. The mass value measured at BIPM, the reference mass of the prototype, should be thus retrieved.

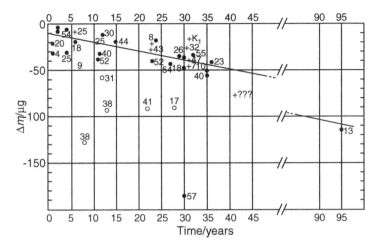

FIGURE 4.1 Mass of a collection of platinum kilograms as function of time.

The effectiveness and reliability of the cleaning and washing process has been assumed to be established by experiments carried out at the BIPM in 1974 on a newly adjusted prototype. About 90% of the surface contamination appears to be removed by a single cleaning and washing. A second treatment appears to remove only a few micrograms, and it was predicted that a third treatment would have no practical effect.

On the basis of experimental results, it was decided that national prototypes sent to BIPM for the third periodic verification of national kilogram prototypes would receive two cleanings and washings.

4.3 Summaries of National Laboratory Studies Related to Cleaning[4]

4.3.1 Cleaning at National Physical Laboratory, United Kingdom (NPL)

The mass stability of kilogram No. 18, the British National Standard made in 1884, has been closely monitored by NPL. After being cleaned and washed at BIPM in 1985, the mass increase of the prototype was monitored for more than 1 year.

After 1990, a best-fit curve fitted to the BIPM data was obtained that expressed the mass change as a function of time:

$$M_t = M_o + 0.356097 t^{0.511678}, \tag{4.1}$$

where

M_t = mass of the prototype at time t after cleaning and washing
M_o = mass of the prototype at the time of cleaning and washing
t = elapsed time in days

Using Eq. (4.1), the mass value of kilogram No. 18 after 6 years since the last cleaning was predicted to within 1.5 μg.

NPL concluded, "if the mass changes after cleaning/washing could be shown to be reproducible, then it should be possible to calculate values by extrapolation to within a few μg for a period of up to ten years provided that the storage conditions can be carefully controlled."

4.3.2   Cleaning at Institut National de Metrologie, France (INM)

INM started a program to verify the efficiency of the BIPM cleaning/washing procedure. The mass of prototype JM15 (made in 1975) increased 9 µg in 8 years. Prototype No. 13 (made in 1889) was cleaned twice; its mass increased by 19 µg in the 8 years since the last BIPM calibration.

4.3.3   Cleaning at National Research Laboratory of Metrology, Japan (NRLM)

NRLM investigated cleaning and contamination of specimens diamond-cut from prototype material. On the surfaces of as-received specimens, gas contaminants (carbon, oxygen, nitrogen) and metal contaminants (copper, mercury) were detected.

Argon-ion sputtering, steam-jet cleaning, and ultrasonic cleaning with solvents were compared. Ultrasonic cleaning by solvents was found to be more effective than steam-jet cleaning in reducing carbon deposits and the adsorption of water. The cleaner the surfaces, the higher the rate of contamination; contamination levels converged to a common value after a 6-month exposure period. Contamination proceeded much faster in the first month than in the succeeding 5 months.

Contaminants as hydrocarbons from the ambient air were identified. The origin of mercury and copper surface contamination was identified to be cutting oil. Oxidation of platinum was not observed; a certain oxidation of iridium was observed, but the oxidized metal was not dissolved by steam.

NPL also carried out a study of the contamination of Pt-Ir by mercury.

4.4   Cleaning of Stainless Steel Mass Standards

4.4.1   Cleaning Procedures Investigated by Weighing and Ellipsometry

To ascertain the most effective cleaning procedure for 1-kg stainless steel mass standards, Schwartz and Glaser[5] of the Physikalisch-Technische Bundesanstalt (PTB) carried out comparative studies of various cleaning procedures. Two independent analytical techniques, weighing and ellipsometry, were used simultaneously.

Five mass standards made of austenitic stainless steel (1.4539/X 2 NiCrMoCu 25 20) with a density of 8040 kg/m³ (8.040 g/cm³) were used. The mass standards were produced as weights of the OIML class E_1, with highly polished surfaces. The surface area of each of the standards was about 151 cm². The nominal value of the mass of each mass standard was 1 kg; the mass difference between the cleaned weights was not larger than 0.5 mg.

The weighings were performed on an automatic 1-kg mass comparator that had a standard deviation of about 1 µg. The ellipsometric measurements were performed with an automatic self-nulled ellipsometer. The mass standard MB was chosen to serve as a reference standard for the weighings. The maximum time interval between the weighings and the ellipsometric measurements was about 8 h.

4.4.1.1   Cleaning Procedures

The following cleaning procedures were investigated:

1. Washing in a Soxhlet apparatus
2. BIPM cleaning/washing procedure
3. Ultrasonic cleaning in ethanol
4. Ultrasonic cleaning in bidistilled water

"In order to apply the different procedures several times, the weights were wiped with uncleaned chamois leather (L)." The wiping procedure increased the mass similarly to a mass increase due to a long-term drift.

In addition to the cleaning procedures, two precautionary different methods of drying the weights in a special oven under vacuum conditions (pressure equal to 50 Pa) were investigated:

1. Strong drying at 130°C for 2 h
2. Moderate drying at 50°C for 4 h

4.4.1.2 Results

1. There was a strong correlation between the gravimetric (weighing) results and the ellipsometric ones.
2. Wiping of a clean 1-kg stainless steel standard with uncleaned chamois caused a mass increase of about 85 µg.
3. Wiping uncleaned 1-kg weights caused smaller mass increases of 30 to 70 µg.
4. A careful cleaning of weight polluted by the wiping, using one of the methods (Soxhlet, BIPM, or ultrasonic cleaning in ethanol), caused uniform mass changes of –70 to –100 µg.
5. Ultrasonic cleaning in ethanol was found to be clearly much more effective than ultrasonic cleaning in water.
6. Ultrasonic cleaning in ethanol after a preceding cleaning with the BIPM or Soxhlet procedure caused small further mass reductions of 15 to 20 µg. No further cleaning effect was observed when ultrasonic cleaning in ethanol was followed by the BIPM method.
7. Strong drying of cleaned surfaces in a vacuum oven led to a reversible mass increase of 15 µg.
8. Moderate drying in a vacuum oven left mass unchanged.

4.4.1.3 Conclusions

1. Ultrasonic cleaning in ethanol seemed to be slightly more efficient for cleaning stainless steel mass standards than the BIPM method.
2. The effect of washing with ethanol in a Soxhlet apparatus was found to be comparable to the effect of the BIPM cleaning/washing method.
3. Moderate drying of a stainless steel mass standard, if necessary, would be preferable to strong drying.

4.4.2 Cleaning of Stainless Steel Mass Standards at BIPM[5]

Bonhoure[2] attempted to apply the same cleaning procedures to stainless steel weights that he found to be so effective on platinum-iridium mass standards. After each step of the cleaning procedures, the mass of the artifacts was measured. A loss of about 100 µg was caused by the final steam cleaning although there was no further effect from successive steam cleanings.

Bonhoure found also that single-piece stainless steel weights that had been used in hydrostatic measurements might change their mass value by an appreciable amount after steam cleaning. These mass changes were not permanent, but recovery could take months.

As a result of the above experiences, it is the practice of the BIPM to clean stainless steel weights exactly as platinum-iridium prototypes are cleaned except that steam cleaning is omitted.

4.4.3 Cleaning of Stainless Steel Mass Standards at NIST[6]

NIST (formerly NBS) used vapor degreasing in inhibited 1,1,1-trichloroethane[7] as the final step in cleaning stainless steel weights. The vapor degreasing method and the BIPM cleaning/washing method were used on steel spheres the diameters of which were then measured optically. The standard deviation for the dimensional measurements on spheres that had been vapor-degreased was lower than the standard deviation for spheres that had been cleaned by the BIPM cleaning/washing method. The reason for the difference in standard deviations and for a systematic difference in dimensional measurements for the two cleaning methods was not known.

Vapor degreasing was found to be an acceptable method of cleaning. A set of stainless steel weights (stackable) with nominal mass values of 1 kg was found to be stable under numerous vapor-degreasing operations over a period of 1 year.

The kilogram weight set had double the surface area of a stainless steel weight designated D_2. D_2 and a weight designated CH-1 were single-piece weights, roughly cylindrical in shape except for a lifting knob on the upper surface, manufactured by the Troemner Company of Philadelphia from an austenitic alloy similar to 18-8 stainless steel. D_2 was used extensively in calibration and research work at NIST.

CH-1 (acquired from the Chyo Company of Kyoto, Japan in 1983) was manufactured from an austentic stainless steel alloy with the following composition: 25.1% Ni, 29.9% Cr, 2.2% Mo, 1.45% Mn, 0.53% Si, 0.2% Cu, 0.07% C, and 0.019% P. The metal was vacuum-melted before being machined. Three vapor degreasings of CH-1 did not result in noticeable changes in mass.

References

1. Girard, G., The procedure for cleaning and washing platinum-iridium kilogram prototypes used at the Bureau International des Poids et Mesures, BIPM, Sèvres, France, 1990.
2. Bonhoure, A., *BIPM Proc.-verb. Com. Int. Poids et Mesures*, 20, 171, 1946.
3. Girard, G., Third Periodic Verification of National Prototypes of the Kilogram, BIPM, May, Sèvres, France, 1993.
4. Plassa, M., Working Group "Mass Standards" report to CCM on the activity 1991–1993, CCM/93-7, unpublished report.
5. Schwartz, R. and Glaser, M., Cleaning procedures for stainless steel mass standards, investigated by weighing and ellipsometry, CCM, 1993, unpublished report.
6. Davis, R. S., Recalibration of the U.S. National Prototype Kilogram, *J. Res. Natl. Bur. Stand.* (U.S.), 90, 263, 1985.
7. Bowman, H. A., Schoonover, R. M., and Carroll, C. L., Reevaluation of the densities of the NBS silicon crystal standards, National Bureau of Standards (U.S.), NBSIR 75–768, 1975.

From Balance Observations to Mass Differences

5.1 Introduction

If one views the operation performed on balance observations to derive the mass difference between nominally equal objects, one would conclude there are only two types of balances. From an engineering point of view there is the very old equal-arm-type balance and the spring balance. The latter has only one pan or load hook and the operator views the stretch of the spring, which has been calibrated in mass units. The two-pan equal-arm balance is used for both transposition and substitution weighing, and the spring balance is used for substitution weighing.

From these two very different weighing devices one can generalize about the manipulation of balance observations to obtain the difference in mass between two nominally equal objects. It is the difference in mass that is required by the weighing equation discussed in Chapter 17 and referred to as δ and as Y_i in Chapter 8 dealing with weighing designs.

All modern weighing devices (magnetic force compensation balances, load cells, one-pan-two-knife balances, etc.) can be viewed as either the spring balance or the equal-arm balance or a variant and the observations can be manipulated in the same manner as for these two instruments.

The direct determination of mass (direct weighing) by weighing an object on a modern electronic balance has been treated as a separate matter in Chapter 28. In principle, direct weighing on the spring balance and direct weighing on a modern electronic balance are the same in that a casual observer might wonder where the mass standard is located.

5.2 Determination of Mass Difference

Consider now the obtaining of the mass difference between two objects, A and B, of nominally equal mass, on the so-called equal-arm balance. It would be an extremely rare occurrence if the arms were actually equal; therefore, the weighing solution must take into account unequal balance arms. Figure 5.1 depicts an equal-arm balance.

Six torque equations commonly referred to as a double transposition weighing are now presented and solved for the difference in mass between objects A and B, $(A - B)$:

$$A'L_1g = B'L_2g + k\theta_1g \tag{5.1}$$

$$B'L_1g = A'L_2g + k\theta_2g \left(A \text{ and } B \text{ are transposed}\right) \tag{5.2}$$

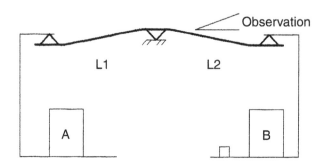

FIGURE 5.1 Illustration of an equal-arm balance.

$$\left(B' + \Delta'\right)L_1 g = A'L_2 g + k\theta_3 g \left(\Delta \text{ added to left pan}\right) \qquad (5.3)$$

$$\left(A' + \Delta'\right)L_1 g = B'L_2 g + k\theta_4 g \left(A \text{ and } B \text{ transposed}\right) \qquad (5.4)$$

$$A'L_1 g = B'L_2 g + k\theta_5 g \left(\Delta \text{ removed}\right) \qquad (5.5)$$

$$A'L_1 g = B'L_2 g + \Delta'L_2 g + k\theta_6 g \left(\Delta \text{ added to right pan}\right), \qquad (5.6)$$

where

A'	$= A[1 - (\rho_a/\rho_A)]$
B'	$= B[1 - (\rho_a/\rho_B)]$
Δ'	$= \Delta[1 - (\rho_a/\rho_\Delta)]$
L_1, L_2	$=$ lengths of the balance arms
$\theta_1, ..., \theta_6$	$=$ balance observations (beam angle to the horizon)
g	$=$ local acceleration due to gravity
k	$=$ a constant of proportionality
Δ	$=$ a small weight of known mass used to calibrate the balance response in mass units, commonly referred to as a sensitivity weight (Δ is unlabeled in Figure 5.1)

Four new equations are now written:

Subtracting Eq. (5.2) from Eq. (5.1):

$$A' = B' = k\left(\theta_1 - \theta_2\right)\Big/\left(L_1 + L_2\right). \qquad (5.7)$$

Subtracting Eq. (5.3) from Eq. (5.2):

$$L_1 = k\left(\theta_3 - \theta_2\right)\Big/\Delta'. \qquad (5.8)$$

Subtracting Eq. (5.3) from Eq. (5.4):

$$A' - B' = k\left(\theta_4 - \theta_3\right)\Big/\left(L_1 + L_2\right). \qquad (5.9)$$

Subtracting Eq. (5.5) from Eq. (5.6):

$$L_2 = k\left(\theta_5 - \theta_6\right)\big/\Delta'. \tag{5.10}$$

Substituting the sum of Eqs. (5.8) and (5.10) for $(L_1 + L_2)$ in Eqs. (5.7) and (5.9) results in

$$A' - B' = \Delta'\left(\theta_1 - \theta_2\right)\big/\left(\theta_3 - \theta_2 + \theta_5 - \theta_6\right) \tag{5.11}$$

and

$$A' - B' = \Delta'\left(\theta_4 - \theta_4\right)\big/\left(\theta_3 - \theta_2 + \theta_5 - \theta_6\right). \tag{5.12}$$

Adding Eqs. (5.10) and (5.11) and dividing by 2, the mean value of $(A' - B')$ is

$$A' - B' = \left(\Delta'/2\right)\left(\theta_1 - \theta_2 + \theta_4 - \theta_3\right)\big/\left(\theta_3 - \theta_2 + \theta_5 - \theta_6\right). \tag{5.13}$$

This solution for $A' - B'$ is called a double-transposition weighing.

Before proceeding, it is useful to consider the above weighing operation in detail.

The balance response is quite nonlinear for angular departures from the gravitational horizon ($\theta_1, \ldots, \theta_6$) greater than 1° (cosine θ error). As a practical matter, the balance operation is restricted to ±1° of angle and consequently Δ must be small in comparison to A and B if the beam displacement is to remain within the imposed limit.

The weight in air of objects A and B (conventional value) are adjusted to be close to each other. This method requires Δ to be placed, in turn, on each balance pan for a total 2Δ beam displacement. An alternative one-pan placement, and therefore a less restrictive approximation method, is discussed later.

When the temporary use of a bubble level on the beam is impractical to determine when the beam is parallel to the horizon, pan swing can be used.[1] Pan swing is induced and balance oscillation is observed for perturbation (motion coupling). Little or no perturbation is observed when the beam is parallel to the horizon and mid-scale should be adjusted to this position. Observing scales are best engraved without negative numbers, i.e., scales engraved as $-10 \ldots 0 \ldots 10$, for example, should be avoided.

Equal-arm balances are functional over a wide load range; unfortunately, the center of gravity and therefore the balance sensitivity change with load (beam flexing). This characteristic might dictate that the mass of Δ be proportional to pan loading. Although pan loading might change the arm ratio, $L_1{:}L_2$, with this method it is of no significant consequence. Although the beam is symmetrical and both arms are loaded equally, changing heat distribution in the beam causes zero drift and this method minimizes the effect of the linear drift,[2] as discussed below.

Linear balance drift is now assumed and the drift units are designated DU. The drift begins at time zero with a balance observation of θ_1 and proceeds uniformly with one drift unit between successive observations. The balance responses are θ_1, $\theta_2 - 1\text{DU}$, $\theta_3 - 2\text{DU}$, $\theta_4 - 3\text{DU}$, $\theta_5 - 4\text{DU}$, and $\theta_6 - 5\text{DU}$. One can include the drift in the above double-transposition solution for $A' - B'$ as follows:

$$A' - B' = \frac{\left(\Delta/2\right)\left[\theta_1 - \left(\theta_2 - 1\text{DU}\right) - \left(\theta_3 - 2\text{DU}\right) + \left(\theta_4 - 3\text{DU}\right)\right]}{\left(\theta_3 - 2\text{DU}\right) - \left(\theta_2 - 1\text{DU}\right) + \left(\theta_5 - 4\text{DU}\right) - \left(\theta_6 - 5\text{DU}\right)} \tag{5.14}$$

For this method, the drift units in the numerator and in the denominator sum to zero, yielding the original expression for $A' - B'$.

Double-transposition weighing has two important virtues. Foremost is that $A - B$ is the average of two measurements, and consequently the balance standard deviation can be divided by the square root

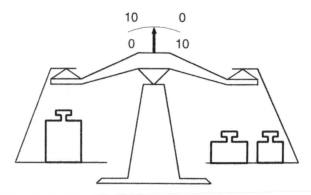

FIGURE 5.2 Illustration of balance sense.

TABLE 5.1   Weighing Operations and Indications

Weighing	Left Pan Load	Right Pan Load	Observation	Lower Scale	Upper Scale
1	A	B	θ_1	4	6
2	B	A	θ_2	6	4
3	$B + \Delta$	A	θ_3	3	7
4	$A + \Delta$	B	θ_4	1	9
5	A	B	θ_5	4	6
6	A	$B + \Delta$	θ_6	7	3

of 2, thereby reducing the measurement uncertainty with little extra work. Second, the present scheme minimizes the effect of balance drift on the measurement.

The above expression for the mass difference $A' - B'$ is all that is required for transposition weighing. However, the intermediate steps have been shown to aid in the description of an approximation method, which follows later, where it is assumed that $L_1 = L_2$, θ_5 and θ_6 are not observed, and Δ is not transposed. Before leaving the accurate method described above, it is useful to perform a thought experiment regarding balance sense; see Figure 5.2.

The balance of Figure 5.2 has two observing scales, the balance arms are equal in length, and the mass of weight A on the left pan is identical to the mass of the aggregate, B, on the right pan. Therefore, the pointer indicates 5 on both observing scales.

Now assume that the mass of A is 1 scale division greater than B. The lower scale under this condition will indicate 4 and the upper scale will indicate 6. The loads are now transposed and the lower scale indicates 6 and the upper scale indicates 4.

Next, a small weight, Δ, is placed on the left pan with B and it is noted that the beam has deflected 3 units on each scale. That is, the lower scale indicates 3 and the upper scale indicates 7. This process is continued until all the weighing operations described above have been performed as indicated in Table 5.1.

Substituting θ_1 through θ_6 for both scales into the equation for $A' - B'$, two equal results are obtained:

$$\text{Lower Scale:} \quad A' - B' = \left(\Delta/2\right)\left[\left(4 - 6 + 1 - 3\right)\big/\left(3 - 6 + 4 - 7\right)\right]$$

$$= \Delta/3$$

$$\text{Upper Scale:} \quad A' - B' = \left(\Delta/2\right)\left[\left(6 - 4 + 9 - 7\right)\big/\left(7 - 4 + 6 - 3\right)\right]$$

$$= \Delta/3$$

For this method it does not matter which scale is used to make the weighing observations. This will not be the case for the approximation method mentioned earlier and now described.

The cardinal features of the approximation method are the conservation of on-scale range, less work, and its asymmetry. That is, it is somewhat less restrictive in load trimming to accommodate Δ without going off scale. Its asymmetry may bias the measured mass difference, $A - B$, when significant zero drift is present. As discussed later, it can be corrected.

If the balance arms are nearly identical in length we can assert that the ratio L_1/L_2 is approximately equal to 1. Let L denote the length of each arm of the balance. Solving Eqs. (5.1) through (5.3) using L in place of L_1 and L_2 in the determination of the mass difference $A - B$, the following approximate relation applies:

$$A' - B' \cong \frac{1}{2}\left(\theta_1 - \theta_2\right)\Delta \big/ \left(\theta_3 - \theta_2\right). \tag{5.15}$$

Depending on in which pan Δ is placed, the result may indicate a decrease in mass (balance sense). This anomaly is prevented by using the absolute value of $(\theta_3 - \theta_2)$, i.e., $|\theta_3 - \theta_2|$.

A similar approximation is obtained using another form of double-transposition weighing in which the objects A and B are transposed but Δ is not, the first four weighings of Table 5.1.

The solution for the four equations is

$$A' - B' \cong \Delta'\left(\theta_4 - \theta_3 + \theta_1 - \theta_2\right) \big/ \left[4\left|\left(\theta_3 - \theta_2\right)\right|\right]. \tag{5.16}$$

Substituting into Eq. (5.16) the first four scale observations for the lower scale:

$$A' - B' \cong \Delta\left(1 - 3 + 4 - 6\right)\big/\left[4\left|\left(\theta_3 - \theta_2\right)\right|\right] = -\Delta/3.$$

Substituting into Eq. (5.16) for the upper scale:

$$A' - B' \cong \Delta\left(9 - 7 + 6 - 4\right)\big/\left[4\left|\left(\theta_3 - \theta_2\right)\right|\right] = \Delta/3.$$

Clearly, the sign of the result changes with the sense of the scale selected. Therefore, the above approximations must be modified to accommodate the balance sense:

$$A' - B' \cong \pm\Delta\left(\theta_1 - \theta_2\right)\big/\left[2\left|\left(\theta_3 - \theta_2\right)\right|\right] \quad \left(\text{Single-transposition approximation}\right)$$

$$A' - B' \cong \pm\Delta\left(\theta_4 - \theta_3 + \theta_1 - \theta_2\right)\big/\left[4\left|\left(\theta_3 - \theta_2\right)\right|\right] \quad \left(\text{Double-transposition approximation}\right)$$

The negative sign must be chosen when increasing mass on the left pan indicates a decrease on the scale, the lower scale in this example.

The accuracy of the approximation methods depends on how close to equal in length are the balance arms. In general use, operators usually fail to account for this error in uncertainty statements. The error related to unequal balance arms can be evaluated by simply weighing objects A and B using the accurate six-observation weighing format. The error resulting from the use of the approximation method is

$$\left[\left(A' - B'\right)\text{ exact} - \left(A' - B'\right)\text{ approx.}\right]\big/\left[\left(A' - B'\right)\text{ exact}\right], \text{ in percent} =$$

$$\pm\left\{\left[\left(\theta_5 - \theta_6\right)\big/\left|\left(\theta_3 - \theta_2\right)\right|\right] - 1\right\}\big/2 \times 100$$

This error is Type B and should be included with other known Type B errors in the weighing process by the method of "root sum square." For most weighings, the other major known Type B error is the uncertainty of the mass standard.

When weighing is performed on a balance of grossly unequal arms, a trimming weight of known mass, T, must be added to either object of mass A or B to restore equilibrium after transposition. If T is added with B after transposing the loads of Eq. (5.2), the solution becomes

$$A' - B' \cong \pm \Delta \left[\left(\theta_4 - \theta_3 + \theta_1 - \theta_2 \right) \right] / \left[4 \left| \left(\theta_3 - \theta_2 \right) \right| \right] + T'/2 \; \left(\text{Double - transposition approximation} \right)$$

T', as are A' and B', is buoyancy-corrected mass.

One also notes that the numerator of the last equation is symmetrical and free of linear balance drift, whereas the denominator is not and requires correction when drift is significant. The equation when corrected for drift becomes

$$A' - B' \cong \pm \Delta \left[\left(\theta_4 - \theta_3 + \theta_1 - \theta_2 \right) \right] / \left[4 \left| \left(3\theta_3 - 3\theta_2 + \theta_1 - \theta_4 \right) \right| \right]. \tag{5.17}$$

The above single-transposition approximation is subject to drift and might also require correction.

Equal-arm balances may be damped such that they rapidly reach the resting equilibrium position. However, they may be used in the free-swinging mode and the θ are estimated from turning points. Turning points, O, are usually observed using a pointer and a scale and noting the minimum and maximum excursions of the pointer.

Typically, either three or five turning points are used to estimate the rest point. θs are determined from Os, for example, $[(O_1 + O_3)/2 + O_2]/2 = \theta_i$. A similar expression can be written for the five turning points.

Furthermore, one can permanently attach a third weight, C, (counterweight), nominally equal to A and B, to the balance arm at the point of rotation of one of the end knives. Doing so creates a so-called one-pan-two-knife balance commonly used from about 1950 to 1990. This type of balance is used in the substitution mode as are the spring balance and the modern force compensation electronic balance. The salient feature of this type of balance is a constant load on the beam regardless of what is being weighed.

Constant loading provides constant sensitivity,[3] whereas the equal-arm balance sensitivity varies with pan loading as a result of variation of beam bending. A convenient feature of the one-pan-two-knife balance is a built-in weight set[3] nearly equal to the range of the balance. Thus, any weights that must be summed with whatever is being weighed to bring the balance into equilibrium are conveniently at hand. The small inequality between A and B is indicated in mass units because the beam center of gravity has been judiciously fixed by the manufacturer.

Figure 5.3 shows a spring balance in use as a mass comparator. The difference in mass between A and B will be determined by substitution weighing. However, as in the equal-arm balance example, the small mass standard Δ is sometimes required to calibrate the balance response. A modern electronic balance may require only periodic calibration with the built-in weight whereas an elastic device (load cell) will require the use of Δ as described above. The choice of Δ is such that it is four times $(A - B)$.[4]

The weighing equations for a single substitution will now be developed. The balance response, O_i, to the following loads consisting of the forces imposed by objects of mass A, B, and Δ on the balance pan are indicated in the following equations:

$$\text{Load} \left(1 \right): \; \left(A - \rho_a V_A \right) g = gkO_1$$

$$\text{Load} \left(2 \right): \; \left(B - \rho_a V_B \right) g = gkO_2$$

$$\text{Load} \left(3 \right): \; \left(B - \rho_a V_B + \Delta - \rho_a V_\Delta \right) g = gkO_3,$$

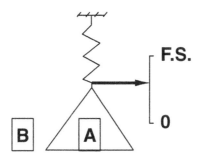

FIGURE 5.3 Spring balance as a mass comparator.

where ρ_a is the density of the air in which the weighings are made and the Vs are the volumes of the objects. Subtracting the equation for load 2 from the equation for load 3 and solving for k yields

$$k = \left(\Delta - \rho_a V_\Delta\right) / \left(O_3 - O_2\right).$$

Subtracting the equation for load 2 from the equation for load 1 and substituting the above quantity for k, the resulting equation is

$$\left(A - \rho_a V_A\right) - \left(B - \rho_a V_B\right) = \left[\left(O_1 - O_2\right)\left(\Delta - \rho_a V_\Delta\right)\right] / \left(O_3 - O_2\right). \tag{5.18}$$

For less-demanding measurements, the buoyancy correction associated with Δ is omitted to yield:

$$\left(A - \rho_a V_A\right) - \left(B - \rho_a V_B\right) = \left[\left(O_1 - O_2\right)\left(\Delta\right)\right] / \left(O_3 - O_2\right), \tag{5.19}$$

and for the modern electronic balance Δ may not be used at all and Eq. (5.18) becomes:

$$\left(A - \rho_a V_A\right) - \left(B - \rho_a V_B\right) = \left(O_1 - O_2\right). \tag{5.20}$$

This variant of Eq. (5.18) assumes that the operator calibrates the balance in the appropriate mass units prior to use.

Whenever significant balance drift is present, one can use the double-substitution weighing method. One merely needs to add an additional load to the ones above:

$$\text{Load}\left(4\right): \quad \left(A - \rho_a V_A + \Delta - \rho_a V_\Delta\right)g = gkO_4.$$

The double-substitution solution is

$$\left(A' - B'\right) = \Delta\left(O_1 - O_2 + O_4 - O_3\right) / \left[2\left(O_3 - O_2\right)\right] - \rho_a\left(V_B - V_A\right). \tag{5.21}$$

A drift correction can be applied to the observations as described earlier for transposition weighing. Doing so yields

$$\left(A' - B'\right) = \Delta\left(O_1 - O_2 + O_4 - O_3\right) / \left[2\left(O_1 - 3O_2 + 3O_3 - O_4\right)\right] - \rho_a\left(V_B - V_A\right) \tag{5.22}$$

For the most stringent measurement requirements, a fifth load[3] is observed:

$$\text{Load } (5): \ \left(A - \rho_a V_A\right)g = gkO_5.$$

This is referred to as the "five-observation-double-substitution weighing format."[3] For this format, Δ need only be 0.86 $(A - B)$.[4]

The drift-free solution is

$$\left(A' - B'\right) = \left(\Delta - \rho_a V_\Delta\right)\left(O_1 - O_2 + O_4 - O_3\right) \Big/ \left[2\left(O_3 - O_2 + O_4 - O_5\right)\right] - \rho_a\left(V_B - V_A\right). \qquad (5.23)$$

Contrary to common belief, most modern electronic force compensation balances do not require leveling. Tilt error is removed with the application of the internal calibration weight. Similarly, there is no gravitational correction when the balance is transported to another elevation and the internal calibration is again performed.

References

1. Cage, M. E. and Davis, R. S., An analysis of read-out perturbations seen on an analytical balance with a swinging pan, *J. Res. Natl. Bur. Stand.* (U.S.), 87, 38, 1982.
2. Cameron, J. M. and Hailes, G. E., Designs for the Calibration of Small Groups of Standards in the presence of Drift, NBS Technical Note 544.
3. Bowman, H. A. and Schoonover, R. M., with an appendix by Jones, M. W., Procedure for high precision density determinations by hydrostatic weighing, *J. Res. Natl. Bur. Stand.* (U.S.), 71, 193, 1967.
4. Davis, R. S., Note on the choice of sensitivity weight in precision weighing, *J. Res. Natl. Bur. Stand.* (U.S.), 92, 239, 1987.

6

Glossary of
Statistical Terms

Central Limit Theorem "Given a population of values with a finite (non-infinite) variance $[\sigma^2]$, if we take independent samples from the population, all of size N, then the population formed by the averages of these samples will tend to have a Gaussian (normal) distribution regardless of what the distribution is of the original population; the larger N, the greater will be this tendency towards 'normality'. In simpler words: The frequency distribution of sample averages approaches normality, and the larger the samples, the closer is the approach."[1]

Concept of a Limiting Mean "The mean of a family of measurements — of a number of measurements for a given quantity carried out by the same apparatus, procedure, and observer — approaches a definite value as the number of measurements is indefinitely increased. Otherwise, they could not properly be called measurements of a given quantity. In the theory of errors, this limiting mean is frequently called the 'true' value although it bears no definite relation to the true quaesitum, to the actual value of the quantity that the observer desires to measure. This has often confused the unwary. Let us call it the limiting mean."[2]

Degrees of Freedom (DF) The number of degrees of freedom in this case is the number of independent deviations.

Deviation A deviation, d, is a value of the measurement minus the mean (in the present context):

$$d_i = M_i - \overline{M}.$$

F Test The F test is a measure of the ratio of two variances, actually two $(SD)^2$, estimates of the population variance, σ^2. In mass measurement, F is the ratio of the square of the observed SD, $(SD_o)^2$, to the square of the long-term SD, $(SD_{lt})^2$:

$$F = \left(SD_o\right)^2 \Big/ \left(SD_{lt}\right)^2.$$

The observed value is determined from the present measurements and the long-term value is determined from a set of measurements made over time. The number of degrees of freedom (DF), in the numerator of the equation for F above is v_1; DF for the denominator is v_2. F provides critical values that will rarely be exceeded if the squares of the two values of SD are estimates of the same σ^2, square of the variance. Table 6.1 is a table of critical values of F at the 5% level. For example, if the calculated value of the ratio F were 5.02 for v_1 of 3 and v_2 of 15, the critical value of F at the 5% level in Table 6.1 of 3.29 is exceeded. Consequently, it is implied that the variances are different. At NIST, a value of F of less than 3.79 was considered to be acceptable. The F test monitors the precision of the measurement process. If in mass measurement the value of F exceeds the critical value, one attempts to improve the performance of the balance by balance maintenance.

TABLE 6.1   Critical Values of F at the 5% Level

v_2						v_1					
	1	2	3	4	5	6	7	8	9	10	12
1	161.0	200.0	216.0	225.0	230.0	234.0	237.0	239.0	241.0	242.0	244.0
2	18.51	19.00	19.16	19.25	19.30	19.33	19.36	19.37	19.38	19.39	19.41
3	10.13	9.55	9.28	9.12	9.01	8.94	8.88	8.84	8.81	8.78	8.74
4	7.71	6.94	6.59	6.39	6.26	6.16	6.09	6.04	6.00	5.96	5.91
5	6.61	5.79	5.41	5.19	5.05	4.95	4.88	4.82	4.78	4.74	4.68
6	5.99	5.14	4.76	4.53	4.39	4.28	4.21	4.15	4.10	4.06	4.00
7	5.59	4.74	4.35	4.12	3.97	3.87	3.79	3.73	3.68	3.63	3.57
8	5.32	4.46	4.07	3.84	3.69	3.58	3.50	3.44	3.39	3.34	3.28
9	5.12	4.26	3.86	3.63	3.48	3.37	3.29	3.23	3.18	3.13	3.07
10	4.96	4.10	3.71	3.48	3.33	3.22	3.14	3.07	3.02	2.97	2.91
11	4.84	3.98	3.59	3.36	3.20	3.09	3.01	2.95	2.90	2.86	2.79
12	4.75	3.88	3.49	3.26	3.11	3.00	2.92	2.85	2.80	2.76	2.69
13	4.67	3.80	3.41	3.18	3.02	2.92	2.84	2.77	2.72	2.67	2.60
14	4.60	3.74	3.34	3.11	2.96	2.85	2.77	2.70	2.65	2.60	2.53
15	4.54	3.68	3.29	3.06	2.90	2.79	2.70	2.64	2.59	2.55	2.48
16	4.49	3.63	3.24	3.01	2.85	2.74	2.66	2.59	2.54	2.49	2.42
17	4.45	3.59	3.20	2.96	2.81	2.70	2.62	2.55	2.50	2.45	2.38
18	4.41	3.55	3.16	2.93	2.77	2.66	2.58	2.51	2.46	2.41	2.34
19	4.38	3.52	3.13	2.90	2.74	2.63	2.55	2.48	2.43	2.38	2.31
20	4.35	3.49	3.10	2.87	2.71	2.60	2.52	2.45	2.40	2.35	2.28
21	4.32	3.47	3.07	2.84	2.68	2.57	2.49	2.42	2.37	2.32	2.25
22	4.30	3.44	3.05	2.82	2.66	2.55	2.47	2.40	2.35	2.30	2.23
23	4.28	3.42	3.03	2.80	2.64	2.53	2.45	2.38	2.32	2.28	2.20
24	4.26	3.40	3.01	2.78	2.62	2.51	2.43	2.36	2.30	2.26	2.18
25	4.24	3.38	2.99	2.76	2.60	2.49	2.41	2.34	2.28	2.24	2.16
26	4.22	3.37	2.98	2.74	2.59	2.47	2.39	2.32	2.27	2.22	2.15
27	4.21	3.35	2.96	2.73	2.57	2.46	2.37	2.30	2.25	2.20	2.13
28	4.20	3.34	2.95	2.71	2.56	2.44	2.36	2.29	2.24	2.19	2.12
29	4.18	3.33	2.93	2.70	2.54	2.43	2.35	2.28	2.22	2.18	2.10
30	4.17	3.32	2.92	2.69	2.53	2.42	2.34	2.27	2.21	2.16	2.09
40	4.08	3.23	2.84	2.61	2.45	2.34	2.25	2.18	2.12	2.07	2.00
50	4.03	3.18	2.79	2.56	2.40	2.29	2.20	2.13	2.07	2.02	1.95
60	4.00	3.15	2.76	2.52	2.37	2.25	2.17	2.10	2.04	1.99	1.92
120	3.92	3.07	2.68	2.45	2.29	2.18	2.09	2.02	1.96	1.91	1.83
∞	3.84	2.99	2.60	2.37	2.21	2.09	2.01	1.94	1.88	1.83	1.75

After balance maintenance, a new value of observed standard deviation, SD_o is determined. Values in Tables 6.1 and 6.2 were taken from Ref. 3.

Mean, Arithmetic Mean, Average   The mean (arithmetic mean, average) of n measurements of mass, $\overline{M}$, is the sum of the n values of M divided by n; which can be expressed as

$$\left(\Sigma_i M_i\right)/n = \overline{M}$$

where Σ indicates sum and the subscript i runs from 1 to n.

Measurement   Measurement is the assignment of numbers to a property.[2] In this book, the property of interest is the mass of an object (a weight, for example) or other related property.

Normal Distribution of Measurements   A general rule relating the frequency of occurrence of a measurement to the deviation of a measurement from the population mean is known as the "normal law of error," that is, measurements are usually "normally distributed."

    68.27% of normally distributed measurements would fall within ±1 SD of the mean.

    95.45% would fall within ±2 SD of the mean.

    99.73% would fall within ±3 SD of the mean.

TABLE 6.2 Critical Values of t (percent probability level)

DF	50	40	30	20	10	5	1
1	1.000	1.376	1.963	3.078	6.314	12.706	63.657
2	0.816	1.061	1.386	1.886	2.920	4.303	9.925
3	0.765	0.978	1.250	1.638	2.353	3.182	5.841
4	0.741	0.941	1.190	1.533	2.132	2.776	4.604
5	0.727	0.920	1.156	1.476	2.015	2.571	4.032
6	0.718	0.906	1.134	1.440	1.943	2.447	3.707
7	0.711	0.896	1.119	1.415	1.895	2.365	3.499
8	0.706	0.889	1.108	1.397	1.860	2.306	3.355
9	0.703	0.883	1.100	1.383	1.833	2.262	3.250
10	0.700	0.879	1.093	1.372	1.812	2.228	3.169
11	0.697	0.876	1.088	1.363	1.796	2.201	3.106
12	0.695	0.873	1.083	1.356	1.782	2.179	3.055
13	0.694	0.870	1.079	1.350	1.771	2.160	3.012
14	0.692	0.868	1.076	1.345	1.761	2.145	2.977
15	0.691	0.866	1.074	1.341	1.753	2.131	2.947
16	0.690	0.865	1.071	1.337	1.746	2.120	2.921
17	0.689	0.863	1.069	1.333	1.740	2.110	2.898
18	0.688	0.862	1.067	1.330	1.734	2.101	2.878
19	0.688	0.861	1.066	1.328	1.729	2.093	2.861
20	0.687	0.860	1.064	1.325	1.725	2.086	2.845
21	0.686	0.859	1.063	1.323	1.721	2.080	2.831
22	0.686	0.858	1.061	1.321	1.717	2.074	2.819
23	0.685	0.858	1.060	1.319	1.714	2.069	2.807
24	0.685	0.857	1.059	1.318	1.711	2.064	2.797
25	0.684	0.856	1.058	1.316	1.708	2.060	2.787
26	0.684	0.856	1.058	1.315	1.706	2.056	2.779
27	0.684	0.855	1.057	1.314	1.703	2.052	2.771
28	0.683	0.855	1.056	1.313	1.701	2.048	2.763
29	0.683	0.854	1.055	1.311	1.699	2.045	2.756
30	0.683	0.854	1.055	1.310	1.697	2.042	2.750
40	0.681	0.851	1.050	1.303	1.684	2.021	2.704
50	0.680	0.849	1.048	1.299	1.676	2.008	2.678
60	0.679	0.848	1.046	1.296	1.671	2.000	2.660
120	0.677	0.845	1.041	1.289	1.658	1.980	2.617
∞	0.674	0.842	1.036	1.282	1.645	1.960	2.576

One measurement in 20 from a sample of measurements may be expected to deviate from the sample mean by more than 2 SD.

Population A statistical population or simply "population" is the totality of all possible measurements.

Population Standard Deviation, σ The standard deviation (SD) for the sample is an estimate of the population standard deviation (σ).

Random Sample A sample selected from a population by a random process.

Sample A statistical sample or simply "sample" is a selection of a number of measurements from the population of measurements.

Standard Deviation (SD) The standard deviation is the square root of the sum of the squares of the deviations divided by $(n - 1)$:

$$SD = \sqrt{\Sigma_i d^2 / (n-1)},$$

where $(n - 1)$ is the number of degrees of freedom.

Student's t In mass measurement, student's t-test is used to compare an observed value (of mass) to the accepted value.

$$t = \left[\left(\overline{M} - \mu\right)\sqrt{n}\right]\Big/ \text{SD},$$

where $\overline{M}$ is the observed mean of n measurements, μ is the accepted value and SD is the standard deviation. Table 6.2 is a table of critical values of t. At NIST, a value of t of less than 3.0 was considered to be acceptable. Thirty-two measurements (n) are about all one needs for a good estimate of long-term SD. For n greater than 32, one can use t values for an infinite (∞) number of degrees of freedom (DF). In practice, in applying the t-test to measurements of the mass of a check standard, if the critical value of t is exceeded and there is no error in weighings, a new set of measurements should be made to determine whether, or to confirm that, the mass has shifted from the accepted value. After enough work has been done to assure the result, a new value of mass can be assigned to the check standard. Or the check standard can be sent to a standards laboratory for assignment of a value.

Variance The variance, σ^2, is the square of the population standard deviation.

References

1. Mandel, J., *The Statistical Analysis of Experimental Data,* Dover, New York, 1964, 64.
2. Dorsey, N. E., The velocity of light, *Trans. Am. Phys. Soc.,* 34, 4, 1944.
3. Youden, W. J., *Statistical Methods for Chemists,* John Wiley & Sons, New York, 1951, appendix.

7

Measurement Uncertainty

7.1 Introduction

A measurement of a quantity, mass, for example, is incomplete without a quantitative statement of the uncertainty of the measurement. In the past, the uncertainty of a measurement (the result of the application of a measurement process) had been considered to consist of random and systematic components.

The random component has generally been considered to be a measure of precision (or *im*precision) of the measurement process as applied to the specific measurement.

> The precision, or more correctly, the imprecision of a measurement process is ordinarily summarized by the standard deviation of the process, which expresses the characteristic disagreement of repeated measurements of a single quantity by the process concerned, and thus serves to indicate how much a particular measurement is likely to differ from other values that the same measurement process might have provided in this instance, or might yield on remeasurement of the same quantity on another occasion."[1]

Eisenhart[1] has defined systematic error thusly:

> The systematic error, or bias, of a measurement process refers to its tendency to measure something other than what was intended; and is determined by the magnitude of the difference $\mu - \tau$ between the process average or limiting mean μ associated with measurement of a particular quantity by the measurement process concerned and the true value τ of the magnitude of this quantity.[1]

In this chapter, we discuss and apply the National Institute of Standards and Technology (NIST) guidelines for evaluating and expressing the uncertainty of measurement results.

7.2 NIST Guidelines

In October 1992, NIST instituted a new policy on expressing measurement uncertainty. The new policy was based on the approach to expressing uncertainty in measurement recommended by the International Committee for Weights and Measures (CIPM) and on the elaboration of that approach given in the *Guide to the Expression of Uncertainty in Measurement*[2] (hereafter referred to as the *Guide*). The *Guide* was prepared by individuals nominated by the International Bureau of Weights and Measures (Bureau International des Poids et Mesures, BIPM), the International Electrochemical Commission (IEC), the International Organization for Standardization (ISO), or the International Organization of Legal Metrology (OIML). The CIPM approach is founded on Recommendation INC-1 (1980) of the Working Group on the Statement of Uncertainties.

NIST prepared a Technical Note, "Guidelines for Evaluating and Expressing the Uncertainty of NIST Measurement Results,"[3] to assist in putting the policy into practice.

Because this treatment of uncertainty is in use in the United States and internationally, it is discussed here.

7.2.1 Classification of Components of Uncertainty

In the CIPM and NIST approach, the several components of the uncertainty of the result of a measurement may be grouped into two categories according to the method used to estimate the numerical values of the components:

Category A. Those components that are evaluated by statistical methods
Category B. Those components that are evaluated by other than statistical methods

Superficially, category A and B components resemble random and systematic components. The NIST Guidelines[3] point out that this simple correspondence does not always exist, and that an alternative nomenclature to the terms *random uncertainty* and *systematic uncertainty* might be:

"Component of uncertainty arising from a random effect"
"Component of uncertainty arising from a systematic effect"

7.2.2 Standard Uncertainty

Each component of uncertainty is represented by an estimated standard deviation, referred to as a *standard uncertainty*. The standard uncertainty for a category A component is here given the symbol u_A. The evaluation of uncertainty by the statistical analysis of a series of observations is termed a *Type A evaluation*.

A category B component of uncertainty, which may be considered an approximation to the corresponding standard deviation is here given the symbol u_B. The evaluation of uncertainty by means other than statistical analysis of series of observations is termed a *Type B evaluation*.

7.2.3 Type A Evaluation of Standard Uncertainty

Type A evaluation of standard uncertainty may be based on any valid statistical method for treating data.

7.2.4 Type B Evaluation of Standard Uncertainty

The NIST Guidelines[3] states that a Type B evaluation of standard uncertainty is usually:

Based on scientific judgment using all the relevant information available, which may include

- previous measurement data,

- experience with, or general knowledge of, the behavior and property of relevant materials and instruments,

- manufacturer's specifications,

- data provided in calibration and other reports, and

- uncertainties assigned to reference data taken from handbooks.[3]

The Guidelines gives examples and models for Type B evaluations. In several of these, the measured quantity in question is modeled by a normal distribution with lower and upper limits $-a$ and $+b$. For such a model "almost all" of the measured values of the quantity lie within plus and minus 3 standard deviations of the mean. Then $a/3$ can be used as an approximation of the desired standard deviation, u_B.

7.2.5 Combined Standard Uncertainty

The *combined standard uncertainty*, with the symbol, u_C, of a measurement result is taken to represent the estimated standard deviation of the result. The combined standard uncertainty is obtained by taking the square root of the sum of the squares of the individual standard uncertainties. That is,

$$u_c = \sqrt{\left(u_A\right)^2 + \left(u_B\right)^2} . \tag{7.1}$$

7.2.6 Expanded Uncertainty

To define the interval about the measurement result within which the value of the quantity being measured is confidently believed to lie, the *expanded uncertainty* (with the symbol U) is intended to meet this requirement.

The expanded uncertainty is obtained by multiplying u_C by a *coverage factor*, with the symbol k. Thus,

$$U = k u_c . \tag{7.2}$$

Typically, k is in the range 2 to 3.

To be consistent with international practice, the value used by NIST for calculating U is, by convention, $k = 2$.

7.2.7 Relative Uncertainties

The *relative standard uncertainty* is the ratio of the standard uncertainty, u_A or u_B, to the absolute value of the quantity measured.

The *relative combined uncertainty* is the ratio of the combined standard uncertainty, u_C, to the absolute value of the quantity measured.

The *relative expanded uncertainty* is the ratio of the expanded uncertainty, U, to the absolute value of the quantity measured.

Relative uncertainties can be expressed as percent or as decimals.

7.3 Example of Determination of Uncertainty

For the calibration of 1-kg mass standards, the estimate of the random uncertainty, expressed as the estimate of the standard deviation, was 0.01571 mg. The estimate of the systematic uncertainty was 0.03000 mg, which is taken to be the upper bound for the total bias with a standard uncertainty of 0.03000/3 mg.

Formerly, an estimate of combined uncertainty would be

$$\left(3 \times 0.01571\right) + 0.03000 = 0.077 \text{ mg}$$

Using the NIST Guidelines and the same figures,

$$u_A = 0.01571 \text{ mg}$$

$$u_B = 0.03000/3 = 0.01000 \text{ mg}$$

$$u_C = \sqrt{\left(0.01571\right)^2 + \left(0.01000\right)^2} = 0.01862 \text{ mg} = 1.862 \times 10^{-6} \%$$

For a coverage factor of $k = 2$,

$$U = 0.037 \text{ mg}$$

We note that for this example the combined uncertainty is approximately one half of the previously conventional estimate of combined uncertainty.

References

1. Eisenhart, C., Realistic evaluation of the precision and accuracy of instrument calibration systems, *J. Res. Natl. Bur. Stand.* (U.S.), 67C, 161, 1963.
2. *Guide to the Expression of Uncertainty in Measurement,* International Organization for Standards, Geneva, Switzerland, 1992.
3. Taylor, B. N. and Kuyett, C. E., Guidelines for evaluating and expressing the uncertainty of NIST measurement results, NIST Technical Note 1297, 1994.

8

Weighing Designs

8.1 Introduction

Weighing designs are treated in detail in the literature. In the present chapter, material from the excellent and authoritative NBS Technical Note 952[1] by Cameron, Croarkin, and Raybold and NBS Special Publication 700-1 by Davis and Jaeger[2] will be used. The interested reader is referred to these publications for more detail. Also, the least-squares method can be studied in a number of other sources.

The fundamental relationship for the comparison of a weight of mass X with a standard weight of mass S can be expressed by the following equation:

$$\left(X'-S'\right)g = \delta g, \tag{8.1}$$

where X' and S' are:

$$X' = X\left[1-\left(\rho_a/\rho_X\right)\right],$$

$$S' = S\left[1-\left(\rho_a/\rho_S\right)\right].$$

ρ_a is the density of the air in which the comparison is made, ρ_X is the density of the weight of mass X, ρ_S is the density of the standard of mass S, g is the local acceleration due to gravity, and δ is mass difference indicated by the balance on which the comparison is made.

Rearranging Eq. (8.1),

$$X = \left\{S\left[1-\left(\rho_a/\rho_S\right)\right]+\delta\right\}\Big/\left[1-\left(\rho_a/\rho_X\right)\right]. \tag{8.2}$$

The mass X is then determined from the mass of the standard and the mass difference determined from the balance indication.

In this discussion it is assumed that weighings are made by the substitution method and the balance indications are given in scale divisions. In single-substitution weighing, a weight of mass X is compared with a standard weight of mass S. A small calibration or sensitivity weight of mass Δ is also required (see Chapter 5).

The steps in single substitution weighing are the following:

1. Place the weight of mass X on the balance pan and record the balance indication, i_1.
2. Remove the weight of mass X from the balance pan.
3. Place the standard weight of mass S on the balance pan and record the balance indication, i_2.
4. Add the sensitivity weight of mass Δ to the balance pan with the standard weight of mass S; and record the balance indication, i_3. Modern electronic balances are customarily calibrated prior to use and it is usually feasible to omit the use of a sensitivity weight.

The following is a schematic of the weighings:

Pan Load	Balance Indication
X	i_1
S	i_2
$S + \Delta$	i_3

The sensitivity weight of mass Δ and the balance indications are used to determine δ. From the schematic:

$$\left(X'-S'\right)=\left(i_1-i_2\right)\left[\Delta\left[1-\left(\rho_a/\rho_\Delta\right)\right]\big/\left(i_3-i_2\right)\right]=\delta. \tag{8.3}$$

Thus, the value of one scale division of the balance indication is equal to $[\Delta \ [1 - (\rho_a/\rho_\Delta)]/(i_3 - i_2)]$ and it is used to convert $(i_1 - i_2)$ to δ in mass units.

Although many mass measurements are made by comparing an object or weight the mass of which is to be determined with a weight or standard of known mass, in many instances it is neither practicable nor desirable to compare each weight in a set with a standard of known nearly equal mass. For those instances, weighing designs are developed.

The weighing designs are such that:

1. The mass values of individual weights can be determined by comparison of a selected group of weights from the set with a weight of known mass.
2. Weights in the set are intercompared with each other.
3. No additional weights of known mass are required, except for the sensitivity weight or a balance calibration weight.

A weighing design will now be illustrated with examples. In these examples, it is assumed that the conversion from scale divisions to mass units has been made or that the balance is direct-reading in mass units. Thus, indicated differences are in mass units.

A combination of two or more weights of unknown mass can be as easily compared with a weight of known mass (hereafter referred to as a standard, S) as can a single weight. The mass value of S is the basis on which the values of the individual weights are fixed.

The case of a combination of two weights is illustrated using Example I:

1. In this example, the unknown masses of two weights, A and B, are nominally equal to each other. The standard is designated S_2.

$$A + B \approx S_2$$

The mass difference between $(A' + B')$ and S_2', a, is determined by weighing and is expressed as

$$\left(A'+B'\right)-S_2'=a, \tag{8.4}$$

where the superscript $'$ indicate buoyancy-corrected masses.

The individual mass values of A and B can be determined by making another measurement, comparing A with B. The relationship between the buoyancy-corrected mass values for A and B can be expressed as

$$A' - B' = b. \tag{8.5}$$

The buoyancy-corrected mass values for A and B in terms of the buoyancy-corrected mass value for S_2 can be found by rearranging Eq. (8.4):

$$A' + B' = S'_2 + a \tag{8.6}$$

and adding Eq. (8.6) to Eq. (8.5), resulting in

$$2A' = S'_2 + a + b, \tag{8.7}$$

from which:

$$A = \left\{ S_2 \left[1 - \left(\rho_a / \rho_{S2} \right) \right] + a + b \right\} \Big/ \left\{ 2 \left[1 - \rho_a / \rho_A \right] \right\}. \tag{8.8}$$

Subtracting Eq. (8.5) from Eq. (8.4),

$$2B' = S'_2 + a - b, \tag{8.9}$$

and

$$B = \left\{ S_2 \left[1 - \left(\rho_a / \rho_{S2} \right) \right] + a - b \right\} \Big/ \left\{ 2 \left[1 - \rho_a / \rho_A \right] \right\}. \tag{8.10}$$

Substituting the mass values of S_2, a, b, and the various densities in Eqs. (9.9) and (8.10), the mass values of A and B are determined. This method is called the "sum and difference" method.

2. It is now assumed that weight B rather than being a single weight is two weights, B_1 and B_2, of mass nominally equal to each other, and the sum of the mass values of which is equal to the mass value of B.

Eqs. (8.4) and (8.5) now take the forms:

$$A' + B'_1 + B'_2 - S'_2 = a, \tag{8.11}$$

$$A' - B'_1 - B'_2 = b. \tag{8.12}$$

Making a weighing comparing B_1 with B_2, the relationship between B_1 and B_2 may be expressed as

$$B'_1 - B'_2 = c. \tag{8.13}$$

Adding Eqs. (8.11) and (8.12),

$$2A' = S'_2 + a + b, \tag{8.14}$$

and

$$A' = \left(S'_2 + a + b \right) \Big/ 2. \tag{8.15}$$

Subtracting Eq. (8.12) from Eq. (8.11),

$$2 \left(B'_1 + B'_2 \right) = S'_2 + a - b, \tag{8.16}$$

and

$$\left(B'_1 + B'_2 \right) = \left(S'_2 + a - b \right) \Big/ 2. \tag{8.17}$$

The values of A' and $(B_1' + B_2')$ have been expressed in terms of S_2' in Eqs. (8.15) and (8.17), respectively. The sum $(B_1' + B_2')$ is now separated to determine B_1' and B_2' separately.

Adding Eqs. (8.13) and (8.17),

$$2B_1' = \left(S_2' + a - b\right)/2 + c, \qquad (8.18)$$

and

$$B_1 = \left\{S_2\left[1 - \left(\rho_a/\rho_{S2}\right)\right] + a - b + 2c\right\}/\left\{4\left[1 - \left(\rho_a/\rho_{B1}\right)\right]\right\}. \qquad (8.19)$$

Subtracting Eq. (8.18) from Eq. (8.17),

$$B_2' = \left(S_2' + a - b\right)/2 - \left(S_2' + a - b + 2c\right)/4, \qquad (8.20)$$

and

$$B_2 = \left\{S_2\left[1 - \left(\rho_a/\rho_{S2}\right)\right] + a - b - 2c\right\}/\left\{4\left[1 - \left(\rho_a/\rho_{B2}\right)\right]\right\}. \qquad (8.21)$$

The mass values for weights A, B_1, and B_2 have been determined in terms of the mass of S_2 without comparing any of the weights individually with a weight of known mass.

The mass differences a, b, and c are used in the above development as uncorrected for buoyancy. In the most accurate measurement of mass, the small buoyancy corrections would be applied and the correct representations would be a', b', and c'.

8.2 Least Squares[3]

8.2.1 Best Fit

An important part of the mathematics of measurement deals with getting the best fit of a line or curve to a set of points.

The procedure that is frequently used to determine the best value or the best equation of a line or curve for a set of data is called the "method of least squares" — least squares because the method minimizes the sum of the squares of the deviations or residuals.

Residual is defined as

Residual = Observed – Calculated

The principle of least squares requires the minimizing of the sum of the weighted squares of the residuals, S. This sum of squares may be written as

$$S = \Sigma\left(w\,\text{res}^2\right), \qquad (8.22)$$

where res is residual and w is weight (not an object).

In the application of the principle of least squares here, observations are made of the mass differences between objects or weights and corresponding values of mass are calculated. Because all the observations are drawn from the same pool of observations, in simple cases all weights (w) are unity.

Thus,

$$S = \Sigma \text{res}^2. \tag{8.23}$$

The summation (denoted by Σ) of the squares of the residuals is taken over all observations that are subject to error.

The *principle* of least squares is the minimizing of S; the *method* of least squares is a rule or set of rules for proceeding with the actual computation.

8.2.2 Simplest Example

Let a *single* sample of n random observations of the mass of an object be taken. An adjusted value can be derived from these observations.

One looks now at the problem as one of fitting the curve:

$$x = a \tag{8.24}$$

to the n observations. This is the simplest of all "curves" because it is a horizontal line. Only one adjustable constant or parameter, a, is to be determined.

The problem is then to minimize S, the sum of the squares of the residuals. For each observation x there is a corresponding residual, $(x - a)$. The sum of the squares of the residuals,

$$S \equiv \Sigma \left(x - a\right)^2, \tag{8.25}$$

is the quantity to be minimized.

The only variable in Eq. (8.25) is the adjustable parameter a. The minimum in S will occur when

$$dS/da = -2\Sigma\left(x - a\right) = 0, \tag{8.26}$$

that is, when

$$\Sigma\left(x - a\right) = 0, \tag{8.27}$$

or

$$\Sigma x = na. \tag{8.28}$$

The least-squares value of a is then

$$a = \left(\Sigma x\right)/n = \text{the mean of the values of } x. \tag{8.29}$$

The average, or the arithmetic mean, or the mean is the best single number to represent a collection of data.

The difference between a measurement and the mean is the deviation. The deviations are squared and added. The minimum of the sum of the squares of the deviations is obtained when the mean is used as the representation of the collection of data. Thus, the mean is the best fit.

8.2.3   Equation of a Line

If a line of form:

$$Y = a + bX \tag{8.30}$$

represents the relationship between the mass X of an object (a weight, for example) and an observation Y, what is sought is values of a and b such that the resulting equation is the best fit for the data.

If the sum of squares of the residuals, observed Y – calculated Y, is a minimum, the equation best represents the collection of data.

The sum of the squares of the residuals of the points (X,Y) from the line represented by the equation is S, where S is given by

$$S = \Sigma Y^2 - \left(\Sigma Y\right)^2 / n - \left[\Sigma\left(XY\right) - \left(\Sigma X \Sigma Y / n\right)^2\right] / \left[\left(\Sigma X^2 - \left(\Sigma X\right)^2 / n\right)\right], \tag{8.31}$$

where Σ represents the sum. The sum of the squares of the residuals, $\Sigma(Y_i - y_i)^2$, is a minimum, where Y_i is the value predicted by Eq. (8.30) for $X = x_i$ and corresponding measured $Y = y_i$.

The application of the principle of least squares provides the best values for a and b. The resulting values of a and b can be expressed as

$$a = \left(\Sigma y\right) / n - b\left(\Sigma x\right) / n, \tag{8.32}$$

$$b = \left(n\Sigma xy - \Sigma x \Sigma y\right) / \left[n\Sigma x^2 - \left(\Sigma x\right)^2\right]. \tag{8.33}$$

8.3   Sequences

Weighing designs can be used for many different kinds of measurements and are particularly used by laboratories that perform a large number of routine mass calibration measurements of laboratory weights. Weighing designs provide a least-squares solution to an over-determined set of weighing equations. Unlike the simple sum-and-difference weighing, designs yield information that is required to determine mass values. This extra information is useful statistical data.

The two prominent sequences used for weight sets are the 5,2,2,1, summation 1 and the 5,3,2,1, summation 1 progressions. Normally, these progressions begin at 20 kg for the first sequence given and at 30 kg for the second sequence and both sequences continue downward to 1 mg, the last weight in the sequences. Therefore, a 5,2,2,1 sequence weight set could be comprised of a 20 kg, 10 kg$_1$, 10 kg$_2$, 5 kg, 2 kg$_1$, 2 kg$_2$, 1 kg, 500 g and so forth until 1 mg is reached. The grouping of the 5 kg, 2 kg$_1$, 2 kg$_2$ and 1 kg weights is referred to as the summation 10 kg (Σ10 kg). Similar groupings of the other weight decades are likewise possible.

Weight sets can begin and end at any mass value in the above sequences and the techniques discussed here will apply. However, the use of weighing designs are most convenient for the above denomination range and for laboratory weights that are defined as those weights that meet the OIML R111 specification[4] for weight classes E$_1$ and E$_2$. There are other similar specifications in use.

The use of weighing designs entails somewhat more work than a simple one-to-one comparison but has additional benefits that will be discussed later.

Because the majority of weight sets are presented as one of the two sequences described above, one need only consider the weighing designs appropriate to them. In Ref. 1, many other designs are given and the application is similar to that described here.

The sequences mentioned above are usually modified at the end of each decade to include an extra 1. This weight is not part of the set but is owned by the calibration agency and is referred to as the check standard. Therefore, the weighing design sequence will be different from the weight set sequences described above to accommodate these check standards.

The complete 5,2,2,1,Σ1 weight sequence can now be efficiently assigned mass values using a 5,2,2,1,Σ1, check standard 1 (Chk. Std. 1), weighing design and similarly for the 5,3,2,1,Σ1 sequence the weighing design sequence is 5,3,2,1,Σ1, and Chk. Std. 1.

Although the calibration can begin at any decade, it begins at 1 kg for weight sets containing a 1-kg weight. Well-characterized check standards provide starting standards for abbreviated weights sets, that is, the 100-g check standard is the starting standard for a weight set beginning at 100 g. The starting weighing design sequence is standard (std.) 1-kg$_1$, std. 1-kg$_2$, 1-kg, and Σ1-kg, i.e., 1,1,1,1 or 1,1,1,1,1, if the set contains two 1-kg weights. The 1,1,1,1 design would also be used when starting at 100 g, 10 g, etc. The mass assigned to Σ1-kg then serves as the standard (restraint) for the first design application of the 5,3,2,1,Σ1, and Chk. Std. 1 sequence going downward.

Each successive decade contains a summation except for the final decade of the weight set. In place of the missing weight summation of the last decade an additional 1 weight is added or a different weighing design is chosen. Either approach is appropriate but here the addition of 1 to the last decade is chosen to simplify the following discussion.

A complete set of weighing data for a 5,2,2,1,Σ1, Chk. Std. 1 sequence from 5 kg to 1 mg is provided here. These data have been analyzed using a version of the National Institute of Standards and Technology (NIST) software. A calculation has been provided as a guide for those wishing to perform these calculations themselves. There are small differences between the hand-calculation and the computer software, but in comparison to the measurement uncertainties these are minor.

The demonstration weight set contains only one 1-kg weight, so one begins with a 1,1,1,1 sequence design. Data for a 1,1,1,1,1 design is not given, but Ref. 1 gives this design along with many others. The data induction process is the same as are the many intermediate calculations.

For each design given in Ref. 1, one has a choice of where to place the restraint. The choices made here are for working upward and downward from the 1-kg level and that two 1-kg standards are available. The mass difference between these 1-kg standards serves as the check standard for this measurement sequence. All other weighing sequences have a donated weight acting as the check standard.

For each design there is a set of parameter multipliers, weight standard deviation factors, and deviation multipliers for computing the least-squares-fit residuals and standard deviation along with other information. The required weighing designs for this weight set and all ancillary information are given below.

Although other designs could be used, the ones discussed have been chosen as a good compromise between the number of weighings required and the resulting standard deviation for each weight assigned a mass value and, hence, measurement uncertainty.

8.3.1 Design A.1.2

Working downward from the kilograms begins here. There are four nominally equal 1-kg balance loads, six mass difference observations, and three degrees of freedom (see Chapter 6). The Σ1-kg is the restraint for the next series below.

Balance Observations	Measured Balance Differences			
	Std 1-kg$_1$	Std 1-kg$_2$	1-kg	Σ1-kg
$Y(1)$	+	−		
$Y(2)$	+		−	
$Y(3)$	+			−
$Y(4)$		+	−	
$Y(5)$		+		−
$Y(6)$			+	−
Restraint, m	+	+		
Restraint for the next descending series				+

Restraint means that one knows the mass, density (volume), and thermal coefficent of expansion of the two weights used as the standards. The mass values assigned the other weights in the design depend on this information.

Restraint for the next series down is the mass value assigned by analysis of this weighing design to $\Sigma 1$-kg. The $\Sigma 1$-kg is the weight aggregate: 500 g + 200 g_1 + 200 g_2 + 100 g, and knowledge of its thermal coefficient of expansion and density are also required.

The $\Sigma 1$-kg now serves as the restaint for the next series (descending) that follows. The next design will be repeated for each decade of the weights (below 1 kg) which is to be calibrated. The last decade will require a substitution of 1 mg for the $\Sigma 1$-mg weight as the set does not contain a $\Sigma 1$-mg weight. The Σ of each decade has the same restraint function for the decade below it as described above.

The following design, C.10, is repeated six times until all the weights from 500 g to 1 mg have been calibrated. The same balance might not be used for all of the applications of this design. The usual case is to select a balance based on capacity and standard deviation that will yield the smallest uncertainty. This means that several balances may be required for the complete weight set calibration. However, balances are usually not mixed within a decade.

Buoyancy is accounted for in two steps. The restraint is adjusted for buoyancy and the computed mass values are adjusted for their respective buoyancy.

8.3.2 Design C.10

$5(--), 2(--)_1, 2(--)_2, 1(--), \Sigma 1(--)$, chk Std $1(--)$

Balance Observations	Measured Balance Differences					
	$5(--)$	$2(--)_1$	$2(--)_2$	$1(--)$	$\Sigma 1(--)$	Chk Std $1(--)$
$Y(1)$	+	−	−	−	−	+
$Y(2)$	+	−	−	−	+	−
$Y(3)$	+	−	−	+	−	−
$Y(4)$	+	−		−	−	−
$Y(5)$	+		−	−	−	−
$Y(6)$		+	−	+	−	
$Y(7)$		+	−	−		+
$Y(8)$		+	−		+	−
Restraint, m	+	+	+	+		
Next series restraint					+	

The above design is also used to work upward from 1-kg to complete the weight set calibration. The starting series begins with the same standard kilograms used for working downward. However, the restraint changes as does the design name; see Design 16 below. If the weight set continues upward beyond 5 kg, then the $\Sigma 10$ kg becomes the restraint for the next series upward. In this way, weight sets that contain a 20 kg weight may be calibrated. This design with this particular restraint is not in Ref. 1 but was provided by Richard S. Davis[5] of BIPM, Sèvres, France and it is given now.

8.3.3 Design 16

5-kg(--), 2-kg$_1$(--), 2-kg$_2$(--), 1-kg(--), Std 1-kg$_1$(--), Std 1-kg$_2$(--)

Balance Observations	Measured Mass Differences					
	5-kg	2-kg$_1$	2-kg$_2$	1-kg	Std 1-kg$_1$	Std 1-kg$_2$
$Y(1)$	+	−	−	−	−	+
$Y(2)$	+	−	−	−	+	−
$Y(3)$	+	−	−	+	−	−
$Y(4)$	+	−		−	−	−
$Y(5)$	+		−	−	−	−
$Y(6)$		+	−	+	−	
$Y(7)$		+	−	−		+
$Y(8)$		+	−		+	−
Restraint					+	+
Next series restraint	+	+	+	+		

When Design 16 is repeated a second time or more to work further upward in a weight set, then the Σ5, 2$_1$, 2$_2$, 1 becomes the restraint for each subsequent application of the design.

The parameters required for computation using these designs, a sample data set, pertinent portions of the solution generated by NIST software, and a hand computation with explanation follow.

8.4 Observation Multipliers for Determining Mass Values and Deviations

8.4.1 Design A.1.2

Four equal weights, $K = 4$, six comparisons (weighings), $N = 6$, and three degrees of freedom, DF = 3.

Parameter Values, Divisor = 4

Observations	Std 1-kg$_1$	Std 1-kg$_2$	1-kg	Σ1-kg
$Y(1)$	2	−2	0	0
$Y(2)$	1	−1	−3	−1
$Y(3)$	1	−1	−1	−3
$Y(4)$	−1	1	−3	−1
$Y(5)$	−1	1	−1	−3
$Y(6)$	0	0	2	−2
Restraint, m	4	4	4	4

Deviations, Divisor = 4

Observations	1	2	3	4	5	6
$Y(1)$	2	−1	−1	1	1	0
$Y(2)$	−1	2	−1	−1	0	1
$Y(3)$	−1	−1	2	0	−1	−1
$Y(4)$	1	−1	0	2	−1	1
$Y(5)$	1	0	−1	−1	2	−1
$Y(6)$	0	1	−1	1	−1	2

8.4.2   Design C.10

$K = 6, N = 8, DF = 3$

Parameter values, Divisor = 70

Observations	5	2_1	2_2	1	Σ1	Chk Std 1
Y(1)	15	−8	−8	1	1	21
Y(2)	15	−8	−8	1	21	1
Y(3)	5	−12	−12	19	−1	−1
Y(4)	0	2	12	−14	−14	−14
Y(5)	0	12	2	−14	−14	−14
Y(6)	−5	8	−12	9	−11	−1
Y(7)	5	12	−8	−9	1	11
Y(8)	0	10	−10	0	10	−10
m	35	14	14	7	7	7

Deviations, Divisor = 7

Observations	1	2	3	4	5	6	7	8
Y(1)	2	−1	−1	0	0	0	−2	2
Y(2)	−1	2	−1	0	0	2	0	−2
Y(3)	−1	−1	2	0	0	−2	2	0
Y(4)	0	0	0	3	−3	1	1	1
Y(5)	0	0	0	−3	3	−1	−1	−1
Y(6)	0	2	−2	1	−1	3	−1	−1
Y(7)	−2	0	2	1	−1	−1	3	−1
Y(8)	2	−2	0	1	−1	−1	−1	3

8.4.3   Design 16

$K = 6, N = 8, DF = 3$

Parameter Values, Divisor = 14

Observations	5	2_1	2_2	1	Std_1	Std_2
Y(1)	−8	−6	−6	−2	−2	2
Y(2)	−8	−6	−6	−2	2	−2
Y(3)	2	−2	−2	4	0	0
Y(4)	14	6	8	0	0	0
Y(5)	14	8	6	0	0	0
Y(6)	5	4	0	3	−1	1
Y(7)	−5	0	−4	−3	−1	1
Y(8)	0	2	−2	0	2	−2
m	35	14	14	7	7	7

Deviations, Divisor = 7

Observations	1	2	3	4	5	6	7	8
Y(1)	2	−1	−1	0	0	0	−2	2
Y(2)	−1	2	−1	0	0	2	0	−2
Y(3)	−1	−1	2	0	0	−2	2	0
Y(4)	0	0	0	3	−3	1	1	1
Y(5)	0	0	0	−3	3	−1	−1	−1
Y(6)	0	2	−2	1	−1	3	−1	−1
Y(7)	−2	0	2	1	−1	−1	3	−1
Y(8)	2	−2	0	1	−1	−1	−1	3

8.5 Factors for Computing Weight Standard Deviations Needed for Uncertainty Calculations

8.5.1 Design A.1.2

Factor	Weight			
	Std 1	Std 2	1-kg	Σ1-kg
0.6124				+
0.6124			+	
0.3536		+		
0.3536	+			
0.5000	+	– (Chk Std)		

Note: The factor is assigned to the weight above the + sign.

8.5.2 Design C.10

Factor	Weight					
	5	2_1	2_2	1	Σ1	Chk Std 1
0.4546						+
0.4546				+		
0.4326			+			
0.3854			+			
0.3854		+				
0.3273	+					

8.5.3 Design 16

Factor	Weight					
	5	2_1	2_2	1	Std 1	Std 2
1.7110	+					
1.0000		+				
1.0000			+			
0.4629				+		
0.2673					+	
0.2673						+
0.5346			(Chk Std)		+	–

8.6 Sample Data Sets and Intermediate Calculations

8.6.1 Design A.1.2

Sample data sets for Design A.1.2 and intermediate calculations: Observed mass differences and restraint, *m*, with buoyancy correction, in milligrams, begin with the weight set calibration working downward:

$Y(1)$	−0.32121
$Y(2)$	−1.05310
$Y(3)$	−0.57339
$Y(4)$	−0.65019
$Y(5)$	−0.36716
$Y(6)$	0.33209
m	−296.6353

Mass correction to nominal values of standards in milligrams: Std_1 = 0.326, Std_2 = 0.680
Densities at 20°C: 8.0017 and 8.0018 g/cm³, respectively
Cubical thermal coefficient of expansion for the standards: 0.000045
Mean air temperature during the measurement: 23.28°C.
Mean air density during the measurement: 1.19065 mg/cm³

Intermediate calculations for design A.1.2:

Volume of standards at 23.28°C ignoring corrections to the nominal mass, (mass/ρ) $[1 + 3\alpha(t - 20)]$, where α is the linear coefficient of thermal expansion:

$$Std_1 = \left(1000 \text{ g}\big/8.0017 \text{ g}\big/\text{cm}^3\right) \times \left[1 + 0.000045\left(23.28°C - 20°C\right)\right] = 124.99189 \text{ cm}^3.$$

$$Std_2 = \left(1000 \text{ g}\big/8.0018 \text{ g}\big/\text{cm}^3\right) \times \left[1 + 0.000045\left(23.28°C - 20°C\right)\right] = 124.99033 \text{ cm}^3.$$

Restraint, M, equals mass correction to the standards minus buoyancy imposed on the standards:

$$M = 0.32600 \text{ mg} - \left(1.19065 \text{ mg}\big/\text{cm}^3 \times 124.99189 \text{ cm}^3\right) + 0.6800 \text{ mg} -$$

$$\left(1.19065 \text{ mg}\big/\text{cm}^3 \times 124.990327 \text{ cm}^3\right).$$

$$M = -296.6353.$$

The 1-kg and Σ1-kg volumes at 23.28°C are 127.5704 cm³ based on an assumed density of 7.84 g/cm³ at 20°C.

8.6.2 Design C.10

For weighing design C.10, working downward from Σ1-kg: Observed mass differences and restraint, m, that includes buoyancy, in milligrams:

$Y(1)$	0.35568
$Y(2)$	−0.64506
$Y(3)$	−0.71092
$Y(4)$	−0.68702
$Y(5)$	−0.38708
$Y(6)$	0.21062
$Y(7)$	0.84962
$Y(8)$	−0.17726
m	−147.74530

Density of all weights undergoing calibration at 20°C: 7.84 g/cm³
Cubical thermal coefficient of expansion: 0.000045
Mean air temperature during the measurement: 23.28°C
Mean air density during the measurement: 1.18995 mg/cm³

Intermediate calculations for Design C.10:

Volumes of standards at 23.28°C including corrections to the nominal mass:

$$\left(1000.0040665 \text{ g}/7.84 \text{ g}/\text{cm}^3\right) \times \left[1 + 0.000045\left(23.28°C - 20°C\right)\right] = 127.5704 \text{ cm}^3.$$

M, the sum of the mass correction and buoyancy for $\Sigma1000$ g:

$$M = 4.0565 \text{ mg} - \left(1.18995 \text{ mg}/\text{cm}^3 \times 127.5704 \text{ cm}^3\right) = -147.7459 \text{ mg}$$

Check standard accepted correction = 0.85169 mg
Density at 20°C: 7.8704 g/cm³
Cubical thermal coefficient expansion: 0.000045

8.6.3 Design 16

For Design 16: Observed mass differences and restraint, m, with buoyancy, in milligrams:

$Y(1)$	−8.83716
$Y(2)$	−10.52204
$Y(3)$	15.44278
$Y(4)$	−5.25467
$Y(5)$	−13.39291
$Y(6)$	3.88068
$Y(7)$	−18.39074
$Y(8)$	−7.54031
m	−289.0835

Density at 20°C: 7.84 g/cm³
Cubical thermal coefficient of expansion: 0.000045
Mean air temperature during the measurement: 23.15°C
Mean air density during the measurement: 1.16045 mg/cm³

Intermediate calculations for Design 16:

Volumes of standards at 23.15°C: 249.9802 cm³

$$M = 1.006 \text{ mg} - \left(1.16045 \text{ mg}/\text{cm}^3 \times 249.9802 \text{ cm}^3\right) = -289.0835 \text{ mg}$$

The above data sets can now be combined with the least-squares solution parameter and deviation multipliers and weight standard deviation multipliers to obtain the following: mass values for each weight in the sequence, standard deviation of the fitted data (balance standard deviation), standard deviation for each weight, mass uncertainty, F-test, and t-test.

One begins with Design 16 and calibration of the 5-kg, 2-kg$_1$, 2-kg$_2$, and 1-kg weights based on two 1-kg mass standards. One first obtains the products of the calculated balance differences and restraint, in milligrams, with the appropriate parameter multipliers; sums the products; divides the sum by the divisor; and adds the appropriate weight buoyancy correction.

Buoyancy correction for the standards was placed in the restraint under the data heading of intermediate calculations. For the reader's convenience the observed balance differences, $Y_{i's}$, and the restraint are listed with these data. The multipliers are shown as subscripts.

8.7   Calculations of Various Values Associated with Design 16 and the 5-kg, 2-kg$_1$, 2-kg$_2$, and 1-kg Weights

Y	5-kg	2-kg$_1$	2-kg$_2$	1-kg	S1-kg$_1$	S1-kg$_2$
(1) −8.83716	70.6973$_{-8}$	53.0230$_{-6}$	53.0230$_{-6}$	17.6743$_{-2}$	17.6743$_{-2}$	−17.6743$_{+2}$
(2) −10.52204	84.1763$_{-8}$	63.1322$_{-6}$	63.1322$_{-6}$	21.0441$_{-2}$	−21.0441$_{+2}$	21.0441$_{-2}$
(3) 15.44278	30.8856$_{+2}$	−30.8856$_{-2}$	−30.8856$_{-2}$	61.7711$_{+4}$	0$_{+0}$	0$_{+0}$
(4) −5.25467	−73.5654$_{+14}$	−31.5280$_{+6}$	−42.0374$_{+8}$	0$_{+0}$	0$_{+0}$	0$_{+0}$
(5) −13.39291	−187.5007$_{+14}$	−107.1433$_{+8}$	−80.3575$_{+6}$	0$_{+0}$	0$_{+0}$	0$_{+0}$
(6) 3.88068	19.4034$_{+5}$	15.5227$_{+4}$	0$_{+0}$	11.6420$_{+3}$	−3.88068$_{-1}$	3.88068$_{+1}$
(7) −18.39074	91.9537$_{-5}$	0$_{+0}$	73.5630$_{-4}$	55.1722$_{-3}$	18.39074$_{-1}$	−18.39074$_{+1}$
(8) −7.54031	0$_{+0}$	−15.0806$_{+2}$	15.0806$_{-2}$	0$_{+0}$	−15.0806$_{+2}$	15.0806$_{-2}$
m −289.0835	−10117.9225$_{35}$	−4047.1690$_{14}$	−4047.1690$_{14}$	−2023.5845$_{+7}$	−2023.5845$_{+7}$	−2023.5845$_{+7}$
Sum	−10081.8723	−4100.1296	−3995.6517	−1856.2808	−2027.5248	−2019.6442
Sum ÷ 14	−720.1337	−292.8664	−285.4037	−132.5915	−144.8232	−144.2603
Buoyancy	740.1909	296.0756	296.0767	147.8456	145.0460	145.0435
Sum = Mass Corr.	20.0572 mg	3.2092 mg	10.6730 mg	15.2541 mg	0.2228 mg	0.7832 mg
NIST values	20.05696 mg	3.20917 mg	10.67299 mg	15.25407 mg	0.22279 mg	0.78319 mg

Deviations:

Y	1	2	3	4	5	6	7	8
(1)	−17.6743$_{+2}$	8.83716$_{-1}$	8.83716$_{-1}$	0$_{+0}$	0$_{+0}$	0$_{+0}$	17.6743$_{-2}$	−17.6743$_{+2}$
(2)	10.5220$_{-1}$	−21.0441$_{+8}$	10.5220$_{-1}$	0$_{+0}$	0$_{+0}$	−21.0441$_{+2}$	0$_{+0}$	21.0441$_{-2}$
(3)	−15.4428$_{-1}$	−15.4428$_{-1}$	30.8856$_{+2}$	0$_{+0}$	0$_{+0}$	−30.8856$_{-2}$	30.8856$_{+2}$	0$_{+0}$
(4)	0$_{+0}$	0$_{+0}$	0$_{+0}$	−15.7640$_{+3}$	15.7640$_{-3}$	−5.2547$_{+1}$	−5.2547$_{+1}$	−5.2547$_{+1}$
(5)	0$_{+0}$	0$_{+0}$	0$_{+0}$	40.1787$_{-3}$	−40.1787$_{+3}$	13.3929$_{-1}$	13.3929$_{-1}$	13.3929$_{-1}$
(6)	0$_{+0}$	7.7614$_{+2}$	−7.7614$_{-2}$	3.8807$_{+1}$	−3.8807$_{-1}$	11.6420$_{+3}$	−3.8807$_{-1}$	−3.8807$_{-1}$
(7)	36.7815$_{-2}$	0$_{+0}$	−36.7815$_{+2}$	−18.3907$_{+1}$	18.3907$_{-1}$	18.3907$_{-1}$	−55.1722$_{-3}$	18.3907$_{-1}$
(8)	−15.0806$_{+2}$	15.0806$_{-2}$	0$_{+0}$	−7.5403$_{+1}$	7.5403$_{-1}$	7.5403$_{-1}$	7.5403$_{-1}$	−22.6209$_{+3}$
Sum	−0.8941	−4.8078	−5.7019	2.3644	−2.3643	−.2185	5.1852	3.3971

Sum ÷ 7 = Deviations, D_i

D1	D2	D3	D4	D5	D6	D7	D8
−0.1277 mg	−0.6868 mg	0.8146 mg	0.3378 mg	−0.3378 mg	−0.8884 mg	0.7407 mg	0.4853 mg

From the deviations the standard deviation of the fitted data, also called the balance standard deviation, is now computed.

In general SD $= [\Sigma\,(D_i)^2/3]^{1/2}$

SD $= \{[(-0.1277)^2 + (-0.6868)^2 + (0.8146)^2 + (0.3378)^2 + (-0.3378)^2 + (-0.8884)^2 + (0.7407)^2 + (0.4853)^2]/3\}^{1/2}$

SD $= 0.99217$ mg, F ratio $=$ observed SD^2/accepted SD^2 $= 0.99217^2/2.3900^2 = 0.1723$

The F ratio, 0.1723, is less than 3.79 and therefore the standard deviation is in control.

t value $=$ (observed correction of check std − accepted)/SD of observed $= [-0.5604$ mg − $(-0.354$ mg)]/(0.5346 × 2.390 mg) $= t = -0.16$

$t < 3$, therefore the check standard is in control.

Usually, in repetitive calibrations, many standard deviations are collected for similar measurements for a balance and are pooled for the so-called long-term estimate. For the balance used in this example the pooled or accepted SD is 2.39 mg. The pooled SD, when available, is used to compute the uncertainties for each of the above assigned mass values.

One begins with the subcalculation of the type A uncertainty, which is the product of the standard deviation factor for the appropriate weight and the long-term standard deviation for the balance. The type A uncertainties are given in the table below.

			Uncertainties			
Nominal	SD Factor	Pooled SD	Type A	Combined	Expanded	NIST Report
5 kg	1.711	2.39 mg	4.089 mg	4.090 mg	8.180 mg	8.18250
2 kg$_1$	1	2.39	2.390	2.390	4.781	4.78072
2 kg$_2$	1	2.39	2.390	2.390	4.781	4.78072
1 kg	0.4629	2.39	1.106	1.106	2.212	2.21320
S 1 kg$_1$	0.2673	2.39	0.639	0.639	1.278	1.27813
S 1 kg$_2$	0.2673	2.39	0.639	0.639	1.278	1.27813

The type B uncertainty for each weight is simply one half of the type B uncertainty of the restraint (S1 kg$_1$ + S1 kg$_2$), 0.036/2 mg multiplied by the nominal value of each unknown weight. Therefore, the combined uncertainties for these weights are $[(\text{type A})^2 + (\text{type B})^2]^{1/2}$. The combined uncertainty multiplied by 2 is the expanded uncertainty. These values are given in the above table as are the values generated by NIST software. The type B was negligible and not shown explicitly.

In the above uncertainty calculation, NIST treated the component arising from uncertainty in the weight density as zero. This, of course, is not true and the uncertainty components from air density parameters and weight density must be accounted for as shown in Chapter 17. The between-time component for this calibration process is believed to be insignificant, i.e., zero. NIST software-generated values for the observed standard deviaton, F ratio, and t value are 0.99216 mg, 0.172, and −0.16, repectively.

Note: In the above example the type A error in the restraint is zero and is not always the case. In the first series (C.10), below the 1-kg level (500 g to 100 g) the restraint is comprised of both type A and type B uncertainty and each component must be proportioned to the nominal value of each weight undergoing a mass assignment in the sequence. The type A uncertainty associated with the restraint is combined by "root sum squaring" with the type A contribution attributed to the measurement process (balance) and then likewise root sum squared with the type B component coming from the restraint.

8.8 Calculations of Various Values Associated with the A.1.2 Design Solution for the 1-kg and Σ1-kg Weights and 500 g through Σ100 g

Mass Corrections to Nominal Values:

Y	S 1 kg$_1$	S 1 kg$_2$	1kg	Σ1 kg
(1) −0.32121	−0.64242$_{+2}$	0.64242$_{-2}$	0$_{+0}$	0$_{+0}$
(2) −1.05310	−1.05310$_{+1}$	1.0540$_{-1}$	3.315930$_{-3}$	1.05310$_{-1}$
(3) −0.57339	−0.57339$_{+1}$	0.57339$_{-1}$	0.57339$_{-1}$	1.72017$_{-3}$
(4) −0.65019	0.65019$_{-1}$	−0.65019$_{+1}$	1.95057$_{-3}$	0.65019$_{-1}$
(5) −0.36716	0.36716$_{-1}$	−0.36716$_{+1}$	0.36716$_{-1}$	1.10148$_{-3}$
(6) 0.33209	0$_{+0}$	0$_{+0}$	0.66418$_{+2}$	−0.66418$_{-2}$
m −296.6355	−1186.5420$_{+4}$	−1186.5420$_{+4}$	−1186.5420$_{+4}$	−1186.5420$_{+4}$
Sum =	−1187.7936	−1185.2904	−1179.8274	−1182.6812
Sum ÷ 8	−148.47420	−148.16130	−147.4784	−147.8352
Buoyancy	148.82159	148.81973	151.8917	151.8917
Sum = mass corr., mg	0.34739	0.65842	4.4133	4.0565

Deviations:

Y	1	2	3	4	5	6
(1)	−0.64242$_{+2}$	0.32121$_{-1}$	0.32121$_{-1}$	−0.32121$_{+1}$	−0.32121$_{+1}$	0$_{+0}$
(2)	1.05310$_{-1}$	−2.10620$_{+2}$	1.05310$_{-1}$	1.05310$_{-1}$	0$_{+0}$	−1.05310$_{+1}$
(3)	0.57339$_{-1}$	0.57339$_{-1}$	−1.14678$_{+2}$	0$_{+0}$	0.57339$_{-1}$	0.57339$_{-1}$
(4)	−0.65019$_{+1}$	0.65019$_{-1}$	0$_{+0}$	−1.30032$_{+2}$	0.65019$_{-1}$	−0.65019$_{+1}$
(5)	−0.36716$_{+1}$	0$_{+0}$	0.36716$_{-1}$	0.36716$_{-1}$	−0.73432$_{+2}$	0.36716$_{-1}$
(6)	0$_{+0}$	0.33209$_{+1}$	0.33209$_{-1}$	0.33209$_{+1}$	−0.33209$_{-1}$	0.66418$_{+2}$
Sum =	−0.03328	−0.22932	0.26260	0.13082	−0.16404	−0.09856
Sum ÷ 4 =						
D_i =	−0.00832 mg	−0.05733 mg	0.06565 mg	0.03270 mg	−0.04101 mg	−0.02464 mg
SD = $[\Sigma(D_i)^2/3]^{1/2}$ = 0.06062 mg						

F ratio = accepted SD^2/observed SD^2 = $(0.0606)^2/(0.0429)^2$ = 1.997 <3.79; therefore SD is acceptable.

t value = (observed correction of check std − accepted)/SD of observed = [−0.31026 mg − (−0.354 mg)]/(0.5 × 0.0606 mg) = t = 1.44

The t value, 1.44 < 3; therefore check standard value is in control.

Uncertainty (k = 1) for 1 kg and Σ1 kg (restraint for the next series below) is $[(0.018)^2 + (0.6124 \times 0.0429)^2]^{1/2}$ = 0.0318 mg

NIST software yielded nearly identical values for all of the above results.

One now proceeds with the calibration of the first decade of weights below 1 kg by application of Design C.10. Therefore, from above, the Σ1-kg nominal mass correction (4.0565 mg) is used as the restraint, with buoyancy added, for the first application of the C.10 design and all necessary data for assignment of mass values are presented as before.

Intermediate Calculations:

Weight volumes at 23.28°C: 500 g = 63.78492 cm^3, 200 g (both weights) = 25.51397 cm^3, 100 g and Σ100 g = 12.7570 cm^3 and the check 100 g = 12.70778 cm^3.

Average air density = 1.18995 mg/cm^3.

Restraint, m, with buoyancy.

$m = 4.0565 - (1.18995 \text{ mg/cm}^3 \times 127.57036) = -147.7453$ mg.

Y	500g	200 g_1	200 g_2	100 g	Σ100 g	Check 100 g
(1) 0.35568	5.3352_{+15}	-2.8454_{-8}	-2.8454_{-8}	0.35568_{+1}	0.35568_{+1}	7.4693_{+21}
(2) -0.64506	-9.6759_{+15}	5.1605_{-8}	5.1605_{-8}	-0.64506_{+1}	-13.5463_{+21}	-0.6451_{+1}
(3) -0.71092	-3.5546_{+5}	8.5310_{-12}	8.5310_{-12}	-13.5075_{+19}	0.7109_{-1}	0.7109_{-1}
(4) -0.68702	0_{+0}	-1.3740_{+2}	-8.2442_{+12}	9.6183_{-14}	9.6183_{-14}	9.6183_{-14}
(5) -0.38708	0_{+0}	-4.6450_{12}	-0.7742_{+2}	5.4191_{-14}	5.4191_{-14}	5.4191_{-14}
(6) 0.21062	-1.0531_{-5}	1.6805_{+8}	-2.5274_{-12}	1.8956_{+9}	-2.3168_{-11}	-0.2106_{-1}
(7) 0.84962	4.2481_{+5}	10.1954_{+12}	-6.7970_{-8}	-7.6466_{-9}	0.8496_{+1}	9.3458_{+11}
(8) -0.17726	0_{+0}	-1.7726_{+10}	1.7726_{-10}	0_{+0}	-1.7726_{+10}	1.7726_{-10}
m -147.74530	-5171.0855_{+35}	-2068.4342_{+14}	-2068.4342_{+14}	-1034.2171_{+7}	-1034.2171_{+7}	-1034.2171_{+7}
Sum	-5175.7858	-2053.5027	-2074.1583	-1038.7276	-1034.8992	-1000.7368
Sum						
$\div$ 70	-73.9398	-29.3358	-29.6308	-14.8390	-14.7843	-14.2962
+ Buoyancy	75.9009	30.3603	30.3603	15.1802	15.1802	15.1216

(sum $\div$ 70) + buoyancy = mass correction

Mass corr. =	1.9611 mg	1.0246 mg	0.7295 mg	0.3412 mg	0.3959[a] mg	0.8254 mg
NIST values	1.9613 mg	1.0248 mg	0.7296 mg	0.3413 mg	0.3959 mg	0.8254 mg

[a] Next series restraint.

Deviations:

Y	1	2	3	4	5	6	7	8
(1)	0.71136_{+2}	-0.35568_{-1}	-0.35568_{-1}	0_{+0}	0_{+0}	0_{+0}	-0.71136_{-2}	0.71136_{+2}
(2)	0.64506_{-1}	-1.29012_{+2}	0.64506_{-1}	0_{+0}	0_{+0}	-1.29012_{+2}	0_{+0}	1.29012_{-2}
(3)	0.71092_{-1}	0.71092_{-1}	-1.42184_{+2}	0_{+0}	0_{+0}	1.42184_{-2}	-1.42184_{+2}	0_{+0}
(4)	0_{+0}	0_{+0}	0_{+0}	-2.06106_{+3}	2.06106_{-3}	-0.68702_{+1}	-0.68702_{+1}	-0.68702_{+1}
(5)	0_{+0}	0_{+0}	0_{+0}	1.16124_{-3}	-1.16124_{+3}	0.38708_{-1}	0.38708_{-1}	0.38708_{-1}
(6)	0_{+0}	0.42124_{+2}	-0.42124_{-2}	0.21062_{+1}	-0.21062_{-1}	0.63186_{+3}	-0.21062_{-1}	-0.21062_{-1}
(7)	-1.69924_{-2}	0_{+0}	1.69924_{+2}	0.84962_{+1}	-0.84962_{-1}	-0.84962_{-1}	2.54886_{3}	-0.84962_{-1}
(8)	-0.35452_{+2}	0.35452_{-2}	0_{+0}	-0.17726_{+1}	0.17726_{-1}	0.17726_{-1}	0.17726_{-1}	-0.53178_{+3}
Sum	0.01358	-0.15912	0.14554	-0.01684	0.01684	-0.20872	0.08236	0.10952

Sum $\div$ 7 = deviations D_i

D1	D2	D3	D4	D5	D6	D7	D8
0.00194 mg	-0.02273 mg	0.02079 mg	-0.00240 mg	0.00240 mg	-0.02982 mg	0.01177 mg	0.01565 mg

NIST computed deviations:

0.00195 mg	-0.02272 mg	0.02079 mg	-0.00240 mg	0.00240 mg	-0.02981 mg	0.01176 mg	0.01564 mg

From the deviations, the standard deviation of the fitted data, also called the balance standard deviation, is computed.

In general, $\text{SD} = [\Sigma(D_i)^2/3]^{1/2}$

$\text{SD} = \{[(0.00194)^2 + (-0.02273)^2 + (0.02079)^2 + (-0.00240)^2 + (0.00240)^2 + (-0.02982)^2 + (0.01177)^2 + (0.01565)^2]/3\}^{1/2}$

$\text{SD} = 0.0273$ mg

F ratio = accepted SD^2/observed SD^2 = 0.0273^2/0.0429^2 = 0.405, NIST value 0.405

The *F* ratio, 0.405, is less than 3.79, and therefore the standard deviation is in control.

t value = (observed correction of check std − accepted)/SD of observed = [-0.8254 mg − (−0.8517 mg)]/(0.4645 × 0.0429 mg) = t = −1.32, NIST −1.31

t < 3, therefore the check standard is in control.

		Uncertainties		
	Type B, mg	Type A, mg[a]	*k* = 1 (mg)	NIST *k* = 1 (mg)
500 g	0.018/2	(0.3273^2) × (0.0429^2) + (1/2)^2 × (0.02672^2)$^{1/2}$	0.0214	0.0212
200 g$_1$	0.018/5	(0.3854^2) × (0.0429^2) + (1/5)^2 × (0.02672^2)$^{1/2}$	0.0177	0.0177
200 g$_2$	0.018/5	(0.3854^2) × (0.0429^2) + (1/5)^2 × (0.02672^2)$^{1/2}$	0.0177	0.0177
100 g	0.018/10	(0.4326^2) × (0.0429^2) + (1/10)^2 × (0.02672^2)$^{1/2}$	0.0188	0.0188
Σ100 g	0.018/10	(0.4645^2) × (0.0429^2) + (1/10)^2 × (0.02672^2)$^{1/2}$	0.0202	0.0202
Chk 100 g	0.018/10	(0.4645^2) × (0.0429)^2 + (1/10)^2 × (0.02672^2)$^{1/2}$	0.0202	0.0202

[a] Proportion Σ1-kg restraint.

The uncertainty for weight combinations can likewise be computed. For example, to compute the uncertainty of the 500 g + 200 g$_1$ + 200g$_2$, refer to NBS Technical Note 952, Ref. 1, for design C.10 standard deviation factor for weight 9 (same as 900 g in this example). It is found to be 0.4326 when the restraint is the summation 1 kg, as is this example, and proceed as above.

		Uncertainties		
	Type B	Type A	*k* = 1	NIST *k* = 1
900 g	0.018/(9/10) mg	[0.4326^2 × 0.0429^2 mg + (9/10)^2 × 0.02672^2 mg]$^{1/2}$	0.0344 mg	0.0341 mg

All the remaining lower decades of this weight set are likewise calibrated by repeated application of the C.10 weighing design. At each decade the weight designated as the summation (5-- + 2$_1$-- + 2$_2$-- + 1--) becomes the restraint for the solution of the individual mass values that comprise the summation, the check standard, and the weight summation for the decade that follows. The mass values of the remaining weights of the set, 50 g to 1 mg, are determined in the same way. All of the necessary data and the mass values and statistical data generated by the NIST software for the remaining weights of the set are provided. As above, there may be slight differences between hand calculations (not provided) and the software values. However, the weighing data and NIST analysis provides a complete set for those wishing to test similar software.

8.8.1   50 g – 10 g NIST Data

50 g – Σ10 g, density: 7.84 g/cm^3 at 20°C, cubical thermal coefficient: 0.000045
10-g check standard, density: 8.3406 g/cm^3 at 20°C
Cubical thermal coefficient: 0.000040
Mass correction: −0.42701 mg
Air density: 1.18345 mg/cm^3 at 22.83°C, 100,849 Pa, and 31% relative humidity

Observed balance differences, 1 to 8, in milligrams (y_i):

1.	−0.51593	**5.**	0.27974
2.	0.40783	**6.**	−0.06402
3.	0.27025	**7.**	−0.35460
4.	0.26502	**8.**	0.44324

Calculated Results:

Item	Correction,	Volume,	Uncertainties, mg		
Weight	mg	cm³ @ T	Type B	Type A	Expanded
50 g	0.17335	6.3784	0.00090	0.0113	0.0227
20 g₁	0.09596	2.5514	0.00036	0.0073	0.0146
20 g₂	0.08682	2.5514	0.00036	0.0073	0.0146
10 g	0.03981	1.2757	0.00018	0.00712	0.0143
Σ10 gᵃ	0.11009	1.2757	0.00018	0.00761	0.0152
Chk 10 g	−0.43100	1.1990	0.00018	0.00761	0.0152

ᵃ Restraint for the next series.

Observed standard deviation of the process: 0.01790 mg
Accepted standard deviation of the process: 0.01580 mg
F ratio 1.283
t value −0.52
Accepted mass correction of check standard: −0.42701 mg
Observed correction of check standard: −0.43100 mg

8.8.2   5 g – 1 g NIST Data

5 g – 1 g, density: 7.84 g/cm³ at 20°C
Cubical thermal coefficient: 0.000045
Σ1 g, density: 8.000 g/cm³ at 20°C
Cubical thermal coefficient: 0.000045
1-g check standard, density: 8.3397 g/cm³ at 20°C
Cubical thermal coefficient: 0.000040
Mass correction: −0.07005 mg
Air density: 1.1831 mg/cm³, at 23.05°C, 100,903 Pa, and 31.7% relative humidity

Observed balance differences, 1 to 8, in milligrams (y_i):

1. −0.037
2. −0.046
3. 0.130
4. 0.109
5. 0.107
6. 0.086
7. −0.089
8. −0.014

Calculated Results:

Item	Correction,	Volume,	Uncertainties, mg		
Weight	mg	cm³ @ T	Type B	Type A	Expanded
5 g	0.03404	0.63785	0.00090	0.00387	0.00775
2 g₁	0.02325	0.25514	0.00004	0.00174	0.00348
2 g₂	0.02839	0.25514	0.00004	0.00174	0.00348
1 g	0.02441	0.12757	0.00002	0.00122	0.00244
Σ1 gᵃ	−0.06819	0.12501	0.00002	0.00127	0.00255
Chk 1 g	−0.06808	0.11991	0.00002	0.00127	0.00255

ᵃ Restraint for the next series.

Observed standard deviation of the process: 0.0027 mg
Accepted standard deviation of the process: 0.0022 mg

F ratio 1.505
t value 1.55
Accepted mass correction of check standard: –0.07005 mg
Observed mass correction of check standard: –0.06808 mg

8.8.3   0.5 g – 0.1 g NIST Data

0.5 g – 0.1 g, density: 8.00 g/cm^3 at 20°C
Cubical thermal coefficient: 0.000045
Σ0.1 g, density: 4.0404 g/cm^3 at 20°C
Cubical thermal coefficient: 0.000063
0.1-g check standard, density: 8.41 g/cm^3 at 20°C
Cubical thermal coefficient: 0.000039
Mass correction: –0.0102 mg
Air density: 1.1860 mg/cm^3, at 22.425°C, 100,923 Pa, and 31.7% relative humidity

Observed balance differences, 1 to 8, in milligrams (y_i).

1.	0.0475	**5.**	0.0182
2.	0.0405	**6.**	–0.0528
3.	0.0385	**7.**	–0.0465
4.	0.0704	**8.**	–0.0549

Calculated Results:

Item Weight	Correction, mg	Volume, cm^3 @ *T*	Uncertainties, mg		
			Type B	Type A	Expanded
0.5 g	–0.01204	0.06251	0.00001	0.00067	0.00133
0.2 g$_1$	–0.04702	0.02500	0.00000	0.00034	0.00068
0.2 g$_2$	0.00450	0.02500	0.00000	0.00034	0.00068
0.1 g	–0.01364	0.01250	0.00000	0.00029	0.00057
Σ0.1 g^a	0.00206	0.02475	0.00000	0.00030	0.00060
Chk 0.1 g	–0.00969	0.01189	0.00000	0.00030	0.00060

^{a} Restraint for the next series.

Observed standard deviation of the process: 0.00039 mg
Accepted standard deviation of the process: 0.00059 mg
F ratio 0.438
t value 1.67
Accepted mass correction of check standard: –0.01020 mg
Observed mass correction of check standard: –0.00969 mg

8.8.4   0.05 g – 0.01 g NIST Data

0.05 g, density: 8.00 g/cm^3 at 20°C
Cubical thermal coefficient: 0.000045
0.020 g – Σ0.01 g, 2.7 g/cm^3 at 20°C
Cubical thermal coefficient: 0.000069
0.01 g check standard, density: 8.41 g/cm^3 at 20°C
Cubical thermal coefficient: 0.000039
Mass correction: –0.00039 mg
Air density: 1.18435 mg/cm^3, at 22.475°C, 100,803 Pa, and 31.7% relative humidity

Observed balance differences, 1 to 8, in milligrams (y_i):

1. −0.0176   5. −0.0173
2. 0.0247   6. −0.0200
3. −0.0210   7. 0.0046
4. −0.0196   8. 0.0258

Calculated Results:

Item Weight	Correction, mg	Volume, cm³ @ T	Uncetainties, mg		
			Type B	Type A	Expanded
0.05 g	−0.00445	0.00625	0.00000	0.00025	0.00049
0.02 g_1	0.00478	0.00741	0.00000	0.00024	0.00047
0.02 g_2	0.00148	0.00741	0.00000	0.00024	0.00047
0.01 g	0.00028	0.00370	0.00000	0.00026	0.00051
Σ0.01 g^a	0.02340	0.00371	0.00000	0.00028	0.00055
Chk 10 mg	−0.00124	0.00119	0.00000	0.00028	0.00055

     [a] Restraint for the next series.

Observed standard deviation of the process: 0.00077 mg
Accepted standard deviation of the process: 0.00059 mg
F ratio 1.709
t value −3.09 (failed)
Accepted mass correction of check standard: −0.00039 mg
Observed mass correction of check standard: −0.00124 mg

8.8.5 0.005 g – 0.001 g NIST Data

0.005 g – 0.001 g, density: 2.70 g/cm³ at 20°C
Cubical thermal coefficient: 0.000069
0.001 g check standard, density: 8.50 g/cm³ at 20°C
Cubical thermal coefficient: 0.000039
Mass correction: −0.00247 mg
Air density: 1.1834 mg/cm³, at 22.50°C, 100,730 Pa, and 31.7% relative humidity

Observed balance differences, 1 to 8, in milligrams (y_i):

1. −0.0351   5. −0.0240
2. −0.0191   6. −0.0015
3. −0.0151   7. −0.0154
4. −0.0208   8. 0.0032

Item Weight	Correction, mg	Volume, cm³ @ T	Uncertainties, mg		
			Type B	Type A	Expanded
0.005 g	−0.00199	0.00185	0.00000	0.00024	0.00047
0.002 g_1	0.00640	0.00074	0.00000	0.00023	0.00047
0.002 g_2	0.01078	0.00074	0.00000	0.00023	0.00047
0.001 g	0.00822	0.00037	0.00000	0.00026	0.00051
0.001 g	Filler weight, not part of the set, there is no summation 1 mg				
Chk 1 mg	−0.00244	0.00012	0.00000	0.00028	0.00055

Observed standard deviation of the process: 0.00079 mg
Accepted standard deviation of the process: 0.00059 mg

F ratio 1.799
t value 0.10
Accepted mass correction of check standard: −0.00247 mg
Observed mass correction of check standard: −0.00244 mg
End of sample data set.

8.9   Commentary

The check standards used in these examples were single weights and may not represent the behavior of the weight summations used as restraints in the solution of all but the A.1.2 series. It is suggested that a wiser course would be to use check standards some of the time that are aggregates similar to weight summations of the restraints, i.e., $\Sigma100$ g, etc.

Analysis of these check values may show a larger between-time component than the single weight checks. If significant, the uncertainty estimates may require revision (see Chapter 22).

The observed balance differences in milligrams referred to here as Y_i are obtained from balance readings. The method used to obtain these differences will vary with the type of weighing instrument used and the observation format, i.e., single substitution, double substitution, drift elimination technique, etc. These weighing techniques are used in other weighing applications and are discussed in Chapter 5.

Weighing designs as illustrated here are quite useful for routine laboratory weight calibration. The extraordinary time required compared to a simple one-to-one calibration yields invaluable statistical data that are necessary to assign realistic uncertainty estimates. However, it is unwise to expend this amount of effort if the densities of the weights have not been determined. Without weight density information, the buoyant force uncertainty contribution to the overall weight calibration uncertainty is omitted and, of course, the uncertainty is incorrect.

The NIST data used to illustrate weighing designs used assumed weight densities and the uncertainty statements do not mention this omission. We show the inclusion of the buoyant force uncertainty contribution as well as all other known sources in Chapter 17. The reader can correct this omission if using erroneous software by simply adjusting the calculated uncertainty. When writing new software or when the source code is available, the code should be written to account for the buoyant force uncertainty.

There are weighing designs applications where very high precision balances are used in hopes of obtaining very small calibration uncertainty estimates. When one goes to such extremes, the use of a weighing series average air density will often cause an increase of the least-squares standard deviation in the weighing (balance) contribution to the overall uncertainty. In addition, the estimated mass values will be slightly different. Usually, an improvement to the least-squares standard deviation is obtained by measuring the air density parameters for each observed balance difference (Y_i) in place of a series average. When this technique is used, the buoyancy is applied to each balance difference before fitting the weighing data. This technique is demonstrated using the data of the beginning kilogram series above, A.1.2.

Previously, each observed balance difference was expressed in milligrams and the buoyancy correction was applied to the restraint before the fitting and for the weights after the least-squares adjustment using the series average. One can use the same data and simulate a new data set by assuming a linear air density drift rate, and create a temperature and air density for each Y_i. With these new data one can calculate and apply the buoyancy to each Y_i before fitting the data. Therefore, the difference between any two weights, A and B, will now be expressed as $Y_i + \rho_a(V_A - V_B)$ in place of Y_i. The new data set for the above A.1.2 series is as follows:

Y	Temperature,°C	Air Density, mg/cm³
(1)	23.43	1.1897
(2)	23.37	1.1901
(3)	23.31	1.1905
(4)	23.25	1.1908
(5)	23.19	1.1912
(6)	23.13	1.1916

The weight volumes corresponding to the above temperatures are computed from the given densities at 20°C using the appropriate thermal coefficient of expansion. Furthermore, the *Y*s are adjusted to include the buoyancy correction, as shown below.

Y	S1 kg$_1$	S1 kg$_2$	1 kg	Σ1 kg
(1) −0.32493	−0.64986$_{+2}$	0.64986$_{−2}$	0$_{+0}$	0$_{+0}$
(2) −4.12114	−4.12114$_{+1}$	4.12114$_{−1}$	12.36342$_{−3}$	4.12114$_{−1}$
(3) −3.64245	−3.64245$_{+1}$	3.64245$_{−1}$	3.64245$_{−1}$	10.92735$_{−3}$
(4) −3.72291	3.72291$_{−1}$	−3.72291$_{+1}$	11.16873$_{−3}$	3.72291$_{−1}$
(5) −3.44090	3.44090$_{−1}$	−3.44090$_{+1}$	3.44090$_{−1}$	10.32270$_{−3}$
(6) 0.33209	0$_{+0}$	0$_{+0}$	0.66418$_{+2}$	−0.66418$_{−2}$
m 1.006	4.0240$_{+4}$	4.0240$_{+4}$	4.0240$_{+4}$	4.0240$_{+4}$
Sum =	2.7744	5.2736	35.30368	32.45392
Sum ÷ 8	0.3468	0.6592	4.4130	4.0567
Sum ÷ 8	equals mass correction in mg			

Deviations Calculation:

Y	1	2	3	4	5	6
(1)	−0.64986$_{+2}$	0.32493$_{−1}$	0.32493$_{−1}$	−0.32493$_{+1}$	−0.32493$_{+1}$	0$_{+0}$
(2)	4.12114$_{−1}$	−8.24228$_{+2}$	4.12114$_{−1}$	4.12114$_{−1}$	0$_{+0}$	−4.12114$_{+1}$
(3)	3.64245$_{−1}$	3.64245$_{−1}$	−7.28490$_{+2}$	0$_{+0}$	3.64245$_{−1}$	3.64245$_{−1}$
(4)	−3.72291$_{+1}$	3.72291$_{−1}$	0$_{+0}$	−7.44582$_{+2}$	3.72291$_{−1}$	−3.72291$_{+1}$
(5)	−3.4409$_{+1}$	0$_{+0}$	3.44090$_{−1}$	3.44090$_{−1}$	−6.88180$_{+2}$	3.44090$_{−1}$
(6)	0$_{+0}$	0.33209$_{+1}$	0.33209$_{−1}$	0.33209$_{+1}$	−0.33209$_{−1}$	0.66418$_{+2}$
Sum =	−0.05008	−0.21990	0.26998	0.123380	−0.173460	−0.09652
Sum ÷ 4 = D_i	−0.01252 mg	−0.05497 mg	0.067495 mg	0.030845 mg	−0.043365 mg	−0.02413 mg

$SD = [\Sigma(D_i)^2/3]^{1/2} = 0.061$ mg

F ratio = accepted SD^2/observed SD^2 = $(0.061)^2/(0.0429)^2 = 2.02$, <3.79; therefore SD is acceptable

t value = observed correction of check std − accepted/SD of observed = [-0.3124 mg − (−0.354 mg)]/(0.5 × 0.061 mg) = t = 1.36

t value, 1.36 < 3; therefore check standard value is in control

The above simulated data cannot improve on the precision of the balance used in this example but do yield slightly different mass values. If the balance used had an accepted standard deviation of 1 µg instead of 42 µg, these changes would be meaningful in terms of a much-reduced measurement uncertainty. In this case, the t value was improved while the standard deviation was unchanged. Had a better balance been used with appropriate sensors to measure air temperature, pressure, and relative humidity, a significant improvement in uncertainty could have been attained.

For example, the standard deviation of the fit using a 1-µg standard deviation balance may have been 12 µg using an average air density. This could be reduced to 2 µg or less had the air density been determined for each Y_i.

A weakness of the weighing design method lies in the use of the mass difference between the starting 1-kg standards as the check standard. With use, the kilogram standards wear and simultaneously accumulate contamination together and at about the same rate. The effects result in undetectable drifts. Therefore, starting kilograms need maintenance, cleaning, and recalibration.

References

1. Cameron, J. M., Croarkin, M. C., and Raybold, R. C., Designs for the Calibration of Standards of Mass, NBS Technical Note 952, 1977.
2. Jaeger, K. B. and Davis, R. S., A Primer of Mass Megrology, NBS Special Publication 770-1, Industrial Measurement Series, 1984.
3. Deming, W. Edwards, *Statistical Adjustment of Data*, John Wiley & Sons, New York, 1938, 14.
4. Organisation International de Metrologie Legale, OIML R111, Weights of Classes E_1, E_2, F_1, F_2, M_1, M_2, M_3, 1994.
5. Davis, R. S., Private communication.

9

Calibration of the Screen and the Built-in Weights of a Direct-Reading Analytical Balance

9.1 Calibration of the Screen

When using a one-pan two-knife-edge, it is necessary to calibrate that part of the balance indication that does not relate to the built-in balance weights, i.e., the "screen" indication. We distinguish between the calibration of a small part of the screen to arrive at K_C in the section on the calibration of the built-in weights of a single-pan direct-reading analytical balance and the calibration of the entire screen for use in normal weighing on the balance. The screen can be calibrated in the same manner as the calibration of built-in weights.

It is convenient to determine a single calibration factor, K_S, to be applied to the entire screen. For greatest accuracy, the lack of linearity of the screen indication must be taken into account. The nonlinearity can be determined by observing the change in screen indication with the application of standard weights at various points in the range of the screen.

If the nonlinearity is of consequence to the user, a series of such observations of change in screen indication can be fitted to an empirical formula that can be applied in subsequent use of the balance. This procedure is discussed in detail in Ref. 1. In the present section we shall be concerned with the determination of the screen calibration factor, K_S.

The screen can be calibrated in the same manner that will be discussed in a later section on the calibration of built-in weights.

The balance is first zeroed and then a standard weight of mass equal to the nominal full-scale screen indication is placed on the balance pan. The screen indication is then observed. The calibration factor, K_S, is equal to the ratio of the true mass of the standard weight, corrected for air buoyancy, to the screen indication.

The value of K_S can also be determined by calibrating the screen using the least increment of the built-in weights. Sufficient mass ("tare") is placed on the balance pan to bring the scale indication to near full scale at a particular dial setting, D_1. The screen indication, O_1, is noted. The dial setting is increased by one least increment and the new dial setting, D_2, and the new screen indication, O_2 (near zero) are noted. K_S is then calculated to be

$$K_S = \left(D_2 - D_1\right)_t \left[1 - \left(\rho_a/\rho_b\right)\right] / \left(O_1 - O_2\right), \tag{9.1}$$

where ρ_a is air density, ρ_b is the density of the built-in weight, and $(D_2 - D_1)_t$ is the mass difference corresponding to the two dial settings.

9.2   Calibration of the Built-in Weights

The built-in weights of a single-pan direct-reading analytical balance are usually adjusted by the manufacturer to one of the "apparent mass" scales (see Chapter 15). The "apparent mass" of an object is the mass that, under specified ambient conditions (20°C and air density of 0.0012 g/cm⁻³), exerts the same force on a balance as the same mass of a reference material of specified hypothetical density.

The hypothetical densities corresponding to the two mass scales in current use are 8.0 g/cm⁻³ (approximately the density of stainless steel) and 8.39039 g/cm⁻³ (approximately the density of brass).

It is necessary to convert from the apparent mass to the approximate true mass of the built-in weights by using the equation:

$$M_b = M_r\left[1-\left(0.0012/\rho_r\right)\right]\Big/\left[1-\left(0.0012/\rho_b\right)\right]$$
$$= M_r Q,$$

(9.2)

where M_b is the approximate mass of the built-in weight, M_r is the mass of the hypothetical reference material (that is, the dial reading of the balance), ρ_b is the density of the built-in weight, and ρ_r is the hypothetical density[1] (see Chapter 15).

The values of ρ_b and ρ_r are supplied by the manufacturer. Values of the ratio on the right-hand side of Eq. (9.2), Q, are listed in Table 9.1.

TABLE 9.1   Examples of Values of Q,
$[1 - (0.0012/\rho_r)]/[1 - (0.0012/\rho_b)]$

ρ_b	ρ_r = 8.0 g/cm³	ρ_r = 8.3909 g/cm³
7.76	1.000 004 6	1.000 001 6
7.80	1.000 003 8	1.000 010 8
7.90	1.000 001 9	1.000 008 9
8.00	1.000 000 0	1.000 007 0

Source: Schoonover, R. M. and Jones, F. E.,
Anal. Chem., 53, 900, 1981.

The value of M_b calculated by use of Eq. (9.2) may differ from the true mass value of the built-in weight, although it might be sufficiently close to the true mass value for the particular measurement the analyst wishes to make.

For more accurate measurements it is necessary to calibrate the built-in weights using standard weights. Since the built-in weights are used in combination in the operation of the balance, it is necessary only to calibrate the individual weights and add the results to arrive at the values for the calibrations.

In the calibration of a built-in weight,[1] a standard weight of mass S (from a set of standard weights covering the mass range of the built-in weights) approximating the mass, D_i ($D_i = \Sigma_o - \Sigma_i$), of the built-in weight, and a small weight of mass Δ approximately one fourth of the range of the screen are used on the balance pan to generate the following balance equations:

$$L_o\left[1-\left(\rho_a/\rho_{L_o}\right)\right]=\Sigma_o\left[1-\left(\rho_a/\rho_b\right)\right]-O_1K_C,$$

(9.3)

$$L_o\left[1-\left(\rho_a/\rho_{L_o}\right)\right]=S\left[1-\left(\rho_a/\rho_S\right)\right]+\Sigma_i\left[1-\left(\rho_a/\rho_b\right)\right]-O_2K_C,$$

(9.4)

$$L_o\left[1-\left(\rho_a/\rho_{L_o}\right)\right]=S\left[1-\left(\rho_a/\rho_S\right)\right]+\Sigma_i\left[1-\left(\rho_a/\rho_b\right)\right]+\Delta\left[1-\left(\rho_a/\rho_\Delta\right)\right]-O_3K_C,\qquad(9.5)$$

$$L_o\left[1-\left(\rho_a/\rho_{L_o}\right)\right]=\Sigma_o\left[1-\left(\rho_a/\rho_b\right)\right]+\Delta\left[1-\left(\rho_a/\rho_\Delta\right)\right]-O_4K_C,\qquad(9.6)$$

where Σ_o is the mass of all of the built-in weights; Σ_i is the mass of all the built-in weights less the mass of the weight removed for calibration; $(\Sigma_o-\Sigma_i)$ corresponds to the dial reading when the standard weight is on the balance pan; L_o is the balance tare; the O terms are the screen observations; K_C is a factor that calibrates the screen over about one fourth the range; ρ_a is the density of air; ρ_{L_o}, ρ_S, ρ_b, and ρ_Δ are the densities of the tare, standard weight, built-in weight, and small weight, respectively. The screen reading of the unloaded pan, O_1, must be near zero. The density of air is calculated from measurements of pressure, temperature, and relative humidity by using an air density equation (see Chapter 12).

From the above set of balance equations,

$$D_i=\left\{S\left[1-\left(\rho_a/\rho_S\right)\right]-\left[\left(O_2+O_3-O_1-O_4\right)K_C/2\right]\right\}\Big/\left[1-\left(\rho_a/\rho_b\right)\right]\qquad(9.7)$$

and

$$K_C=\Delta\left[1-\left(\rho_a/\rho_\Delta\right)\right]\Big/\left(O_3-O_2\right).\qquad(9.8)$$

The sequence of measurements that generated the mass balance equations is repeated for each dial reading that represents a discrete weight; the remaining dial readings correspond to combinations of the discrete weights and are readily assigned mass values by summing the appropriate mass values.

K_C and K_S, mentioned earlier, have approximately the same value for a given balance.

Reference

1. Schoonover, R. M. and Jones, F. E., Air buoyancy correction in high-accuracy weighing on analytical balances, *Anal. Chem.*, 53, 900, 1981.

10

A Look at the Electronic Balance*

10.1 Introduction

If one were to wander through the nation's industrial facilities and laboratories, one would find electronic balances being used for everything from counting batches of resistors to adjusting the component ratio of epoxy mixtures. Many of these balances are suitable for the most demanding analytical work, whereas others are less precise but serve many purposes well.

In the following discussion, the analytical balance is defined as an instrument with capacity ranging from 1 g to a few kilograms and with a precision of at least one part in 10^5 of maximum capacity. Many modern electronic balances have a precision of one part in 10^7 at full capacity, and the accuracy is usually comparable.

The inspiration for the modern analytical balance came from a weighing method change suggested by Borda.[2] To overcome the difficulties of unequal arm lengths inherent with the two-pan equal-arm balance, Borda suggested a method known as substitution weighing to be used in place of transposition weighing.

In 1886, a balance was designed and built specifically for substitution weighing, although the modern one-pan substitution balance did not become commonplace until the 1950s.

Although these balances were completely mechanical in operation, the optical readout system was usually assisted with a light bulb.

10.2 The Analytical Balance and the Mass Unit

Before looking in detail at electronic balances, it is worthwhile to consider how mass is determined from a balance weighing and to look briefly at the inner workings of a modern mechanical substitution balance.

The reader should keep in mind that of the many forces the balance can respond to, we are interested in only the gravitational and buoyant forces and would like to exclude all others.

First, we must recognize that the balance reading is not the mass of the sample being weighed and therefore is not the desired result. The balance manufacturer has built the balance to indicate so-called *apparent mass* (see Chapter 15) of the material being weighed.

In essence, if the material being weighed has a density of 8.0 g/cm³ at 20°C and the air density is 0.0012 g/cm³, then and only then does the balance indicate the mass of the object being weighed. Obviously, these conditions are rarely met, and the balance reading must be corrected to obtain the desired mass.[3]

The important point of the above discussion is that all analytical balances are calibrated by the manufacturers to indicate what is known as "apparent mass vs. 8.0 g/cm³."

---

*Chapter is based on Ref. 1.

In some instances, the balance may actually contain a single weight or a set of weights, or a weight may be supplied separately by the balance manufacturer for calibration of the balance.

It is from this initial calibration and subsequent recalibrations that the user is tied to the mass unit, directly for standard conditions and by computation for all other conditions.

For many years, another apparent mass scale based on this principle had been in use. The basis of this scale was a brass weight the density of which was specified to be 8.3909 g/cm³ at 20°C with an air density of 0.0012 g/cm³. This scale is generally referred to as the "apparent mass vs. brass" scale.

10.3 Balance Principles

Today, there are two dominant types of electronic balances in use — the hybrid and the electromagnetic force balance. The hybrid balance uses a mix of mechanical and electronically generated forces, whereas the electromagnetic force balance uses electronically generated forces entirely.

A brief review of mechanical balance principles is worthwhile before the hybrid and electromagnetic force balances are discussed.

In all cases, a null indicator is used to determine when the internal force balances the force generated by the sample.

10.3.1 The Mechanical Balance

The modern mechanical balance is a one-pan two-knife balance with the force on the pan exerted by the object being weighed and a collection of built-in weights nearly counterbalanced by a fixed weight built into the balance beam. Any residual inequality of forces causes an angular displacement of the beam. Figure 10.1 is a sketch of a modern one-pan mechanical analytical balance.

The balance readout is attained by summing the weight dial indications with a beam displacement indicator such as a projected optical scale.

Whatever means are used to indicate the angular displacement of the balance beam, the balance is manufactured and calibrated to indicate the same apparent mass scale as that of the built-in weights.

The prominent features of the one-pan two-knife mechanical balance are the built-in weights (i.e., mass standards) and beam damping. However, of significance is the constant loading of the balance beam

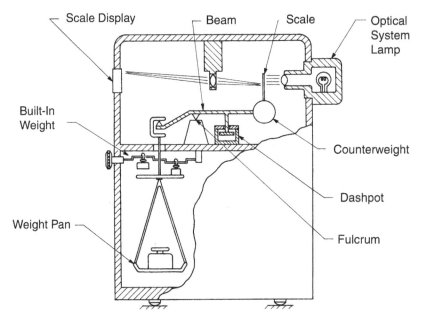

FIGURE 10.1 A modern one-pan mechanical analytical balance.

regardless of the sample weight on the pan. This feature provides a constant balance sensitivity; that is, the angular beam displacement in response to a small force change remains constant regardless of pan loading, a characteristic not found in equal-arm balances.

10.3.2 The Hybrid Balance

The hybrid balance is identical to the mechanical balance just described except that the balance beam is never allowed to swing through large angular displacements when the applied loading changes. Instead, the motion is very limited and when in equilibrium the beam is always restored to a predetermined reference position by a servo-controlled electromagnetic force applied to the beam.

10.3.3 The Electromotive Force Balance

The more recent development of the electromagnetic force balance is a radical departure from the past in several ways. First, an electromagnetic force balances the entire load either by direct levitation or through a fixed-ratio lever system. Second, the loading on the electromechanical mechanism that constitutes the balance is not constant, but varies directly with applied load. Finally, the sensitivity and response are no longer dominated by the dynamics of the balance beam but are largely controlled by servo system characteristics.

10.3.4 The Servo System

Servo systems differ in classification and in design, but a simple explanation of how a balance servo works in principle is given here. The details of the circuitry are omitted since there are many different means to achieve the desired result.

There are two distinct electronic approaches to the servo system. In one method there is a continuous current through the servomotor coil and in the other method the current is pulsed. The latter technique has the advantage of simpler coupling to digital readout indicators although both systems work well and, from the user's standpoint, there are many advantages available from the electronic nature of the mechanism regardless of its internal operation.

In an electromagnetic servo system (Figure 10.2), the force associated with the sample being weighed is mechanically coupled to a servomotor that generates the opposing magnetic force. When the two forces are

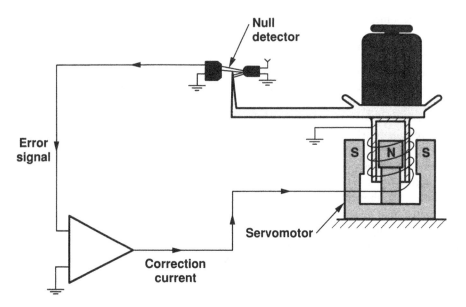

FIGURE 10.2 Simplified electromagnetic servo system.

in equilibrium, the error detector is at the reference position and the average electric current in the servomotor coil is proportional to the resultant force that is holding the mechanism at the reference position.

When the applied load changes, a slight motion occurs between the fixed and moving portions of the error-detector components, resulting in a very rapid change in current through the coil. The direction and magnitude of this current change are such that the equilibrium condition is restored.

The reference or "null" position is usually not arbitrarily chosen, but is selected to allow the flexure pivots to remain in a relaxed state with the beam parallel to the gravitational horizon.

10.4   A Closer Look at Electronic Balances

10.4.1 The Hybrid Balance

When the mechanical balance is modified by the addition of servo control and other electronics, several beneficial things occur.

Functionally, the most important change occurs when the balance beam is servoed to a reference or null position. Doing this not only results in an electronic output but also makes knife adjustments and edge quality less critical.

Furthermore, other forms of pivots, such as flexural pivots, taut fibers, etc., which are more rugged and economical to manufacture, can then be used in place of knife edges.

Like the purely mechanical balance, the balance output from a hybrid balance is a summation of the built-in weights being used for a particular weighing and the restoring force required to hold the beam at the null position. In some balances the output reading indicates both of these in a single digital display, the necessary summing having taken place electronically.

The most salient features that distinguish the electronic hybrid from the electromagnetic force balance are the still-recognizable balance beam and the built-in weights. Equilibrium is closely approximated by the selected built-in weights so that the forces generated by the servomotor are small.

If industry trends continue, the hybrid balance will eventually disappear at capacities much above 1 g, but will probably survive for some time to come at the lower capacities. At these small capacities there is at present a problem of balance precision surpassing the expressed uncertainty of the built-in weights and any external weight sets that may be required in the use of a balance. Normally, the manufacturer will calibrate these weights to a tolerance comparable to the balance precision. When this is not possible, these weights are sometimes viewed merely as tare weights.

10.4.2 The Force Balance

Although the top-loading electronic balance is a familiar sight in many situations, its principle of operation is least understood by most operators. In recent years the force balance principle has also been incorporated into classical enclosures, similar to those used for the early mechanical balances.

For this reason, it is convenient to refer to the electromechanical heart of the balance as the electromagnetic force-balance cell or, simply, the cell. The logic behind having two distinct instrument configurations will become apparent later in the discussion.

Like the hybrid balance, one of the outstanding features of the cell is the principle of servoing the mechanical mechanism to a null position. In either configuration, the forces generated by the servo motor are much larger than those required for an equivalent-capacity hybrid balance.

Of course, the balance beam, if any, no longer resembles its predecessor, and it is always tethered by a parallelogram loading constraint (guides).

Unlike the mechanical and hybrid balances that have a counterweight and therefore a compensating buoyant force, the higher-precision cells are more reflective of air density changes and may require frequent calibration.

In the top-loading configuration of the weighing cell, the sample to be weighed is loaded on a weight pan and the loading guides prevent torsional forces, caused by off-center loading, from perturbing the alignment of the balance mechanism.

These same guides may serve to stabilize the moving portion of the servo motor, or an additional restraint may perform this task.

As previously described, an electromagnetic force is generated to oppose the net gravitational and buoyant force imposed by the mass being weighed. The balance readout is proportional to the amount of current passing through the coil when the equilibrium position is established. The constant of proportionality provides the conversion from current units to apparent mass units.

Calibration is performed by the application of a calibrating weight and adjustment of the circuitry to indicate the apparent mass of the calibrating weight.

Higher precision can be obtained if the weighing pan is placed below the cell rather than above it. This is due to better axial alignment at the pan hook (minimum off-center loading) and a reduction in servomotor force with a corresponding drop in capacity.

Both configurations of the cell require air-draft shielding for high levels of precision.

The industry trend at this writing is toward the complete elimination of the hybrid balance at the very highest levels of precision and toward increasing capacity.

10.5 Benefits and Idiosyncrasies of Electronic Balances

Because the modern analytical balance is electronic in nature, it is very often found with a microprocessor as part of the package. The level of sophistication of these instruments may be very high, and there are many balance functions and idiosyncrasies that are unfamiliar to those who are acquainted only with mechanical balances. Let us examine the electronic capabilities that are inherent in the basic balance and those that may be optional.

10.5.1 Benefits

It is assumed that all balances have the a digital display to indicate "weight" and a means to provide a zero indication when no load is applied to the weigh pan. Beyond these basic features we are apt to find taring control, dual capacity and precision, selectable sampling period, etc. A brief summary of these features is given below.

10.5.1.1 Taring Control

A taring control is a means to ignore the indication of a weighing constant, such as a weighing boat, by forcing a zero indication when the boat is placed on the balance pan. Most balances permit taring to 100% of capacity.

10.5.1.2 Dual Capacity and Precision

This feature allows the balance capacity to be decreased by a predetermined amount with a comparable gain in precision, i.e., a two-for-one balance. The choice of where to place the smaller, more precise range can vary from one balance to another.

10.5.1.3 Selectable Sampling Period

The time required for a balance to obtain a reliable indication under ambient conditions may vary from one location to another; therefore, provision is made so that the user can vary the integration period by a simple circuit change. As the sampling period is lengthened, the overall balance weighing cycle is likewise lengthened.

10.5.1.4 Filters

Some manufacturers provide electronic filters that eliminate certain portions of the noise spectrum from the servo loop.

10.5.1.5 Computer Compatibility

A means is provided to access the balance by an external computer that may range from a simple data transfer such as a binary coded decimal (BCD) output to complete functional control as offered by an RS-232C or IEEE-488 interface bus.

10.5.1.6 Computation

A balance option is provided that performs computations such as counting, standard deviation calculations, or user-programmable computations.

10.5.1.7 Environmental Weighing Delay

Some balances have a circuit that protects the user from collecting poor data, due to unusually strong air currents or vibrations, by not displaying the weighing results during these periods.

Some balances may have an override capability provided by the manufacturer for the acquisition of data during such periods.

10.5.2 Idiosyncrasies

Some idiosyncrasies of the electronic balance follow. They may present problems in some applications.

10.5.2.1 Weighing Ferromagnetic Materials

Ferromagnetic materials may perturb the magnetic field associated with the servomotor, leading to systematic errors in the weighing.

This can be checked by moving the material in and around the pan area while looking for large changes in the balance zero reading. An easy remedy is to weigh below the balance pan because magnetic forces fall off rapidly with distance.

The above effects would be even more pronounced if weighing magnetic materials were necessary.

10.5.2.2 Electromagnetic Radiation

Electronic balances may malfunction in the presence of a strong electromagnetic field. This effect may be checked by keying a handheld radio transmitter near the balance and looking for a change in balance indication.

10.5.2.3 Dust Susceptibility

Dust migration (from a dirty environment) into the gap between the pole pieces associated with the permanent magnet and moving coil of the servomotor can cause insidious changes in the balance precision and calibration. When these dust particles are ferromagnetic, the balance may be rendered inoperable. Such environments should be avoided.

There may also exist a severe explosion potential in this type of environment, mandating the use of an entirely mechanical balance.

10.6 Black Box Comparison

In selecting a particular balance for a task, the following are the familiar specifications to consider along with what has been covered above in detail: capacity, accuracy (agreement with mass scale), precision (reproducibility of a measurement), zero stability, linearity, built-in weight tolerance or calibration accuracy, speed of response, pan size (top-loading balances), temperature coefficient, resolution (least significant figure in the display), and power requirements.

Because all balance manufacturers calibrate their products to the 8.0 apparent mass scale, any potential user can test these specifications with the appropriate 8.0 laboratory weights.

It should be kept in mind that the manufacturer has tested the product under the best laboratory conditions and if a laboratory is not as conducive to weighing this will be reflected in the test results.

The analyst should keep in mind that the balance precision attained in the laboratory will usually dominate the measurement error. However, precision errors must be combined with linearity errors and weight calibration errors in determining the measurement uncertainty.

Repeating, if the density of the material being weighed is not close to 8.0 g/cm^3, there is a buoyancy correction that, when omitted, can be a significant systematic error.

10.7 The Future

If the past can be used as a guide to the future, we can expect improvements in balance structure that will allow for increasing applications. Thus, we can look forward to force-measuring transducers that require no moving parts or expensive mechanical adjustments and that are more rugged than present instruments. A few such balances are at this writing beginning to make their appearance in the market.

One recent new application is the immersed balance for the density determination of solid objects[4]; others will certainly follow.

References

1. Schoonover, R. M., A look at the electronic analytical balance, *Anal. Chem.*, 54, 973A, 1982.
2. Miller, W. H., On the construction of the new Imperial pound; and on the comparison of the Imperial standard pound with the Kilogramme des Archives, *Philos. Trans. R. Soc. London*, 146, 753, 1846.
3. Schoonover, R. M. and Jones, F. E., Air buoyancy correction in high accuracy weighing on analytical balance, *Anal. Chem.*, 53, 900, 1981.
4. Schoonover, R. M. and Davis, R. S., Quick and accurate density determination of laboratory weights, In *Weighing Technology*, 8th Conference of the IMEKO Technical Committee TC3, Krakow, Poland, Sept. 9–11, 1980 (Polish Federation of Engineering Associations).

11

Examples of Buoyancy
Corrections in Weighing

11.1 Introduction

Weighing on a balance essentially involves the balancing of forces. The action of the acceleration due to gravity on the weighed object generates a vertical force on the balance pan. The action of the acceleration due to gravity on the built-in weight or weights in the balance or on an external standard weight generates an offsetting force. In addition to these gravitational forces, the various objects and weights are partially supported by buoyant forces.

For accurate weighing, corrections accounting for the buoyant forces must be applied. It is the objective of this chapter to discuss buoyancy corrections and the application of buoyancy corrections to mass determination.

11.2 Buoyant Force and Buoyancy Correction

The downward vertical gravitational force exerted on a balance by an object is

$$\mathbf{F}_g = Mg, \tag{11.1}$$

where $\mathbf{F}_g$ is the gravitational force, M is the mass of the object, and g is the local acceleration due to gravity.

The upward vertical buoyant force exerted on the body by the air in which the weighing is made is

$$\mathbf{F}_b = \rho_a g\left(M/\rho_m\right) = M\left(\rho_a/\rho_m\right)g, \tag{11.2}$$

where $\mathbf{F}_b$ is the bouyant force, ρ_a is the density of the air, and ρ_m is the density of the object weighed. This net upward vertical force can be called the buoyant force.

The overall net vertical force, $\mathbf{F}$, is

$$\mathbf{F} = \mathbf{F}_g - \mathbf{F}_b = M\left[1 - \left(\rho_a/\rho_m\right)\right]g. \tag{11.3}$$

The quantity $[1 - \rho_a/\rho_m)]$ when applied to weighing is a buoyancy correction factor. The net force exerted on the balance is thus less than the gravitational force.

For a reference air density (see Chapter 15), ρ_a, of 0.0012 g/cm^3 and an object density, ρ_m, of 8.0 g/cm^3 (approximating the density of stainless steel),

$$\left(\rho_a/\rho_m\right) = 0.00015$$

and the buoyancy correction factor is

$$\left[1-\left(\rho_a/\rho_m\right)\right]=0.99985.$$

The buoyancy correction, $(\rho_a/\rho_m)M$, $0.00015M$, corresponds to 150 parts per million (ppm) of M or 0.015% of M.

If water, with an approximate density of 1.0 g/cm³, were being weighed, the buoyancy correction would be 1200 ppm of M or 0.12% of M.

For the nominal stainless steel example, the correction for a mass of 1 kg would be 150 mg; for a mass of 100 g, the correction would be 15 mg. For the water case, the correction for a mass of 1 kg would be 1.2 g; for a mass of 100 g, the correction would be 120 mg.

For a reference air density, ρ_o, of 0.0012 g/cm³ and substance densities from 0.7 to 22 g/cm³, the values of the buoyancy correction factors range from 0.9982857 to 0.9999455, and the values of the buoyancy correction range from 1714.3 to 54.5 ppm. Values of the buoyancy correction factor and the ratio ρ_a/ρ_m are tabulated in Table 11.1.

The significance of these corrections depends on the desired accuracy for the particular substance in the particular situation and on the precision of the balance, among other things.

TABLE 11.1   Buoyancy Correction Factors and Ratios, $A = [1 - (\rho_a/\rho_m)]$, $B = (\rho_a/\rho_m)$

			B	
g/cm³	A	%	ppm	mg/100 g
0.7	0.998286	0.1714	1714	171.4
1.0	0.9988	0.12	1200	120.0
1.5	0.9992	0.08	800	80.0
2.0	0.9994	0.06	600	60.0
3.0	0.9996	0.04	400	40.0
4.0	0.9997	0.03	300	30.0
5.0	0.99976	0.024	240	24.0
6.0	0.9998	0.020	200	20.0
7.0	0.999829	0.0171	171	17.1
8.0	0.99985	0.015	150	15.0
9.0	0.999867	0.0133	133	13.3
10.0	0.99988	0.012	120	12.0
11.0	0.999891	0.0109	109	10.9
12.0	0.9999	0.01	100	10.0
13.0	0.999908	0.0092	92	9.2
14.0	0.999914	0.0086	86	8.6
15.0	0.99992	0.0080	80	8.0
16.0	0.999925	0.0075	75	7.5
16.5	0.999927	0.0073	73	7.3
17.0	0.999929	0.0071	71	7.1
18.0	0.999933	0.0067	67	6.7
19.0	0.999937	0.0063	63	6.3
20.0	0.999940	0.0060	60	6.0
21.0	0.999943	0.0057	57	5.7
22.0	0.999945	0.0055	55	5.5

11.3 Application of the Simple Buoyancy Correction Factor to Weighing on a Single-Pan Two-Knife Analytical Balance

In a simple case using a single-pan analytical balance, the unknown mass of an object, M_x, is balanced by built-in weights of total mass, M_s. The balance indication, U_s, is equal to M_s. It is assumed that the built-in weights exactly balance M_x.

The force exerted on the balance pan by the *object X* of mass M_x is

$$F_x = M_x\left[1-\left(\rho_a/\rho_x\right)\right]g, \tag{11.4}$$

where ρ_a is the density of air and ρ_x is the density of the weight (object X).

The force exerted on the balance pan by an *assemblage of built-in weights* of total mass M_s is

$$F_s = M_s\left[1-\left(\rho_a/\rho_s\right)\right]g, \tag{11.5}$$

where ρ_s is the density of the assemblage of built-in weights.

When these forces are equal, that is, the object is balanced by the built-in weights,

$$F_x = F_s, \tag{11.6}$$

$$M_x\left[1-\left(\rho_a/\rho_x\right)\right] = M_s\left[1-\left(\rho_a/\rho_s\right)\right], \tag{11.7}$$

$$M_x = M_s\left[1-\left(\rho_a/\rho_s\right)\right]\Big/\left[1-\left(\rho_a/\rho_x\right)\right]. \tag{11.8}$$

We now make calculations of M_x for several values of ρ_x and for the following fixed values:

$$\rho_a = 0.0012 \ \text{g}/\text{cm}^3$$

$$t = t_o = 20°\text{C}$$

$$\rho_s = 8.0 \ \text{g}/\text{cm}^3$$

$$M_s = 1000 \ \text{g}$$

For $\rho_x = 1.0 \ g/cm^3$, the approximate density of water,

$$M_x = 1000\left[1-\left(0.0012/8.0\right)\right]\Big/\left[1-\left(0.0012/1.0\right)\right]$$

$$M_x = 1001.051 \ \text{g}.$$

That is, under these conditions, *1001.051 g of weight X of density 1.0 g/cm³* would balance 1000 g of S weights of density 8.0 g/cm³. The difference, 1.051 g, between the masses of X and the S weights is due to the difference in buoyant forces acting on the object and the weights. This, of course, is due to the difference in density of the object and the weights and, consequently, to the difference in volume of air displaced by the object and the weights.

For $\rho_x = 2.7\ g/cm^3$ (density of aluminum),

$$M_x = 1000\left[1-\left(0.0012/8.0\right)\right]\big/\left[1-\left(0.0012/2.7\right)\right],$$

$$M_x = 1000.295\ \text{g}.$$

For $\rho_x = 16.6\ g/cm^3$ (density of tantalum),

$$M_x = 1000\left[1-\left(0.0012/8.0\right)\right]\big/\left[1-\left(0.0012/16.6\right)\right],$$

$$M_x = 999.922\ \text{g}.$$

For $\rho_x = 21.5\ g/cm^3$ (density of platinum),

$$M_x = 1000\left[1-\left(0.0012/8.0\right)\right]\big/\left[1-\left(0.0012/21.5\right)\right],$$

$$M_x = 999.906\ \text{g}.$$

We note that, for these latter two cases, M_x is less than 1000 g. This is, of course, because ρ_x is greater than 8.0 g/cm³ in these two cases.

For $\rho_x = 8.0\ g/cm^3$ (approximate density of stainless steel alloys),

$$M_x = 1000\left[1-\left(0.0012/8.0\right)\right]\big/\left[1-\left(0.0012/8.0\right)\right],$$

$$M_x = 1000\ \text{g}.$$

For $\rho_x = 8.0$ g/cm³, we have the conditions that define apparent mass (see Chapter 15).

11.4   The Electronic Analytical Balance

In an electronic force balance (see Chapter 10):

1. An electronic force is generated to oppose the net gravitational and buoyant force imposed by the object being weighed.
2. The readout of the balance is proportional to the current in a servomotor coil.
3. In calibration of the balance, a built-in calibrating weight is used and the electronic circuitry is adjusted so that the readout indicates the approximate *apparent mass* of the calibrating weight.

11.4.1   Electronic Balance Calibration and Use

The calibration and performance of an electronic balance with a built-in calibrating weight is now investigated.

The *true mass* (which should be referred to as mass) of the built-in calibrating weight is assumed to be *100 g*. Throughout this discussion, M with no superscript refers to true mass. The density of the calibrating weight, ρ_b, is assumed to be *8.0 g/cm³*, ρ_r. The temperature at which the balance is calibrated at the factory is assumed to be *20°C*, and the air density at the factory is assumed to be *0.0012 g/cm³*.

The defining equation for *apparent mass* is

$$^A M_x = {}^T M_r = {}^T M_x\left[1-\left(\rho_o/\rho_x\right)\right]\big/\left[1-\left(\rho_o/\rho_r\right)\right], \tag{11.9}$$

where ^AM_x is the apparent mass of the object of interest; ^TM_r is the true mass of a quantity of reference material of density 8.0 g/cm³; ^TM_x is the true mass of the object; ρ_o is the reference air density, 0.0012 g/cm³; ρ_x is the density of the object; and ρ_r is the density of the reference material, 8.0 g/cm³.

Under the above conditions,

$$^TM_x = 100 \text{ g} = {}^TM_b$$

$$\rho_o = 0.0012 \text{ g/cm}^3$$

$$\rho_r = 8.0 \text{ g/cm}^3 = \rho_b$$

$$^TM_r = 100 \text{ g}$$

$$\rho_x = 8.0 \text{ g/cm}^3,$$

where ^TM_b and ρ_b are the true mass and density, respectively, of the calibrating or built-in weight.

Note that "mass" is a *property* and "weight" is an *object*. Thus,

$$^AM_b = 100\left[1-\left(0.0012/8.0\right)\right]\Big/\left[1-\left(0.0012/8.0\right)\right] = {}^TM_b = 100 \text{ g}. \tag{11.10}$$

Therefore, at the factory under the above conditions, the apparent mass of the calibrating weight is equal to the true mass of the calibrating weight.

At the factory, the balance is calibrated (in this case, using the built-in calibrating weight) to indicate apparent mass or the conventional value of weighing in air (see Chapter 15).

In the balance, an electromotive force, F, is generated to equal and oppose the net force impressed on the balance pan by the gravitational force minus the buoyant force. The electromotive force, F, is generated by the current, I, passing through the coil of an electromotive force cell. F is proportional to I. The indication of the balance, U, is proportional to I at equilibrium. Thus,

$$F = kI, \tag{11.11}$$

$$U = cI, \tag{11.12}$$

where k and c are constants of proportionality.

$$I = F/k = U/c, \tag{11.13}$$

$$U = \left(c/k\right)F = KF, \tag{11.14}$$

where $(c/k) = K$.

Again, at the factory,

$$\rho_a = \rho_o = 0.0012 \text{ g/cm}^3$$

$$g = g_f = \text{the acceleration due to gravity at the balance location in the factory}$$

$$\rho_b = \rho_r = 8.0 \text{ g/cm}^3$$

$$t = t_o = 20°C$$

$$1/K = K_o.$$

The force, F_x, exerted on the balance by an unknown mass, M_x, is

$$F_x = M_x\left[1-\left(\rho_o/\rho_x\right)\right]g_f = U_x K_o. \tag{11.15}$$

The force, F_b, exerted on the balance by the built-in calibrating weight of mass M_b and density 8.0 g/cm³, is

$$F_b = M_b\left[1-\left(\rho_o/\rho_b\right)\right]g_f = U_b K_o. \tag{11.16}$$

At the factory, the electronics are adjusted in such a way that the indication of the balance, U_b, is equal to the apparent mass of the built-in weight (100 g, for example) with the built-in weight introduced to the balance. We shall refer to this operation as the *adjusting* of the balance rather than the *calibration* of the balance.

Thus, under the above conditions,

$$U_b = {}^A M_b = {}^T M_b. \tag{11.17}$$

Then F_b is given by

$$F_b = U_b\left[1-\left(\rho_o/\rho_b\right)\right]g_f = U_b K_o, \tag{11.18}$$

and, thus,

$$K_o = \left[1-\left(\rho_o/\rho_b\right)\right]g_f. \tag{11.19}$$

At the *factory*, for a *standard* weight of true mass M_s, density ρ_s and the conditions:

$$\rho_a = \rho_o = 0.0012 \text{ g/cm}^3$$
$$\rho_s = \rho_s$$
$$\rho_b = 8.0 \text{ g/cm}^3$$
$$t_o = 20°C$$
$$g = g_f$$
$$1/K = K_o,$$

the force exerted on the balance is

$$F_s = M_s\left[1-\left(\rho_o/\rho_s\right)\right]g_f = U_s K_o = U_s\left[1-\rho_o/\rho_b\right]g_f, \tag{11.20}$$

where U_s is the balance indication with M_s on the balance. Then,

$$M_s = U_s\left[1-\left(\rho_o/\rho_b\right)\right]/\left[1-\left(\rho_o/\rho_s\right)\right], \tag{11.21}$$

and

$$U_s = M_s\left[1-\left(\rho_o/\rho_s\right)\right]\big/\left[1-\left(\rho_o/\rho_b\right)\right]. \tag{11.22}$$

This last equation is recognized to be the definition of the apparent mass of the standard weight, M_s, since $\rho_b = \rho_r$. Therefore, the indication of the balance is the apparent mass of the standard weight:

$$U_s = {}^A M_s.$$

Similarly, for a weight of unknown true mass M_x,

$$U_x = M_x\left[1-\left(\rho_o/\rho_x\right)\right]\big/\left[1-\left(\rho_o/\rho_b\right)\right], \tag{11.23}$$

and, again, the balance indication is equal to the apparent mass of the weight on the pan:

$$U_x = {}^A M_x.$$

11.4.2 Usual Case for Which the Air Density Is Not the Reference Value

The more usual case for which the air density, ρ_a, is not equal to the reference value, $\rho_o = 0.0012$ g/cm³, is now considered.

In the laboratory, the balance is adjusted using the built-in weight and the conditions:

$$\rho_a = \rho_a$$
$$t = 20°C$$
$$g = g_L$$
$$\rho_b = 8.0 \text{ g/cm}^3,$$

where g_L is the local acceleration due to gravity in the laboratory at the location of the balance.

The force exerted on the balance by the introduction of the built-in weight of mass M_b is

$$F_b = M_b\left[1-\left(\rho_a/\rho_b\right)\right]g_L = U_b K_L. \tag{11.24}$$

The electronics of the balance are adjusted so that the scale indication, U_b, is equal to the apparent mass of the built-in weight (which is also the true mass). Then,

$$K_L = \left[1-\left(\rho_a/\rho_b\right)\right]g_L. \tag{11.25}$$

For a *standard* weight of true mass M_s on the pan of the balance, the force exerted by the standard weight is

$$F_s = M_s\left[1-\left(\rho_a/\rho_s\right)\right]g_L = U_s\left[1-\left(\rho_a/\rho_b\right)\right]g_L, \tag{11.26}$$

and

$$U_s = M_s\left(1-\rho_a/\rho_s\right)\big/\left(1-\rho_a/\rho_b\right), \tag{11.27}$$

and

$$M_s = U_s \left(1 - \rho_a/\rho_b\right) \big/ \left(1 - \rho_a/\rho_s\right). \tag{11.28}$$

Therefore, if the balance were operating perfectly, with the standard weight on the pan, the balance indication would be equal to the right side of Eq. (11.27). Deviation of the balance indication from this value would represent a weighing error or a random deviation.

For a weight of unknown mass, M_x, on the balance, the force exerted on the balance is

$$F_x = M_x \left[1 - \left(\rho_a/\rho_x\right)\right] g_L = U_x \left[1 - \left(\rho_a/\rho_b\right)\right] g_L, \tag{11.29}$$

and

$$U_x = M_x \left[1 - \left(\rho_a/\rho_x\right)\right] \big/ \left[1 - \left(\rho_a/\rho_b\right)\right], \tag{11.30}$$

and

$$M_x = U_x \left[1 - \left(\rho_a/\rho_b\right)\right] \big/ \left[1 - \left(\rho_a/\rho_x\right)\right]. \tag{11.31}$$

Therefore, if the balance were operating perfectly, with the unknown weight on the pan, the balance indication would be equal to the right side of Eq. (11.30), and the true mass of the unknown would be calculated using Eq. (11.31).

Note: It is the *true mass* of the unknown that is the desired mass quantity, *not* the *indication* of the balance, and *not* the *apparent mass* of the unknown. Even if the apparent mass were measured perfectly, a calculation must be made to determine the true mass.

11.5   Examples of Effects of Failure to Make Buoyancy Corrections

In the calibration of flowmeters for liquids at the National Institute of Standards and Technology (NIST), two liquids were used: (1) water and (2) Stoddard's solvent (Mil. Spec. 7024, Type 2).

The density of Stoddard's solvent at 60°F (15.56°C) is 770.9 kg/m³ or 0.7709 g/cm³. For an air density, ρ_a, of 0.0012 g/cm³; a mass of solvent, M_x, of 1000 g; a density of the balance built-in weight of 8.0 g/cm³; and a density of Stoddard's solvent of 0.7709 g/cm³; the indication of the balance using Eq. (11.30) is:

$$U_x = 1000 \left[1 - \left(0.0012/0.7709\right)\right] \big/ \left[1 - \left(0.0012/8.0\right)\right].$$

U_x is then 998.593 g.

If the indication of the balance, 998.593 g, were taken to be the measurement of the mass of the solvent, it would be in error by 1.407 g (1000 − 998.593) or 0.1407%, a quite significant error. This is the consequence of not making a buoyancy correction.

The true value of the mass of solvent would be calculated by applying the buoyancy correction, that is, by dividing the balance indication by the buoyancy correction factor, [(1 − (0.0012/0.7709)]/[1 − (0.0012/8.0)]:

$$M_x = 998.593 \big/ \left[1 - \left(0.0012/0.7709\right)\right] \big/ \left[1 - \left(0.0012/8.0\right)\right],$$

$$M_x = 1000 \text{ g.}$$

The density of air-saturated water[3] at 60°F (15.56°C) is 999.010 kg/m³ or 0.999010 g/cm³. For the other conditions in the Stoddard's solvent example, Eq. (11.30) becomes

$$U_x = 1000\left[1-\left(0.0012/0.999010\right)\right]\Big/\left[1-\left(0.0012/8.0\right)\right].$$

U_x is then 998.949 g.

If the indication of the balance, 998.949 g, were taken to be the measurement of the mass of the water, it would be in error by 1.051 g (1000 − 998.949) or 0.1051%, again a quite significant error. Again, this is the consequence of not making a buoyancy correction.

The true value of the mass of water would be calculated by dividing the balance indication by the buoyancy correction factor, [(1 − 0.0012/0.999010)]/[1 − (0.0012/8.0)]:

$$M_x = 998.949\Big/\left[1-\left(0.0012/0.999010\right)\right]\Big/\left[1-\left(0.0012/8.0\right)\right],$$

$$M_x = 1000 \text{ g}.$$

11.6 Other Examples of Buoyancy Correction

11.6.1 Weighing of Syringes[1]

The volume of a liquid dispensed from a syringe is determined from a calibration of the syringe. The syringe is calibrated[2] by filling it with water to a particular graduation on the barrel of the syringe, weighing the syringe before and after dispensing the water (in both cases the syringe is open to the atmosphere), and inferring the volume from the mass and density of the water dispensed.

In Ref. 2, it was demonstrated that the volume of water contained at the 0.500 graduation of a 1-ml-capacity syringe can be determined with a standard deviation of the mean (for 15 measurements), Type A, of 1.2 parts in 10,000. The buoyancy effect was 1.05 parts in 1000. Therefore, for this example, it is necessary to include the buoyancy effect in the calculation of syringe volume, or of the mass of water contained.

The mass balance equation for the weighing of the *empty syringe* is

$$M_Y\left[1-\left(\rho_a/\rho_Y\right)\right] = M_B\left[1-\left(\rho_a/\rho_B\right)\right]+\left(O_2-O_1\right)K, \tag{11.32}$$

where M_Y and ρ_Y are the mass and density, respectively, of the syringe; M_B and ρ_B are the mass and density, respectively, of the combination of built-in weights of a direct-reading single-pan analytical balance[1]; O_2 and O_1 are the screen indications of the direct-reading single-pan analytical balance with and without, respectively, the syringe on the balance pan; and K is the calibration factor for the screen. O_1 is usually adjusted to zero.

The mass balance for the weighing of the *syringe containing a mass M_W of water* is

$$M_Y\left[1-\left(\rho_a'/\rho_Y\right)\right] + M_w\left[1-\left(\rho_a'/\rho_w\right)\right] = M_c\left[1-\left(\rho_a'/\rho_B\right)\right]+\left(O_4-O_3\right)K, \tag{11.33}$$

where ρ_a' is the density of air at the time of the weighing; M_C and ρ_B are the mass and density, respectively, of the combination of built-in weights; ρ_w is the density of the water inferred from the temperature of the water; and O_4 and O_3 are the screen indications with and without, respectively, the syringe on the balance pan.

If the two weighings are done sufficiently close in time that the air density does not change significantly, $\rho_a = \rho_a'$. If O_1 and O_3 are set equal to zero, the mass of water, M_w, in the syringe is

$$M_w = \left\{\left(M_C-M_B\right)\left[1-\left(\rho_a/\rho_B\right)\right]+\left(O_4-O_2\right)K\right\}\Big/\left[1-\left(\rho_a/\rho_w\right)\right]. \tag{11.34}$$

11.6.2 Buoyancy Applied to Weighing in Weighing Bottles[1]

For the determination of the mass, M_m, of a solid material (granular, powder, etc.) or liquid of density ρ_m by weighing in a weighing bottle, an equation similar to Eq. (11.34) applies:

$$M_m = \left\{ (M_C - M_B)\left[1 - (\rho_a/\rho_B)\right] + (O_4 - O_2)K \right\} \Big/ \left[1 - (\rho_a/\rho_m)\right]. \qquad (11.35)$$

Application of Eq. (11.35) to the Determination of Mass of Titanium Dioxide Powder[1]

Titanium dioxide (TiO_2) powder was weighed in a weighing bottle on a direct-reading single-pan analytical balance. The values of the various quantities in Eq. (11.35) are listed below:

$$M_C = 92.5 \times 1.0000113 \text{ g}$$

$$M_B = 72.3 \times 1.0000113 \text{ g}$$

$$\rho_a = 0.001173 \text{ g}/\text{cm}^3$$

$$\rho_B = 7.78 \text{ g}/\text{cm}^3$$

$$\rho_m = 4.60 \text{ g}/\text{cm}^3$$

$$O_4 = 2083 \text{ divisions}$$

$$O_2 = 4399 \text{ divisions}$$

$$K = 0.000010 \text{ g}/\text{division}$$

The calculated value of the mass of TiO_2 using Eq. (11.35) and the above values is 20.1792 g.

A "Buoyancy Corrections in Weighing" course is included among the Appendices.

References

1. Kell, G. S., Effects of isotopic composition, temperature, pressure and dissolved gases on the density of liquid water, *J. Phys. Chem. Ref. Data*, 6, 1109, 1977.
2. Schoonover, R. M. and Jones, F. E., Air buoyancy correction in high-accuracy weighing on electronic balances, *Anal. Chem.*, 53, 900, 1981.
3. Jones, F. E., Telescopic viewer in syringe calibration, *Anal. Chem.*, 52, 364, 1980.

12

Air Density Equation

12.1 Introduction

In applying air buoyancy corrections to weighing, an equation to be used in calculating the density of air is required unless one makes a direct determination.

The air density, ρ_a, is typically determined from an equation of state for moist air. "In the 1970s, …, it was appreciated that the equation-of-state itself has great importance and that several such equations were in wide use. Furthermore, it had not been demonstrated experimentally that any of the equations-of-state in use were adequate for actual mass comparisons."[1]

In 1977, Jones made a definitive derivation of "a semi-empirical equation-of-state based on up-to-date data."[1] The equation was developed quickly and was published in 1978.[2]

The equation developed by Jones, "with minor changes,"[1] was endorsed for use in mass metrology by CIPM (Comité International des Poids et Mesures) in 1981.[3] The equation given in Ref. 3 is now referred to as the CIPM-81 equation-of-state for moist air and is used for mass metrology by most national laboratories. Use of CIPM-81 instead of its predecessor[2] makes a negligible change in routine mass calibrations.[1]

The efficacy of CIPM-81 and its predecessor was tested by determining the mass difference between two nominally equal weights with and without reliance on the equation-of-state. The measurement without reliance on the equation-of-state was typically made in vacuum. Results of this type of comparison done at the Physikalisch-Technische Bundesantalt (PTB)[4] agreed to within the expected uncertainty, 1×10^{-4} in air density, ρ_a.

The following is quoted from an excellent paper by Davis:[5]

> Recently, Balhorn [Balhorn, R., Berucksichtigung de Luftdichte durch Wagung beim Massevergleich, Kochsiek, M., ed., Massebestimmung hoher Genauigkeit, *PTB (F.R.G.) Ber.*, Me-60, 65, June 1984] has compared results obtained by using the CIPM equation and direct measurements based on Archimedes' principle (the latter measurements involve vacuum weighing to find mass differences independent of a buoyancy correction). His expected experimental uncertainty is at the level of that estimated by Jones for uncertainties in the equation-of-state. Balhorn finds no unexpected results. This is an important confirmation because Balhorn has measured air density by a buoyancy method. Similar results, but at a somewhat increased uncertainty, have been obtained in other laboratories as well.

Davis[6] pointed out in 1992 that in the 10 or so years since the publication of the BIPM/CIPM equation (referred to as CIPM-81) one of the most important constant parameters in the equation, the molar gas constant, had become better known; in addition, updated values for some constant parameters had become available.

The Comité Consultatif pour las Masse et les grandeurs apparentees (CCM), on the advice of the Working Group on Density, considered it worthwhile in 1991 to amend several of the constant parameters in the CIPM-81 equation; the basic form of the equation and the principles by which it was derived remained unchanged. The CCM recommended that the 1981 equation incorporating the amended parameters be designated as the "1981/1991 equation for the determination of the density of moist air." The CIPM accepted the amended equation in 1991 (abbreviated by Davis[6] as the "1981/1991 equation"). The 1981/1991 equation is valid over the same range of pressure, temperature, relative humidity (or dew-point temperature), and carbon dioxide mole fraction as the 1981 equation.

Davis[6] concluded that:

1. Air densities calculated using the 1981/1991 equation do not differ significantly from those calculated from the 1981 equation.
2. The overall uncertainty of the 1981/1991 equation is not significantly improved over that of the 1981 equation.
3. The reason for making the changes is to ensure that the values for all constant parameters used in the equation were the best currently available.

In this chapter, the development of Jones[2] will be presented in detail as background material. Then, the 1981 equation and the 1981/1991 equation will be presented and discussed. The BIPM equation should be used in practice.

12.2   Development of the Jones Air Density Equation

The ideal gas equation,

$$PV = nRT, \tag{12.1}$$

relates the total pressure, P, the total volume, V, and the absolute temperature, T, of an ideal gas or a mixture of ideal gases. The number of moles of the gas or the mixture of gases is n, and R is the universal gas constant or molar gas constant.

In terms of density, ρ, rather than volume, Eq. (12.1) becomes

$$P = \rho RT / M, \tag{12.2}$$

where M is the molecular weight of the gas or the apparent molecular weight of the mixture.

For a mixture of dry air (indicated by the subscript a) and water vapor (indicated by the subscript w), ρ and M are respectively, the density and apparent molecular weight of the air–water vapor mixture. Because

$$M = m/n = \left(m_a + m_w\right)/\left(n_a + n_w\right), \tag{12.3}$$

where m is the mass of the mixture and n is the number of moles of the mixture,

$$M = \left(n_a M_a + n_w M_w\right)/\left(n_a + n_w\right) = M_a\left[\left(1 + n_w M_w / n_a M_a\right)/\left(1 + n_w / n_a\right)\right]. \tag{12.4}$$

We now introduce the water vapor mixing ratio, r:

$$r = \left(\text{mass of water vapor}/\text{mass of dry air}\right) = n_w M_w / n_a M_a, \tag{12.5}$$

and designate the ratio M_w/M_a by ε, whereby Eq. (12.4) becomes

$$M = M_a\left(1 + r\right)/\left(1 + r/\varepsilon\right). \tag{12.6}$$

We now substitute Eq. (12.6) in Eq. (12.2) and note that the effective water vapor pressure, e', in moist air is defined[7] by

$$e' = rP/(\varepsilon + r).$$ (12.7)

Then

$$P = (\rho RT/M_a)\{1/[1+(\varepsilon-1)e'/P]\}.$$ (12.8)

Eq. (12.8) is the *ideal* gas equation for a mixture of dry air and water vapor with water vapor pressure of e'. If the air–water vapor mixture behaved as a mixture of ideal gases,

$$P/[(\rho RT/M_a)\{1/[1+(\varepsilon-1)e'/P]\}] = Z = 1,$$ (12.9)

where Z is the compressibility factor.

Since a mixture of air and water vapor is not ideal, the magnitude of the nonideality is reflected in the departure of Z from 1. Eq. (12.9) then becomes

$$P = (\rho RTZ/M_a)\{1/[1+(\varepsilon-1)e'/P]\}.$$ (12.10)

Eq. (12.10) is the *real* gas equation for a mixture of dry air and water vapor. By rearrangement of Eq. (12.10), the expression for the air density is

$$\rho = (PM_a/RTZ)[1+(\varepsilon-1)e'/P].$$ (12.11)

12.2.1 Parameters in the Jones Air Density Equation

12.2.1.1 Universal Gas Constant, *R*

The value of the molar gas constant, R, listed in a compilation by Cohen and Taylor,[8] is 8.31441 ± 0.00026 J/K/mol. Quinn et al.[9] made a new determination of R by measuring the speed of sound in argon by means of an acoustic interferometer. Their value was 8315.73 ± 0.17 J/K/kmol. Gammon[10] deduced a value of R from measurements of the speed of sound in helium; his later reported value[11] is 8315.31 ± 0.35 J/K/kmol, which, considering the uncertainties, was in close agreement with the Quinn et al. value. Rowlinson and Tildesley[12] interpreted the experimental measurements of Quinn et al. and arrived at a value of 8314.8 ± 0.3 J/K/kmol, which is in close agreement with the Cohen and Taylor value within the uncertainties attached to the values. All of the uncertainties listed are estimates of standard deviation.

Jones[2] chose to use the Cohen and Taylor value "with the realization that in the future it might be replaced by a new value." Such a new value has appeared. Moldover et al.[13,14] reported a new experimental value of the gas constant of 8.314471 ± 0.000014 J/mol/K, where the uncertainty quoted is the estimate of the standard deviation. This uncertainty is smaller by a factor of 5 than that of earlier values. The Moldover et al. value is the preferred value and will be used throughout this chapter.

12.2.1.2 Apparent Molecular Weight of Air, *M_a*

The apparent molecular weight of dry air, M_a, is calculated as a summation, Σ, using the relationship

$$M_a = \Sigma_i^k M_i x_i,$$ (12.12)

where each M_i is the molecular weight of an individual constituent and x_i is the corresponding mole fraction, the ratio of the number of moles of a constituent to the total number of moles in the mixture.

The molecular weights and typical mole fractions of the constituents of dry air are tabulated in Table 12.1. Other constituents are present in abundances that are negligible for the present application.

The values of the atomic weights of the elements[15] are based on the carbon-12 scale. The molecular weights are taken to be the sums of the atomic weights of the appropriate elements.

The value for the abundance of oxygen is taken from Ref. 16.

The value for the abundance of carbon dioxide is taken from an unpublished compilation of data on *atmospheric* concentration of carbon dioxide at seven locations throughout the world. It must be emphasized that 0.00033 was the mole fraction of CO_2 in the atmosphere and should be considered a *background* value. The mole fraction of CO_2 in laboratories, which is, of course, the value of interest here, is in general greater than 0.00033 and is *variable*. For example, three samples of air taken from a glove box in the Mass Laboratory at the National Bureau of Standards (now NIST) had a mean value of 0.00043, and four samples of laboratory air taken at the National Center for Atmospheric Research in Boulder, CO had a mean value of 0.00080. Clearly, then, the optimum utilization of the air density calculation would necessitate a measurement of CO_2 abundance of an air sample taken at the time of the laboratory measurements of interest.

One of the options one has in dealing with the variability of CO_2 abundance is to select a reference level, for example, 0.00033 or 0.00043, and to provide an adjustment to M_a to account for known departures from the reference level.

Gluekauf,[17] in discussing the variation of the abundance of oxygen in the atmosphere, stated that "all major variations of the O_2 content must result from the combustion of fuel, from the respiratory exchange of organisms, or from the assimilation of CO_2 in plants. The first process does not result in more than local changes of O_2 content, while the latter two processes, though locally altering the CO_2/O_2 ratio, leave their sum unchanged."

The constancy of the sum is expressed by the equation (for convenience, the subscript i has been replaced by the chemical symbol):

$$x_{CO2} + x_{O2} = \text{constant} = 0.20979. \tag{12.13}$$

The contribution of O_2 and CO_2 to the apparent molecular weight of dry air is

$$M_{O2}x_{O2} + M_{CO2}x_{CO2} = 31.9988x_{O2} + 44.0098x_{CO2}. \tag{12.14}$$

From Eq. (12.13),

$$x_{O2} = 0.20979 - x_{CO2}, \tag{12.15}$$

and

$$M_{O2}x_{O2} + M_{CO2}x_{CO2} = 12.011x_{CO2} + 6.7130. \tag{12.16}$$

Therefore,

$$\partial(M_a) = \partial[M_{O2}x_{O2} + M_{CO2} - x_{CO2}]$$
$$= 12.011\delta(x_{CO2}). \tag{12.17}$$

That is, the variation in M_a due to a variation in CO_2 abundance is equal to 12.011 (the atomic weight of carbon) multiplied by the variation in CO_2 abundance.

TABLE 12.1 Composition of Dry Air

Constituent	Abundance (mole fraction)	Molecular Weight
Nitrogen	0.78102	28.0134
Oxygen	0.20946	31.9988
Carbon dioxide	0.00033	44.0098
Argon	0.00916	39.948
Neon	0.00001818	20.179
Helium	0.00000524	4.00260
Krypton	0.00000114	83.80
Xenon	0.000000087	131.30
Hydrogen	0.0000005	2.0158
Methane	0.0000015	16.0426
Nitrous oxide	0.0000003	44.0128

The variation in M_a due to the difference between the reference levels of CO_2 abundance 0.00033 and 0.00043 is thus 0.0012 g/mol, which corresponds to a relative variation of 41 ppm in M_a and a corresponding variation of 41 ppm in air density.

The adjusted M_a accounting for the departure of the CO_2 abundance from the reference level of 0.00033 becomes

$$M_a = M_{a033} + 12.011\left[x_{CO2} - 0.00033\right],$$
(12.18)

where M_{a033} is the apparent molecular weight of dry air with a CO_2 mole fraction of 0.00033.

The value of the abundance of argon in dry air, 0.00916, is that calculated from the mass spectrometric determination of the ratio of argon to argon and nitrogen by Hughes.[18]

The value for the abundance of nitrogen was arrived at by the usual practice of inferring nitrogen abundance to be the difference between unity and the sum of the mole fractions of the other constituents.

The abundances of the constituents neon through nitrous oxide in Table 12.1 were taken to be equal to the parts per volume concentration in the U.S. Standard Atmosphere, 1976.[19]

From the data of Table 12.1, the apparent molecular weight of dry air with a CO_2 mole fraction of 0.00033 is calculated by Eq. (12.12) to be 28.963. For dry air with a CO_2 mole fraction of 0.00043, the apparent molecular weight is calculated to be 28.964.

12.2.1.3 Compressibility Factor, Z

The compressibility factor was computed using the virial equation of state of an air–water vapor mixture expressed as a power series in reciprocal molar volume, $1/v$,

$$Z = Pv/RT = 1 + B_{mix}/v + C_{mix}/v^2 + \ldots,$$
(12.19)

and expressed as a power series in pressure,

$$Z = Pv/RT = 1 + B'_{mix}P + C'_{mix}P^2 + \ldots,$$
(12.20)

where B_{mix} and B'_{mix} are second virial coefficients and C_{mix} and C'_{mix} are third virial coefficients for the mixture.

The virial coefficients of the pressure series are related to the virial coefficients of the volume power series by

$$B'_{mix} = B_{mix}/RT$$
(12.21)

and

$$C'_{mix} = \left(C_{mix} - B_{mix}^2\right) \Big/ \left(RT\right)^2. \tag{12.22}$$

Each mixture virial coefficient is a function of the mole fractions of the individual constituents and the virial coefficients for the constituents. The virial coefficients are functions of temperature only.

Using the virial coefficients provided by Hyland[20] and Wexler,[21] a table of compressibility factor, Z, for CO_2-free air, Table 12.2, has been generated. Table 12.2 is applicable to moist air containing reasonable amounts of CO_2.

Alternatively, Z can be calculated using the following equations:

For P in pascals and t in °C,

$$\begin{aligned}
Z = {} & 0.99999 - 5.8057 \times 10^{-9}\, P + 2.6402 \times 10^{-16}\, P^2 \\
& - 3.3297 \times 10^{-7}\, t + 1.2420 \times 10^{-10}\, Pt \\
& - 2.0158 \times 10^{-18}\, P^2 t + 2.4925 \times 10^{-9}\, t^2 \\
& - 6.2873 \times 10^{-13}\, Pt^2 + 5.4174 \times 10^{-21}\, P^2 t^2 \\
& - 3.5 \times 10^{-7}\, (\mathrm{RH}) - 5.0 \times 10^{-9}\, (\mathrm{RH})^2.
\end{aligned} \tag{12.23}$$

For P in psi (pounds per square inch) and t in °C,

$$\begin{aligned}
Z = {} & 0.99999 - 4.0029 \times 10^{-5}\, P + 1.2551 \times 10^{-8}\, P^2 \\
& - 3.3297 \times 10^{-7}\, t + 8.5633 \times 10^{-7}\, Pt \\
& - 9.5826 \times 10^{-11}\, P^2 t + 2.4925 \times 10^{-9}\, t^2 \\
& - 4.3349 \times 10^{-9}\, Pt^2 + 2.5753 \times 10^{-13}\, P^2 t^2 \\
& - 3.5 \times 10^{-7}\, (U) - 5.0 \times 10^{-9}\, (U)^2.
\end{aligned} \tag{12.24}$$

U is relative humidity (for example, $U = 50$ is equal to 50% relative humidity).

For temperatures and/or pressures outside the range of Table 12.2, the table of compressibility factor of moist air (also CO_2-free) in the *Smithsonian Meteorological Tables*[22] can be used, with some loss of precision since the listing there is to the fourth decimal place.

12.2.1.4 Ratio of the Molecular Weight of Water to the Molecular Weight of Dry Air, ε

The molecular weight of water is 18.0152.[15] The ratio, ε, of the molecular weight of water to that of dry air is, therefore, 0.62201 for dry air with a CO_2 mole fraction of 0.00033. For dry air with a CO_2 mole fraction of 0.00043, the ratio, ε, is 0.62199.

12.2.1.5 Effective Water Vapor Pressure, e′

Since e' is the effective vapor pressure of water in *moist air*, a word of caution with regard to inferring e' from measurements of relative humidity is in order. Relative humidity, U, can be defined[23] by

$$U = \left(e'/e'_s\right) \times 100\%, \tag{12.25}$$

where e'_s is the *effective* saturation vapor pressure of water in moist air. The introduction of a second gas (air in this case) over the surface of water increases the saturation concentration of water vapor above

TABLE 12.2 Compressibility Factor, *Z*, Calculated Using Eq. (12.23)

t (°C)	*P* (Pa)	RH(%) 0	25	50	75	100
19.0	95000	0.99964	0.99963	0.99961	0.99960	0.99957
	100000	0.99962	0.99961	0.99959	0.99958	0.99956
	101325	0.99962	0.99960	0.99959	0.99957	0.99955
	105000	0.99960	0.99959	0.99958	0.99956	0.99954
	110000	0.99958	0.99957	0.99956	0.99954	0.99952
20.0	95000	0.99965	0.99964	0.99962	0.99960	0.99958
	100000	0.99963	0.99962	0.99960	0.99958	0.99956
	101325	0.99963	0.99962	0.99960	0.99958	0.99956
	105000	0.99961	0.99960	0.99958	0.99957	0.99954
	110000	0.99959	0.99958	0.99957	0.99955	0.99953
21.0	95000	0.99966	0.99965	0.99963	0.99961	0.99958
	100000	0.99964	0.99963	0.99961	0.99959	0.99956
	101325	0.99964	0.99962	0.99961	0.99959	0.99956
	105000	0.99962	0.99961	0.99959	0.99957	0.99955
	110000	0.99960	0.99959	0.99958	0.99956	0.99953
22.0	95000	0.99967	0.99965	0.99963	0.99961	0.99958
	100000	0.99965	0.99964	0.99962	0.99960	0.99957
	101325	0.99965	0.99963	0.99961	0.99959	0.99956
	105000	0.99963	0.99962	0.99960	0.99958	0.99955
	110000	0.99962	0.99960	0.99958	0.99956	0.99954
23.0	95000	0.99968	0.99966	0.99964	0.99962	0.00059
	100000	0.99966	0.99964	0.99962	0.99960	0.99957
	101325	0.99965	0.99964	0.99962	0.99960	0.99957
	105000	0.99964	0.99963	0.99961	0.99958	0.99956
	110000	0.99963	0.99961	0.99959	0.99957	0.99954
24.0	95000	0.99968	0.99967	0.99965	0.99962	0.99959
	100000	0.99967	0.99965	0.99963	0.99961	0.99958
	101325	0.99966	0.99965	0.99963	0.99960	0.99957
	105000	0.99965	0.99964	0.99962	0.99959	0.99956
	110000	0.99964	0.99962	0.99960	0.99957	0.99954
25.0	95000	0.99969	0.99968	0.99965	0.99962	0.99959
	100000	0.99968	0.99966	0.99964	0.99961	0.99958
	101325	0.99967	0.99966	0.99963	0.99961	0.99957
	105000	0.99966	0.99964	0.99962	0.99960	0.99956
	110000	0.99965	0.99963	0.99961	0.99958	0.99955
26.0	95000	0.99970	0.99968	0.99966	0.99963	0.99959
	100000	0.99969	0.99967	0.99964	0.99961	0.99958
	101325	0.99968	0.99966	0.99964	0.99961	0.99957
	105000	0.99967	0.99965	0.99963	0.99960	0.99956
	110000	0.99966	0.99964	0.99961	0.99959	0.99955

the surface of the water; the effective saturation vapor pressure of water, e'_s, is greater than the saturation vapor pressure of pure phase (i.e., water vapor without the admixture of air or any other substance), e_s.

12.2.1.6 Enhancement Factor, *f*

This "enhancement" of water vapor pressure is expressed by the enhancement factor, *f*, which is defined by

$$f = e'_s/e_s.$$ (12.26)

A published, experimentally derived value of f^{24} at 20°C and 100,000 Pa, is 1.00400. Therefore, the common practice of inferring e' from measured U and tabulated value of e_s introduces a significant error in e' if f has been ignored. The corresponding relative error in ρ at 20°C, 101,325 Pa, and 50% relative humidity is about 1.7×10^{-5}.

f is a function of temperature and pressure. Jones[2] has fitted Hyland's values of f^{24} to a three-parameter equation in the pressure (P, Pa) and temperature (t, °C) ranges of interest in national standards laboratories. The resulting equation is

$$f = 1.00070 + 3.113 \times 10^{-8} P + 5.4 \times 10^{-7} t^2. \tag{12.27}$$

Over the temperature range 19.0 to 26.0°C and the pressure range 69,994 Pa (10.152 psi) to 110,004 Pa (15.955 psi), f ranges from 1.0031 to 1.0045. The maximum variation of f from a nominal value of 1.0042 is equal to 0.11% of the nominal value. The corresponding relative variation of air density is equal to 0.00040%, which is negligible.

12.2.1.7   Saturation Vapor Pressure of Water, e_s

The expression for e' is found by combining Eqs. (12.25) and (12.26) to be

$$e' = (U/100) f e_s. \tag{12.28}$$

The systematic relative uncertainties in ρ due to the uncertainties assigned to f^{24} and e_s^{25} are approximately $\pm 1 \times 10^{-6}$ and $\pm 2 \times 10^{-7}$, respectively.

The e_s data of Besley and Bottomley[26] in the temperature range 288.15 K (15°C) to 298.04 K (24.89°C) and calculated values for the remainder of the temperature range to 301.15 K (28°C) have been fitted[2] to a two-parameter equation. The resulting equation is

$$e_s = 1.7526 \times 10^{11} e^{(-5315.56/T)}, \tag{12.29}$$

where $e = 2.7182818\ldots$ is the base of Naperian logarithms.

Values calculated using Eq. (12.29) are sufficiently close to experimental and calculated values of e_s, within $\pm 0.1\%$, to be used in the present application.

Calculated values of e_s expressed in mmHg are tabulated in Table 12.3.

12.2.1.8   Carbon Dioxide Abundance, x_{CO2}

In a previous section it was stated that the CO_2 abundance in laboratory air is in general variable. A variation of 0.0001 in CO_2 mole fraction is equivalent to a relative variation of 4×10^{-5} in calculated air density. The CO_2 abundance should be known for optimum utilization of the air density calculation. Eq. (12.18) can be used to adjust M_a for departures of CO_2 abundance from the reference level, 0.00033.

12.2.2   The Jones Air Density Equation[2]

By combining Eqs. (12.11) and (12.28) and substituting M_w/M_a for ε, the air density equation developed by Jones[2] becomes

$$\rho = (PM_a/RTZ)\left[1 - (1 - M_w/M_a)(U/100)(f e_s/P)\right]. \tag{12.30}$$

By substituting the Moldover et al. value of R,[13,14] 8.314471 J/mol/K, and the value of 18.0152 for M_w, Eq. (12.30) becomes

TABLE 12.3 Saturation Vapor Pressure
of Water Calculated Using Eq. 12.29

t (°C)	e_s (mmHg)	t (°C)	e_s (mmHg)
19.0	16.480	22.6	20.565
19.1	16.583	22.7	20.691
19.2	16.686	22.8	20.817
19.3	16.790	22.9	20.943
19.4	16.895		
19.5	17.000	23.0	21.071
19.6	17.106	23.1	21.199
19.7	17.212	23.2	21.327
19.8	17.319	23.3	21.457
19.9	17.427	23.4	21.587
		23.5	21.718
20.0	17.535	23.6	21.849
20.1	17.644	23.7	21.982
20.2	17.753	23.8	22.114
20.3	17.863	23.9	22.248
20.4	17.974		
20.5	18.085	24.0	22.383
20.6	18.197	24.1	22.518
20.7	18.309	24.2	22.653
20.8	18.422	24.3	22.790
20.9	18.536	24.4	22.927
		24.5	23.065
21.0	18.650	24.6	23.204
21.1	18.765	24.7	23.344
21.2	18.880	24.8	23.484
21.3	18.996	24.9	23.625
21.4	19.113		
21.5	19.231	25.0	23.767
21.6	19.349	25.1	23.909
21.7	19.467	25.2	24.052
21.8	19.587	25.3	24.196
21.9	19.707	25.4	24.341
		25.5	24.487
22.0	19.827	25.6	24.633
22.1	19.949	25.7	24.780
22.2	20.071	25.8	24.928
22.3	20.193	25.9	25.077
22.4	20.317		
22.5	20.441	26.0	25.226

$$\rho = 0.000120272 \left(PM_a / TZ \right) \left[1 - \left(1 - 18.0152 / M_a \right) \left(U / 100 \right) \left(f e_s / P \right) \right], \qquad (12.31)$$

where

$$M_a = 28.963 + 12.011 \left(x_{CO2} - 0.00033 \right). \qquad (12.32)$$

For T = 293.15 K (20°C), P = 101,325 Pa, 50% relative humidity, and M_a = 28.963 g/mol, the air density calculated using Eq. (12.31) is 1.1992 kg/m^3 = 0.0011992 g/cm^3 = 1.1992 mg/cm^3.

For M_a of 28.964 g/mol (x_{CO2} = 0.00043) and the same values of the other parameters above, ρ = 1.1993 kg/m^3 = 0.0011993 g/cm^3 = 1.1993 mg/cm^3.

12.2.3 Uncertainties in Air Density Calculations

We now return to the air density equation as expressed by Eq. (12.30).
The uncertainty propagation equation[27] as it applies to air density is

$$\left(SD_\rho\right)^2 = \Sigma_i\left[\left(\partial\rho/\partial Y_i\right)^2\left(SD_i\right)^2\right], \tag{12.33}$$

where SD_ρ is the estimate of standard deviation of ρ, the Y_i are quantities on the right-hand side of Eq. (12.30), and the SD_i are the estimates of standard deviation of the quantities on the right-hand side of Eq. (12.30).

12.2.3.1 Uncertainties in Quantities Other than P, T, U, and x_{CO2}

We first concentrate on the uncertainties in the quantities other than the measurements of P, T, U, and x_{CO2} in order to find the minimum uncertainty in ρ if P, T, U, and x_{CO2} were measured perfectly. This minimum uncertainty can then be considered the limitation on the determination of ρ using the air density equation. We shall refer to these quantities loosely as nonenvironmental quantities.

12.2.3.1.1 Partial Derivatives, $(\partial\rho/\partial Y_i)$, for the Nonenvironmental Quantities

The partial derivatives of ρ with respect to the non-environmental quantities are now taken and evaluated using the following values of the parameters:

$$P = 101325 \text{ Pa}$$
$$M_a = 28.963 \text{ g/mol}$$
$$R = 8314.471 \text{ J/kmol/K}$$
$$T = 293.15 \text{ K}$$
$$Z = 0.99960$$
$$M_w = 18.0152 \text{ g/mol}$$
$$U = 50$$
$$f = 1.0040$$
$$e_s = 2338.80 \text{ Pa}$$
$$x_{CO2} = 0.00033$$

The value of ρ_a calculated using the quantities above is 1.1992 kg/m³.
The partial derivatives for the nonenvironmental quantities are

$$\left(\partial\rho/\partial M_a\right) = \left(1/RTZ\right)\left[P - \left(Ufe_s/100\right)\right] = 4.1106 \times 10^{-2}$$

$$\left(\partial\rho/\partial R\right) = \left(1/R^2TZ\right)\left[\left(M_a - M_w\right)\left(Ufe_s/100\right) - PM_a\right] = -1.4423 \times 10^4$$

$$\left(\partial\rho/\partial Z\right) = \left(1/RTZ^2\right)\left[\left(M_a - M_w\right)\left(Ufe_s/100\right) - PM_a\right] = -1.1997$$

$$\left(\partial\rho/\partial f\right) = \left[Ue_s/\left(100\ RTZ\right)\right]\left(M_w - M_a\right) = -5.2546 \times 10^{-3}$$

$$\left(\partial\rho/\partial e_s\right) = \left[Uf/\left(100\ RTZ\right)\right]\left(M_w - M_a\right) = -2.2557 \times 10^{-6}$$

$$\left(\partial\rho/\partial M_w\right) = \left(1/RTZ\right)\left(Ufe_s/100\right) = 4.8189 \times 10^{-4}$$

For convenience, units have not been given for the partial derivatives.

12.2.3.1.2 Uncertainties in the Nonenvironmental Quantities (SD$_i$)

The estimates of the standard deviations for the nonenvironmental quantities are given below:

$$M_a = 1 \times 10^{-3} \text{ g/mol}$$

$$R = 1.4 \times 10^{-2} \text{ J/kmol/K}$$

$$Z = 5.7 \times 10^{-6}$$

$$f = 7 \times 10^{-5}$$

$$e_s = 3.9 \times 10^{-2} \text{ Pa}$$

$$M_w = 2 \times 10^{-4} \text{ g/mol}$$

12.2.3.1.3 Products of the Partial Derivatives and the Estimates of Standard Deviation, $(\partial\rho/\partial Y_i)\cdot(SD_i)$, for the Nonenvironmental Quantities

The product of the partial derivative and the estimate of standard deviation for each of the nonenvironmental quantities is given below:

$$M_a = 4 \times 10^{-5} \text{ kg/m}^3$$

$$R = -2.0 \times 10^{-6} \text{ kg/m}^3$$

$$Z = -6.8 \times 10^{-6} \text{ kg/m}^3$$

$$f = -4 \times 10^{-7} \text{ kg/m}^3$$

$$e_s = -8.8 \times 10^{-8} \text{ kg/m}^3$$

$$M_w = 1 \times 10^{-7} \text{ kg/m}^3$$

The above products were squared and summed as indicated in Eq. (12.33) and the square root was taken. The resulting estimate of the standard deviation of the air density, ρ_a, is 4×10^{-5} kg/m³. The estimate of standard deviation of ρ_a is dominated by the uncertainty in the apparent molecular weight of dry air, M_a.

Therefore, the minimum uncertainty in the calculation of air density using the air density equation, that which can be considered to be intrinsic to the equation, is an estimate of standard deviation of 4×10^{-5} kg/m³ or 33 ppm in the present example. Note that this does not include uncertainties in the measurements of pressure, temperature, relative humidity, and mole fraction of CO_2.

12.2.3.2 Uncertainties in the Environmental Quantities

12.2.3.2.1 Partial Derivatives, $\partial\rho/\partial Y_i$, for the Environmental Quantities

The partial derivatives for the environmental quantities, P, T, U, and x_{CO2} are

$$\left(\partial\rho/\partial P\right) = \left(M_a/RTZ\right) = 1.1888 \times 10^{-5}$$

$$\left(\partial\rho/\partial T\right) = \left(1/RT^2Z\right)\left[-PM_a + \left(M_a - M_w\right)\left(Ufe_s/100\right)\right]$$

$$= -4.0908 \times 10^{-3}$$

$$\left(\partial\rho/\partial U\right) = \left(1/RTZ\right)\left(fe_s/100\right)\left(M_w - M_a\right) = -1.0551 \times 10^{-4}$$

Again, for convenience, the units for the partial derivatives have not been given.

For atmospheres in which the mole fraction of CO_2, x_{CO2}, is other than the reference value of 0.00033,

$$M_a = 28.963 + 12.011\left(x_{CO2} - 0.00033\right). \tag{12.32}$$

and

$$\left(\partial M_a / \partial x_{CO2}\right) = 12.011.$$

Then,

$$\left(\partial\rho / \partial x_{CO2}\right) = \left(\partial\rho / \partial M_a\right)\left(\partial M_a / \partial x_{CO2}\right) = \left(4.1106 \times 10^{-2}\right)\left(12.011\right)$$

$$= 0.49372.$$

12.2.3.2.2 Uncertainties in the Environmental Quantities (SD_i)

The estimates of the standard deviations of each of the four environmental parameters are

$$P = 20.5 \text{ Pa}$$

$$T = 0.05 \text{ K}$$

$$U = 0.51$$

$$x_{CO2} = 0.00005$$

12.2.3.2.3 Products of the Partial Derivatives and the Estimates of Standard Deviation, ($\partial\rho / \partial Y_i$)·($SD_i$), for the Environmental Quantities

The product of the partial derivative and the estimate of standard deviation for each of the environmental parameters is

$$P = 2.4 \times 10^{-4} \text{ kg}/\text{m}^3$$

$$T = -2.0 \times 10^{-4} \text{ kg}/\text{m}^3$$

$$U = -5.4 \times 10^{-5} \text{ kg}/\text{m}^3$$

$$x_{CO2} = 2.5 \times 10^{-5} \text{ kg}/\text{m}^3$$

The above products were squared and summed, and the square root was taken. The resulting estimate of the standard deviation of the air density, ρ_a, is 3×10^{-4} kg/m³.

The largest uncertainty in ρ is that due to pressure measurement, and the next largest is that due to temperature measurement. The estimates of uncertainty are based on state-of-the-art measurements of the variables.

The overall uncertainty contributed by all of the quantities in the air density equation is 3×10^{-4} kg/m³.

12.2.4 Use of Constant Values of F, Z, and M_a in the Air Density Equation

By considering the expected variations in pressure, temperature, and relative humidity in the laboratory, it might be possible to use constant values of f, Z, and M_a.

For example, in the Mass Laboratory of NIST, constant values of f (1.0042), Z (0.99966), and M_a (28.964) are considered to be adequate. With these values of f, Z, and M_a, the resulting equation for calculating air density is, for P in pascals and absolute temperature $T = 273.15 + t$ (°C),

$$\rho_a = \left(0.0034847/T\right)\left(P - 0.0037960Ue_s\right). \tag{12.34}$$

For P in psi and t in °C,

$$\rho_a = \left[24.026/\left(t + 273.15\right)\right]\left(P - 0.0037960Ue_s\right) \tag{12.35}$$

For $P = 101{,}325$ Pa $= 14.69595$ psi, $T = 293.15$ ($t = 20$°C), $U = 50$, and $e_s = 2338.80$ Pa, $\rho_a = 1.1992$ kg/m³ $= 0.0011992$ g/cm³ $= 1.1992$ mg/cm³ for both Eqs. (12.34) and (12.31).

12.3   CIPM-81 Air Density Equation

The CIPM-81 equation[3] is

$$\rho = pM_a\left[1 - x_v\left(1 - M_v/M_a\right)\right]/ZRT, \tag{12.36}$$

where p is pressure, x_v is the mole fraction of water vapor in moist air, M_v is the molar mass of water vapor in moist air (M_w in the Jones development), and M_a is molar mass of dry air (M_a in the Jones development). The mole fraction of water vapor, x_v, is equal to $(U/100)$ (fe_s/P) and is determined from the relative humidity, U, or the dew-point temperature.

12.4   CIPM 1981/1991 Equation

The CIPM 1981/1991[6] equation is the same as the CIPM-81 equation, Eq. (12.36). Davis[6] has tabulated the amended constant parameters appropriate to the CIPM 1981/1991 equation. Davis stated that air densities calculated from the 1991 parameters are smaller by about 3 parts in 10^5 relative to calculations using the 1981 parameters, and that the overall uncertainty for air density calculated using the 1981/1991 equation is essentially the same as if the 1981 equation were used. Davis[6] noted that ITS-90 should be used.

12.5   Recommendation

The difference between the air density calculated using the CIPM 1981/1991 and that calculated using the CIPM-81 are well within the practical uncertainty. If one prefers, the CIPM 1981/1991 equation can be used.

12.6   Direct Determination of Air Density

12.6.1   Introduction

The air density in a balance case was determined directly by Koch et al.[28] using a simple device that enabled mass comparisons in air without the need to correct for air buoyancy. The device was an evacuated canister that was weighed on a laboratory balance with either of two masses inside.

The experimental method was used to determine the mass difference between two stainless steel weights of widely different densities. By a simple weighing of each of the two objects, the mass difference determination and the volume difference of the two weights were used to determine directly the density of the air in the balance case.

12.6.2 Experimental Procedure

Two stainless steel weights were used. One was hollow and of low density with mass of M_H and exterior volume of V_H; the other was solid and of high density with mass of M_S and external volume of V_S.

The two weights were weighed, successively, within an evacuated enclosure of mass M_E and external volume V_E. For air density, ρ_a, a force equation can be written:

$$\left(M_E + M_H - \rho_a V_E\right)g = \left(M_E + M_S - \rho_a V_E + \Delta M\right)g, \tag{12.37}$$

where g is the local acceleration due to gravity and ΔM represents the small mass difference indicated on the optical scale of the balance, calibrated using platinum weights.

If the volume of the evacuated enclosure, V_E, and g are constant throughout the procedure, the mass difference between the two weights:

$$\Delta M = M_H - M_S \tag{12.38}$$

is obtained without an air buoyancy calculation.

If the mean densities of the two objects are sufficiently different, the weights thus characterized can be used to determine the density of the air in the balance case. If m is the difference in air weight (divided by g) between M_H and M_S, and if

$$\Delta V = V_H - V_S, \tag{12.39}$$

then,

$$m = \left(M_S - \rho_a V_S\right) - \left(M_H - \rho_a V_H\right), \tag{12.40}$$

or

$$\rho_a = \left(\Delta M + m\right)\big/ \Delta V. \tag{12.41}$$

The values of air density so determined could be compared to those calculated using an existing air density algorithm and measurements of barometric pressure, temperature, and relative humidity.

Mass comparisons using the evacuated canister were made on a commercial single-pan balance with air damping and an optical scale. The balance had a capacity of 100 g and a resolution of 10 µg.

The weighings in air of the two mass artifacts for checking the adequacy of the air density algorithm were all made on single-pan balances with optical scales. These balances had a capacity of 20 g and resolution of 1 µg.

12.6.3 Results and Conclusions

The experimental and calculated values of air density agreed throughout to within 1.0 µg/cm³ (where the normal air density was about 1.2 mg/cm³). The uncertainty was 0.08% of the air density, at 1 standard deviation.

The experimental and calculated values of day-to-day fluctuations in air density agreed to within 0.5 µg/cm³.

A mass determination of the following materials will have at most the relative uncertainty indicated (due solely to the air density algorithm) starting from platinum-iridium standards:

Silver 5×10^{-8}
Stainless steel 8×10^{-8}
Water 9×10^{-7}

The measurements showed that the assignment of the mass unit from one kilogram to another can be made with an uncertainty attributable to the air density algorithm of no greater than 0.10 mg/100 cm^3 in the volume difference of the kilograms.

12.7 Experimental Determination of Air Density in Weighing on a 1-kg Balance in Air and in Vacuum

12.7.1 Introduction

Glaser et al.[29] made experimental determinations of air density by comparing two mass artifacts of about the same mass and surface but different well-known volumes on a 1-kg mass comparator in air and in vacuum.

The density of air, ρ_a, was calculated from the differences in "apparent mass" in air and in vacuum. Apparent mass, m', was defined as the difference between the mass of the artifact, m, and the product of the volume, V, of the object and the air density, $(m - V\rho_a)$.

The following formula was used to calculate ρ_a:

$$\rho_a = \left(\Delta m'_v - \Delta m'_a\right)/\Delta V, \tag{12.42}$$

where $\Delta m'_v$ is the difference of apparent mass in vacuum, $(m_1 - m_2)$, of the two artifacts, $\Delta m'_a$ is the difference of apparent mass in air, $\Delta m'_a = (m'_1 - m'_2)_a$, and ΔV is the difference in volume of two artifacts (Nos. 1 and 2).

Surface desorption effects, loss of mass from surfaces of the artifacts during evacuation, can be minimized by designing the artifacts to have as nearly equal surface areas as possible.

In general, the mass, m, of an artifact changes due to the desorption of mass m_s by

$$m = m'_v + m_s. \tag{12.43}$$

Eq. (12.43) is then modified to become

$$\rho_a = \left(\Delta m'_v - \Delta m'_a + \Delta m_s\right)/\Delta V, \tag{12.44}$$

where Δm_s is the difference in desorbed mass.

Desorption influences were estimated by the use of artifact No. 1 and an artifact, No. 3, of equal volume to that of No. 1 but of different surface area.

12.7.2 Results and Conclusions

Weighings were made of the three artifacts in air and in vacuum on a 1-kg mass comparator. The air comparisons were corrected for buoyancy using the BIPM-recommended air density equation.[3]

An estimation was made of the uncertainy of the air density determined by use of the recommended equation and of the air density determined in these experiments.

It was concluded that the relative uncertainty for the experimental measurements (at a level of 1 standard deviation) was 5×10^{-5}; this value was calculated to be smaller by a factor of 2.4 than that associated with the recommended equation. The uncertainties used for the air density equation, however, can be questioned, particularly the uncertainty for the apparent molecular weight of air.

The surface adsorption and desorption effects in vacuum and in air were found to be small compared with the experimental uncertainty.

For an artifact of mass 1 kg of stainless steel compared with one of platinum-iridium, the buoyancy correction is about 94 mg at an air density of 1.2 kg/m^3.

A relative uncertainy of 5×10^{-5} corresponds to an uncertainty of about 0.1 mg in the comparison of a stainless steel kilogram with a platinum-iridium kilogram.

12.8   A Practical Approach to Air Density Determination[30]

12.8.1   Introduction

Mass standards are used directly or indirectly with the electronic balance in many common measurement applications such as mass, density, volumetric capacity, force, and pressure. These applications generally require some knowledge of air density. This section offers the metrologist, bench scientist, and laboratory technician a practical guide to the determination of air density and the associated uncertainty estimate.

The measurement process usually begins with a problem definition and then an error analysis. Many problems are repetitive and set forth in paper standards. Often, the metrologist or experimenter is faced with a measurement problem that will be performed a very limited number of times and for which the metrologist has little experience. Metrologists often perform tasks by rote as passed along by others or follow written prescriptions that are without detail. These circumstances often lead to unnecessary work, the purchase of inappropriate equipment, and inaccurate measurement results and uncertainty estimates, all of which give rise to economic inefficiencies. This section provides a basis for the determination of air density where correction for the buoyant effect of air is necessary.

Sometimes it is unnecessary for the user to determine air density while at other times the user must appeal to an air density equation that requires the measurement of other parameters. This section examines the use of average air density for a given location and the appropriate measurement of temperature, barometric pressure, and relative humidity when one applies an equation. The Organization of Legal Metrology (OIML) document R111[31] sets forth a framework for laboratory mass standard specifications that form a useful basis to examine the problems associated with the determination of air density.

The primary purpose of R111 is the creation of a system for reference weights that in normal usage air density, that is, buoyancy corrections, can be ignored. This is accomplished by relying on a very old concept that in the past was referred to as "apparent mass" and then, in R111, as the "conventional value of the result of weighing in air" (see Chapter 15) as defined by OIML R33.[32]

Unfortunately, the small maximum permissible errors (MPE) for class E_1 and E_2 weights (laboratory mass standards) and the departure of air density from the standard air density of 1.2 mg/cm³ force the user to measure mass (sometimes called "true mass") and compute the conventional value or to make tedious corrections to obtain the correct conventional value.

Warning is given in R33 regarding the incorrectness of conventional values when the weighings are performed at air densities differing from 1.2 mg/cm³ by more than 10%. This warning applies to weights the densities of which are within the domain of R111, and may not be adequate for objects outside this domain.

12.8.2   Air Density

In the use of force machines, and in some instances in the use of dead-weight pressure gauges, the metrologist need only account for the gross effect of air buoyancy, and therefore an average air density for a given location is adequate. In the "conventional value" usage one may not want to depart by more than 10% from the reference air density, 1.2 mg/cm³. These applications are easily accommodated by consulting a weather bureau for a location average air density or by calculating an air density based on the station elevation. One need only consider elevation as nature limits local variations in air density related to variation in air pressure to ±3%. Laboratory temperature and relative humidity (RH) is usually maintained near 25°C and 50% RH, respectively. Normal variations from these values have a negligible effect when making approximate buoyancy corrections.

From the laboratory elevation, h, in meters, one can calculate the standard pressure, P, in millimeters of mercury by application of either of the following two formulas derived from data given in the *Smithsonian Meteorological Tables*[33]:

$$\ln P = 6.6342205 - 0.00012157179h \tag{12.45}$$

$$P = \exp\left(6.6342205 - 0.00012157179h\right). \tag{12.46}$$

TABLE 12.4 Elevation and Air Pressure

Elevation, meters	Air Pressure, mmHg
0 (sea level)	760
300	733.4
600	707.2
900 (10% limit)	681.8
1200	657.4
1500	633.9
1609.3	625.5 (Denver, CO, USA)
1800	611.2
2000	596.5

The pressure uncertainty (at $k = 1$) is 0.44 mmHg. The standard pressure at sea level (0 m) is 760 mmHg at 15°C. The standard pressure is near the upper extreme of the natural pressure variation of ±3%. Therefore, the average station pressure, P_A, calculated from P is $P_A \geq P - 3\%\ P$. If the station elevation is unknown it can be determined from many sources (topographical maps); however, a nearby airport elevation may be adequate. Otherwise, an aircraft altimeter can be set to read the airfield elevation and then transported to the laboratory. The indicated elevation is uncertain (at k = 1) by 25 m. For the reader's convenience, a few calculated air pressures, P, from sea level to 2000 m are provided in Table 12.4.

Having obtained an average station pressure, one can apply it to an equation that expresses the density of air in terms of the air temperature, t, relative humidity, U, and air pressure, P. The international community generally accepts the equation developed by Jones[2] and now slightly modified and known as the CIPM 81/91 air density equation.[6] A variant of this equation is offered here that is simpler to use and easier to follow in discussion[35]:

$$\rho_a = \left[0.0034836/(TZ)\right]\left(P - 0.0037960 U e_s\right), \tag{12.47}$$

where ρ_a is air density in kg/m³, T is the air temperature in kelvin, Z is the compressibility factor for CO_2-free air, P is ambient air pressure in pascals, U is the relative humidity in %, and e_s is the saturation vapor pressure of water in pascals. e_s is obtained by application of the formula:

$$e_s = 1.7526 \times 10^{11} \varepsilon^{\left(-5315.56/T\right)}, \tag{12.48}$$

where ε is the base of natural logarithms. A value for Z is obtained from the following formula:

$$Z = 0.99999 - 5.0857 \times 10^{-9}\ P + 2.6402 \times 10^{-16}\ P^2$$
$$-3.3297 \times 10^{-7}\ t + 1.2420 \times 10^{-10}\ Pt$$
$$-2.0158 \times 10^{-18}\ P^2 t + 2.4925 \times 10^{-9}\ t^2 \tag{12.49}$$
$$-6.2873 \times 10^{-13}\ Pt^2 + 5.4174 \times 10^{-21}\ P^2 t^2$$
$$-3.5 \times 10^{-7}\ U - 5.0 \times 10^{-9}\ U^2,$$

where P is in pascals, t in °C, and U in %. For the reader's convenience, Table 12.5 provides values of Z for a few selected values of P, t, and U.

For many mass and mass-related measurements such as density and volumetric capacity, greater knowledge of the air density is required along with an uncertainty estimate. Eq. (12.47) is appropriate for these applications but requires measurement of air pressure, air temperature, and relative humidity.

TABLE 12.5   Selected Values of Compressibility
Factor, Z

t, °C	P, Pa	Relative Humidity, U		
		25%	50%	75%
20	101,325	0.99962	0.99960	0.99958
	90,000	0.99966	0.99964	0.99962
	80,000	0.99969	0.99967	0.99965
25	101,325	0.99966	0.99963	0.99961
	90,000	0.99969	0.99967	0.99964
	80,000	0.99972	0.99970	0.99967

A background value for CO_2 concentration was assumed to be 430 ppm in the derivation of Eq. (12.47). Each parameter sensor must be calibrated and accompanied with an uncertainty estimate before an error propagation can be performed for the calculated air density. Sensor calibration will now be discussed before proceeding with the error propagation for Eq. (12.47).

12.8.2.1   Temperature

For the most accurate work, thermometers used to measure air temperature should be placed inside the weighing chamber and close to the weight on the weighing pan. Of course, the thermal history 24 h prior to the weighing should be monitored to assure thermal equilibrium between the object undergoing test and the surrounding air.[36] If the data show a change greater than 0.5°C/h, the local environment may not be adequate for applications involving large buoyancy corrections and where high accuracy is required.

Regardless of the type of thermometer used, that is, platinum resistance, thermistor, mercury-in-glass, or thermocouple, it should be periodically tested by the user, *as should the pressure and humidity sensors*. A single point check is not always adequate because it fails to check the response slope of the sensor. Therefore, a more thorough thermometer check requires two widely separated temperature reference points that span the normal laboratory working temperatures. Fortunately, nature has provided three fixed reference temperature sources that are easy and economical to access. They are the melting point of ice (0°C), the triple point of water (0.01°C), and the steam condensation temperature (approximately 100°C). Techniques to implement these fixed points that are quite simple and reliable are now described in detail.

12.8.2.2   Melting Point of Ice

The melting point of ice is achieved by packing finely divided ice (crushed ice) made from distilled water into a dewar flask or common Thermos™ bottle to prevent rapid melting.[37] The thermometer to be checked is not allowed to come into contact with the walls of the flask while ice is packed by firm hand pressure around the thermometer; see Figure 12.1. The thermometer stem must be immersed at least ten diameters (more is better) except for mercury-in-glass types. The latter usually have an immersion line or may require total immersion. Then, distilled water is added to the flask and the excess water is forced out of the flask by adding more crushed ice. The constituents should be in thermal equilibrium within ½ h. Periodically, crushed ice must to be added to displace liquid water created by the melting ice. The amount of liquid water must be minimized.

Thermal equilibrium is confirmed by noting a stable thermometer reading for 10 min. The thermometer correction is

$$\text{thermometer correction} = t_{\text{ref}} - t,$$

where t_{ref} is the reference temperature 0°C and t is the thermometer reading in °C. The reference temperature uncertainty is 0.003°C with a coverage factor $k = 2$.

To achieve a reliable ice point, it is extremely important that everything that comes into contact with the distilled water and ice not impart any impurities. Therefore, the hands of the operator must be

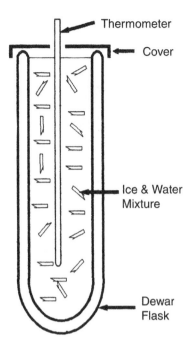

FIGURE 12.1 Melting point of ice setup.

covered with clean protective gloves, and the ice cube trays, the ice crusher, flask, and any other objects that come in contact with the ice or water must be thoroughly cleaned. An easy cleaning method is to scrub the equipment with a glassware detergent and water mixture and then rinse with tap water followed by an ethanol (or methanol) rinse. Then, after air-drying, a distilled water rinse and then steaming by holding the inverted flask over a boiling beaker of distilled water follow. The ice crusher and other objects can be cleaned by pouring boiling distilled water over them or through them.

If the operator dedicates the required equipment to this procedure and keeps it in clean storage when not in use, future use requires less preparation. It is also wise to keep the ice cube trays covered with plastic bags during their storage in the freezer during the ice-making process.

12.8.2.3 Triple Point of Water

The triple point of water is achieved by freezing a triple point of water cell.[38] These cells are available commercially and are easy to use. The reader should refer to Figures 12.2 and 12.3. The cell should be prechilled in a container filled with a mixture of ordinary tap water and ice as should the thermometer to be tested. An ice mantle is formed around the thermometer well by keeping the thermometer well packed with chips of solid CO_2 (dry ice). Once an adequate mantle is formed, the ice adjacent to the thermometer well is melted (inner melt) by filling the well with chilled water from the soaking bath and then introducing a room temperature, loose-fitting solid glass rod into the well. After 15 to 20 s the cell is held vertically and given a brisk quarter-turn rotation. If an inner melt has been obtained, the ice mantle will rotate freely about the thermometer well; if not, the warming process should be repeated until the mantle can rotate. Once the rotation is achieved, the cell is accurate to about 0.0002°C, more than adequate for the purpose here.

The cell can be maintained in this state for many days when stored in a thermally insulated mixture of packed ice and water. The cell user should always be sure that the ice mantle is free of the thermometer well before each use as it can refreeze, and also that the mantle diameter is not allowed to increase to where the cell is destroyed. The operator can prevent the destructive formation of an ice bridge inside the cell (near the upper end) by holding the cell with a bare hand during the mantle-formation procedure.

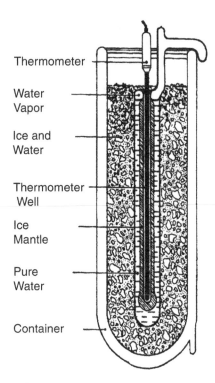

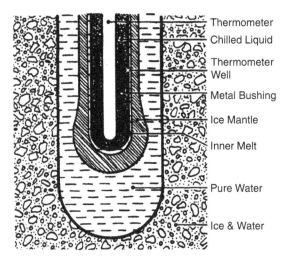

FIGURE 12.2 Triple point of water cell.

FIGURE 12.3 Detail of triple point of water cell.

12.8.2.4 Steam Point

The water vapor above the surface of boiling pure water, when not confined, condenses at a well-characterized temperature near 100°C. This phenomenon is also referred to as the boiling point of water. The relationship between the steam condensation temperature and atmospheric pressure[39] is given in Table 12.6. The air pressure in mmHg must be measured immediately adjacent to the free surface of the water. The resulting temperature is expressed on IPTS 68 and must be converted to ITS 90.[39] The conversion formula is

TABLE 12.6 Condensation Temperature of Steam

P	0	1	2	3	4	5	6	7	8	9
				Pressure in mmHg (standard)						
				Temperature in degrees on the International Practical						
				Temperature Scale of 1968						
550	91.183	.231	.279	.327	.375	.423	.471	.518	.566	.614
560	91.661	.709	.756	.803	.851	.898	.945	.992	*.039	*.086
570	92.133	.179	.226	.274	.320	.367	.413	.460	.506	.552
580	92.598	.644	.690	.736	.782	.828	.874	.920	.965	*.011
590	93.056	.102	.147	.193	.238	.283	.328	.373	.418	.463
600	93.5079	.5527	.5975	.6422	.6868	.7315	.7760	.8204	.8648	.9092
610	93.9534	.9976	*.0419	*.0860	.1300	.1740	.2179	.2617	.3055	.3493
620	94.3940	.4366	.4802	.5237	.5671	.6105	.6538	.6972	.7404	.7835
630	94.8266	.8698	.9126	.9555	.9985	*.0413	*.0841	*.1268	*.1694	*.2120
640	95.2546	.2972	.3396	.3819	.4242	.4665	.5087	.5508	.5929	.6351
650	95.6771	.7190	.7609	.8027	.8445	.8862	.9280	.9696	*.0122	*.0526
660	96.0941	.1355	.1769	.2182	.2595	.3007	.3419	.3830	.4240	.4650
670	96.5060	.5469	.5878	.6286	.6693	.7100	.7506	.7912	.8318	.8724
680	96.9128	.9532	.9936	*.0339	*.0741	*.1143	*.1546	*.1947	*.2347	*.2747
690	97.3147	.3546	.3945	.4343	.4741	.5138	.5535	.5931	.6327	.6722
700	97.7117	.7511	.7906	.8300	.8693	.9085	.9477	.9869	*.0260	*.0651
710	98.1042	.1431	.1821	.2210	.2598	.2986	.3373	.3761	.4148	.4534
720	98.4920	.5305	.5690	.6074	.6458	.6842	.7225	.7608	.7990	.8372
730	98.8753	.9134	.9515	.9896	*.0275	*.0654	*.1033	*.1411	*.1789	*.2107
740	99.2544	.2922	.3298	.3673	.4049	.4424	.4798	.5172	.5546	.5920
750	99.6293	.6666	.7038	.7409	.7780	.8151	.8522	.8893	.9262	.9631
760	100.0000	.0368	.0736	.1104	.1472	.1839	.2205	.2571	.2937	.3302
770	100.3667	.4031	.4395	.4760	.5123	.5486	.5848	.6210	.6572	.6934
780	100.7295	.7656	.8016	.8376	.8736	.9095	.9453	.9811	*.0169	*.0528
790	101.0885	.1242	.1598	.1954	.2310	.2665	.3020	.3376	.3730	.4084

* Indicates change in integer.

$$T_{90} = T_{68} - 0.026°C.$$

The difference between the two temperature scales is nearly constant space for sea level and for an air pressure of 600 mmHg. The steam point temperature has an uncertainty (at $k = 1$) of 0.0015°C when the pressure uncertainty is 0.1 mmHg.

The metrologist can readily access the steam condensation temperature to check thermometers using the apparatus shown in Figure 12.4. The apparatus is constructed from a 3-cm-diameter glass tube about 30 cm long and heated with the smallest possible flame that will sustain boiling and provide a steady thermometer indication. The water must be distilled and kept pure. It is important that a clean boiling bead of glass or Teflon™ be in the water to prevent superheating. A fiber glass insulation sleeve surrounds the glass tube about 2 cm above the water surface. The sensor tip is placed in the center of the tube about 2 cm above the boiling water and will be in thermal equilibrium within a few minutes after the water condensation drop forms on the thermometer tip. The thermometer correction is determined as described above for the ice point.

It is extremely important that the system be open to atmospheric pressure and that the thermometer never touch the water or the container during the calibration process. The air pressure is determined

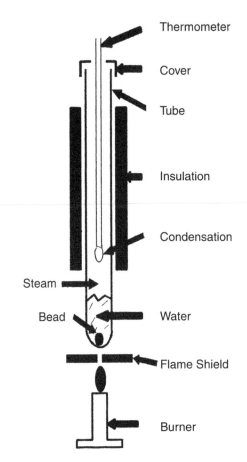

Thermometer

Cover

Tube

Insulation

Condensation

Steam

Bead

Water

Flame Shield

Burner

FIGURE 12.4 Steam condensation temperature apparatus.

adjacent to the apparatus and at the elevation of the water surface. A flame shield prevents overheating of the walls of the glass tube near the thermometer, and radiation heating. However, the flame shield has a small hole to expose the glass to the flame.

Like the ice point, cleanliness of the apparatus is important and the same cleaning techniques are applied here. The thermometer must also be clean.

12.8.2.5 Relative Humidity

For gravimetric measurement, the relative humidity in the laboratory should be maintained between 30 and 60%. Low humidity (20 to 25%) will support the formation of electric charge on the balance, operator's clothing, and nonconducting objects being weighed such as glassware. The resulting force imposed on the balance mechanism results in erroneous balance observations. With high humidities there is the possibility of water condensation on the balance mechanism and the object being weighed. The result of this is a degradation of balance precision or an error in the mass of the object being weighed. These limits are somewhat arbitrary and can be extended when dictated by circumstances; however, the operator should be aware that the above errors might be encountered.

Although many laboratories control relative humidity, others must accept what their environment offers. Rapid fluctuations should be avoided as should the above-mentioned extremes. Fluctuations less than 5% RH per hour is a reasonable limit for high-precision weighing.

The most widely used instruments for measuring relative humidity are transducers sensitive to the amount of water vapor in the air by the use of thin films and require exposure to known relative humidities for calibration. Dew-point meters, accurate to 1% RH or better, are less common, more expensive, and need only have their mirrors cleaned and their internal thermometers calibrated.

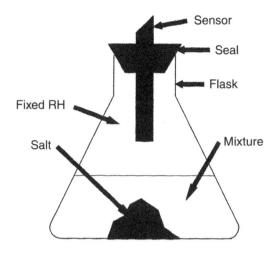

FIGURE 12.5 Apparatus for calibration of humidity sensors.

Wet- and dry-bulb thermometers require thermometer calibration, a source of distilled water, and clean wicks. The wet- and dry-bulb thermometer method is limited to an accuracy of approximately 6% RH for the best instruments and is no longer in common use for mass calibration.

The following calibration method utilizing saturated salt solutions,[40] that is, distilled water and pure chemical compounds, is an excellent means to calibrate the transducers that require exposure to known relative humidity. One or two of the salts given in Table 12.7 are each mixed separately in chemically nonreactive containers (glass or polyethylene) with distilled water. Some of the chemical compound (salt) must remain visible after mixing and given time to create a saturated solution. The container is covered with a lid that allows insertion of the humidity sensor and at the same time be hermetically sealed; see Figure 12.5. A recommended 24 h should be given to establish equilibrium.

A salt solution is chosen from Table 12.7 with a relative humidity near the working relative humidity; otherwise two solutions are selected that closely surround the working range. If required, the entire range of the sensor can be calibrated in this manner. A sensor correction is obtained in the same manner described for the thermometer.

This method depends on pure chemical compounds, American Chemical Society ACS grade, for example, pure distilled water, and a very clean container and sensor. Techniques similar to those described above for the ice point can be used to clean the container. Care must be taken not to immerse the sensor in the liquid solution as it may be damaged. The uncertainty of this method is 1% RH (at $k = 2$). We note that lithium chloride undergoes an exothermic reaction when mixed with water and caution is advised.

12.8.2.6 Pressure

Air pressure (barometric pressure) is easily measured by modern sensors, but like Fortin and aneroid barometers, the sensors require calibration. Unlike mercury barometers and deadweight pressure gauges, they do not define pressure.

Most metrologists are unable to generate the known pressures necessary to perform this calibration in-house and must transport the instrument elsewhere for this expensive calibration. Unfortunately, upon return of the instrument, one can never be certain the calibration remains valid. It has been demonstrated that pressure sensors can be calibrated gravimetrically by weighing two nearly identical known masses of widely differing but known volumes.[35]

The metrologist can choose to calibrate the barometric sensor gravimetrically, simply test the validity of a calibration performed elsewhere, or perform a direct determination of air density. In any case, the method is the same. Adequately calibrated weights of known mass and volume are available commercially. The measurements are easily performed using a 1-kg weight fabricated from aluminum and another

Handbook of Mass Measurement

TABLE 12.7 Relative Humidities Established by Saturated Salt Solutions

Lithium chloride, LiCl·H₂O						
Temp., °C	0.23	9.56	19.22	29.64	39.64	46.76
RH, %	14.7	13.4	12.4	11.8	11.8	11.4
Magnesium chloride, MgCl₂·6H₂O						
Temp., °C	0.42	9.82	19.53	30.18	39.96	48.09
RH, %	35.2	34.1	33.4	33.2	32.7	31.4
Sodium dichromate, Na₂Cr₂O₇·2H₂O						
Temp., °C	0.60	10.14	19.82	30.01	37.36	47.31
RH, %	60.4	57.8	55.3	52.4	50.4	48.0
Magnesium nitrate, Mg(NO₃)₂·6H₂O						
Temp., °C	0.39	9.85	19.57	30.47	40.15	48.10
RH, %	60.7	57.5	55.8	51.6	49.7	46.2
Sodium chloride, NaCl						
Temp., °C	0.92	10.23	20.25	30.25	39.18	48.30
RH, %	75.0	75.3	75.5	75.6	74.6	74.9
Ammonium sulfate, (NH₄)₂SO₄						
Temp., °C	0.39	10.05	20.04	30.86	39.97	47.96
RH, %	83.7	81.8	80.6	80.0	80.1	79.2
Potassium nitrate, KNO₃						
Temp., °C	0.62	10.17	20.01	30.70	40.35	48.12
RH, %	97.0	95.8	93.1	90.6	88.0	85.6
Potassium sulfate, K₂SO₄						
Temp., °C	0.54	10.08	19.81	30.44	39.94	48.06
RH, %	99.0	98.0	97.1	96.8	96.1	96.0

from stainless steel. The mass difference between these two weights when they are compared on a balance and accounting for air buoyancy is

$$SS - AL = \left(O_1 - O_2\right) - \rho_g\left(V_A - V_S\right).$$ (12.50)

Rewriting Eq. (12.50) to express ρ_g, explicitly:

$$\rho_g = \left[\left(SS - AL\right) - \left(O_1 - O_2\right)\right]\big/\left(V_S - V_A\right),$$ (12.51)

where ρ_g (g/cm³) is the gravimetrically determined air density, AL is the mass of a low-density weight such as an aluminum weight, SS is the mass of a stainless steel weight nominally equal to that of the aluminum weight, and V_A and V_S are their respective displacement volumes (cm³). O_1 and O_2 are the respective balance indications when SS and AL are weighed.

Equating the right-hand members of Eq. (12.51) to the right-hand members of Eq. (12.47), one can write an expression for the air pressure at the time of weighing and compare it to the pressure indicated by the pressure sensor during the weighing[35]:

$$P = \left[\left(\rho_g TZ\right)\big/0.0034836\right] + \left(0.0037960Ue_s\right).$$ (12.52)

12.8.2.7 Uncertainty

The partial derivatives for Eq. (12.51) are

$$\partial \rho_g / \partial \mathrm{AL} = 1 / \left(V_A - V_S \right) \tag{12.53}$$

$$\partial \rho_g / \partial \mathrm{SS} = -1 / \left(V_A - V_S \right) \tag{12.54}$$

$$\partial \rho_g / \partial O_1 = 1 / \left(V_A - V_S \right) \tag{12.55}$$

$$\partial \rho_g / \partial O_2 = -1 / \left(V_A - V_S \right) \tag{12.56}$$

$$\partial \rho_g / \partial V_A = -\left[\left(\mathrm{AL} - \mathrm{SS} \right) + \left(O_1 - O_2 \right) \right] / \left(V_A - V_S \right)^2 \tag{12.57}$$

$$\partial \rho_g / \partial V_S = \left[\left(\mathrm{AL} - \mathrm{SS} \right) + \left(O_1 - O_2 \right) \right] / \left(V_A - V_S \right)^2 . \tag{12.58}$$

The partial derivatives[41] for the air density, Eq. (12.47), that utilizes the sensor observations of temperature, pressure, and relative humidity are

$$\partial \rho_a / \partial T = -\rho_a / T \tag{12.59}$$

$$\partial \rho_a / \partial P = \rho_a / P \tag{12.60}$$

$$\partial \rho_a / \partial U = -0.0037960 e_s \left(\rho_a / P \right) . \tag{12.61}$$

Sample data and values for ancillary terms such as e_s and Z with uncertainty estimates can be found in Table 12.8. These data were used in the calculated air densities, the gravimetrically derived air pressure, and error propagation that follows in Tables 12.9 and 12.10.

A guideline for estimating uncertainties is given in Ref. 42. The guideline is abbreviated here:

$$\mathrm{RSS} = \left\{ \Sigma \left[\partial f(x) / \partial \mathrm{var} \right]^2 \left(\mathrm{SD} \right)^2 \right\}^{1/2} \tag{12.62}$$

For Eq. (12.62), RSS (uncertainty at $k = 1$) is the square root of the sum of the products of the square of the partial derivatives and the square of the SDs for each variable (var) of the functional relationship, $f(x)$.

12.8.2.8 Mass Calibration

Having obtained an air density and the associated uncertainty either gravimetrically or by measuring the parameters T, P, and U, it remains to propagate the uncertainty through a typical weight calibration.

The calculated air density and uncertainty found in Table 12.9 have been used to propagate uncertainties for a 1-kg weight of class E_1 when used as a standard to calibrate a 1-kg weight of class E_2. The data provided by Jones and Schoonover[43] (see Chapter 16) with the above air density exception have been

TABLE 12.8   Sample Data Set

Variable	Value	SD
Air Density Calculated from P, T, and U		
T	296.28	0.1 K
	(23.13°C)	(0.1°C)
P	98,917 Pa	3.3 Pa
	(741.94 mmHg)	(0.02 mmHg)
U	26.6	2
ρ_a calculated from Eq. (12.47)		1.1601 kg/m³
Intrinsic uncertainty		0.0001 kg/m³
Combined uncertainty		0.0005 kg/m³
Gravimetric Air Density		
SS	1000.01444 g	0.0001 g
AL	1000.02638 g	0.0001 g
O_1–O_2	−0.00424 g	0.00005 g
V_A	333.736 cm³	0.034 cm³
V_S	126.418 cm³	0.013 cm³
ρ_a calculated from Eq. (12.51)		1.1600 kg/m³
Uncertainty		0.0006 kg/m³
Ancillary Data and Equivalents		
Z	0.99964	1.7×10^{-5}
e_s	2831 Pa	0.03 %

Equivalents: 101,325 Pa = 760 mmHg; 1.2 kg/m³ = 1.2 mg/cm³ = 0.0012 g/cm³.

Air Pressure by Weight, Eq. (12.52) = 98,908 Pa.

TABLE 12.9   Air Density Error Propagation

Variable	Value	SD	$\partial\rho/\partial\text{var.}$
Air Density — Gravimetric			
AL	1000.026 g	0.0001	4.7×10^{-3}
SS	1000.014 g	0.0001	-4.7×10^{-3}
V_A	338 cm³	0.0338	-3.6×10^{-7}
V_S	126 cm³	0.0126	3.6×10^{-7}
O_1	0.000 g	5×10^{-5}	4.7×10^{-3}
O_2	−0.004 g	5×10^{-5}	-4.7×10^{-3}

$\rho_a = 0.00116$ g/cm³
RSS = 0.00000074 g/cm³

Air Density — Calculated from T, P, U			
T	296.28 K	0.1	3.9×10^{-6}
P	98917 Pa	3.3	1.2×10^{-8}
U	26.6	2	1.3×10^{-7}

$\rho_a = 0.00116$ g/cm³
RSS = 0.00000047 g/cm³
With intrinsic = 0.00000048 g/cm³

TABLE 12.10   Error Propagation for
Gravimetrically-Determined Air Pressure

Variable	Value	SD	$\partial P / \partial$var.
ρ_a	1.16 kg/m³	0.0005	85,019
T	298 K	0.01	320
Z	0.99964	1.7×10^{-5}	94,826
U	26	2	10.74
e_s	2831 Pa	85	0.10

RSS = 49 Pa. Pressure Calculated by Eq. (12.52) =
98908 Pa.

TABLE 12.11   Error Propagation
for a Class E_2 1-kg Weight

Variable	$\partial E_2 / \partial$Var.	SD
Mass of E_1	1	83×10^{-6}
Balance observation	-1	50×10^{-6}
Density of E_1	0.014	8×10^{-6}
Density of E_2	-0.0135	5×10^{-4}
Air density	-2.16	5×10^{-7}

RSS = 97×10^{-6} g; mass of E_2 = 999.9994 g.

used. The E_1 weight has mass of 1000.0002 g and a density of 8.067 g/cm³. The E_2 weight has a density of 7.810 g/cm³ and its mass is determined to be 999.9994 g. See Table 12.11 for the uncertainty propagation. As required by R111, the uncertainty of a class E_2 1-kg weight must not exceed 0.5 mg. The uncertainty for the example, $2 \times 97 \times 10^{-6}$ g, meets this requirement.

12.8.2.9   Summary

A practical information source for the determination of air density and its uncertainty (at $k = 1$) along with calibration procedures for the associated instrumentation have been provided.

12.9   Test of Air Density Equation at Differing Altitudes

12.9.1   Introduction

A series of measurements of the mass of aluminum and tantalum artifacts compared against mass standards of stainless steel was undertaken at the National Bureau of Standards (NBS) and the results were published in 1975.[44]

Inconsistencies were reported that seemed to correlate with barometric pressure.

The air density equation of Jones[2] was subsequently developed. Measurements for intercomparison between two objects of different density were made to determine the mass difference between the objects when weighed in air.[45]

The intercomparison measurements could be used to test the Jones air density equation.[2] The agreement between the measurements and the equation was within the experimental uncertainty, 600 ppm in air density.

An experiment was then undertaken to attempt to reconcile the measurements, which were consistent with the equation, with those in Ref. 44.

12.9.2   Experimental Details

A series of measurements with a selection of kilogram artifacts was made at NBS, Gaithersburg, MD. Similar measurements were made with the same artifacts at Sandia Laboratories at Albuquerque, NM.

TABLE 12.12   Characteristics and Designations
of the Ten Weights

Artifact Designation	Nominal Mass (kg)	Volume (cm³) at 20°C	Nominal Surface Area (cm²)
B1	1	127.385	145
D2	1	127.625	145
H1	1	337.381	270
H2	1	337.666	270
R1	1	126.395	270
R2	1	126.392	270
S1	1	126.549	660
S2	1	126.545	660
A	1	359.488	280
T	1	60.027	85

The NBS, Gaithersburg, laboratories are near sea level, the Sandia Laboratories are at about 1600 m (about 5249 ft) above sea level. The artifacts included the aluminum and tantalum kilograms used in Ref. 44, as well as several other weights designed to elucidate surface effects.

At both NBS and Sandia Laboratories, mass measurements were made on commerically available single-pan kilogram balances of conventional design, each with a precision of 25 to 50 μg.

Ten different 1-kg weights were used in the experiment. In Table 12.12, the characteristics and designations of the weights are listed.

B1 and D2, the most conventional weights, were used as standards. They were of single-piece stainless steel construction, of nearly minimum surface area, and with knobs for ease of handling.

H1 and H2, of stainless steel, were designed to have density near that of aluminum. They were hollow, right circular cylinders of minimum surface area (diameter equal to height), with an internal centerpost for rigidity, and filled with helium at about 1 atmosphere pressure.

R1 and R2, constructed as companions to the hollow weights, were thick-walled stainless steel tubes with surface areas nominally equal to those of H1 and H2.

S1 and S2, of solid stainless steel with surface areas roughly equal to twice those of the R weights, were each in the form of two nested stainless steel tubes on a circular stainless steel base. A centerpost welded to the base allowed easy manipulation of the S weights.

A was a single-piece weight of aluminum, constructed of bar stock in the form of a right circular cylinder of minimum surface area.

T was a single-piece weight of tantalum, of nearly minimum surface area, with a knob for ease of handling.

Weights A and T were the same ones used in the experiment reported in Ref. 44.

All of the weights except A and T were steam-cleaned prior to these experiments. The weights with stainless steel surface were also vapor-degreased with 1,1,1-trichloroethane. No further cleaning was attempted; all weights were dusted with a soft, lint-free brush prior to each use.

12.9.3   Calculation of Air Density for Buoyancy Correction

The values of air density for the measurements were calculated using the air density equation of Jones.[2]

12.9.4   Measurements of Parameters in the Air Density Equation

12.9.4.1   Temperature

Temperature was the most elusive of all the measurements made in the course of the experiment. Type E thermocouples were distributed within the weighing chamber. A mercury-in-glass total immersion thermometer was used as a backup to the thermocouples.

12.9.4.2 Pressure

Aneroid barometers were used to read barometric pressure at both NBS and Sandia. One of the aneroid barometers covered air pressures near sea level at NBS, and the other covered atmospheric pressures at an altitude of about 1600 m at Sandia.

The aneroid barometer used at NBS was calibrated at NBS, and it was also checked at ambient pressure twice daily against a cistern-type mercury manometer.

The aneroid barometer used at Sandia was calibrated first by a private calibration laboratory and then *in situ* at Sandia. In addition to the mercury manometer, a piston gauge and a sensitive quartz pressure transducer were taken to Sandia.

The quartz pressure transducer was used to calibrate the aneroid barometer and the transducer was calibrated twice daily against the piston gauge. The calibrated quartz transducer agreed with the mercury manometer to within 30 ppm.

The value of the local acceleration due to gravity, g, which was needed to make accurate pressure measurements using the piston gauge and the mercury manometer, was supplied by personnel of the Sandia Laboratories.

12.9.4.3 Relative Humidity

A Dunmore-type humidity-sensing element was mounted in one door of the balance case to measure relative humidity. The humidity elements used were calibrated at NBS. Periodic checks of calibration were made by immersing the element in air above a saturated salt solution, which produced an atmosphere of 34% relative humidity (RH), close to ambient humidity at both NBS and Sandia.

12.9.4.4 Carbon Dioxide Content

The carbon dioxide content of the air in the balance case was tested twice daily for both NBS and Sandia. Air samples were drawn into evacuated spheres, sealed with vacuum-type stopcocks, and then analyzed at NBS.

12.9.5 Weighings

All of the weighings were made using a so-called four-ones pattern in which four objects of nominally equal mass were intercompared in each of six possible combinations (see Chapter 8).

A least-squares fit of the weighing data resulted in the assignment of mass to each of three of the objects when the mass of the fourth object was known. B1 and D2 were used in every measurement. Since the mass of B1 and D2 was known, their sum was used in the least-squares solution.

12.9.6 Conclusions

Over a period of several months, five groups of measurements of the mass of the aluminum and the tantalum kilogram against stainless steel mass standards were carried out. Four groups of measurements were made at NBS, Gaithersburg and one group was made at the Sandia Laboratories.

The agreement between the measurements and the equation of Jones[2] was within the experimental uncertainty, 600 ppm in air density.

The groups of data exhibited significant differences among them of the type reported in Ref. 44. However, the magnitude of these discrepancies was a factor of 5 less than had been previously observed by the author of Ref. 44. The authors of the experimental work reported in Ref. 45 were unable to reproduce or satisfactorily explain the earlier results.

Surface effects were considered likely to play a role in the discrepancies reported in this later work.

It was hypothesized that this observed behavior was due to weights not being in sufficiently good thermal equilibrium with the balance.

References

1. Davis, R. S., New assignment of mass values and uncertainties to NIST working standards, *J. Res. Natl. Bur. Stand.* (U.S.), 95, 79, 1990.
2. Jones, F. E., An air density equation and the transfer of the mass unit, *J. Res. Natl. Bur. Stand.* (U.S.), 83, 419, 1978.
3. Giacomo, P., Equation for the determination of the density of moist air (1981), *Metrologia*, 18, 33, 1982.
4. Balhorn, R., *PTB Mit.*, 93, 303, 1983.
5. Davis, R. S., Recalibration of the U.S. National Prototype Kilogram, *J. Res. Natl. Bur. Stand.* (U.S.), 90, 263, 1985.
6. Davis, R. S., Equation for the determination of the density of moist air (1981/91), *Metrologia*, 29, 67, 1992.
7. List, R. J., *Smithsonian Meteorological Tables*, 6th rev. ed., Smithsonian Institution, Washington, D.C., 1951, 347.
8. Cohen, E. R. and Taylor, B. N., *J. Phys. Chem. Ref. Data*, 2, 663, 1973.
9. Quinn, T. J., Colclough, A. R., and Chandler, T. R. D., *Philos. Trans. R. Soc. London*, A283, 367, 1976.
10. Gammon, B. E., The velocity of sound with state properties in helium at −175 to 159°C with pressure to 150 atm, *J. Chem. Phys.*, 64, 2556, 1976.
11. Gammon, B. E., Private communication.
12. Rowlinson, J. S. and Tildesley, D. J., *Proc. R. Soc. London*, A358, 281, 1977.
13. Moldover, M. R. et al., Measurement of the universal gas constant R using a spherical acoustic resonator, *Phys. Rev. Lett.*, 60, 249, 1988.
14. Moldover, M. R. et al., Measurement of the universal gas constant R using a spherical acoustic resonator, *J. Res. Natl. Bur. Stand.* (U.S.), 93, 85, 1988.
15. International Union of Pure and Applied Chemistry, Inorganic Chemistry Division, Commission on Atomic Weights, *Pure Appl. Chem.* 47, 75, 1976.
16. Machta, L. and Hughes, E., Atmospheric oxygen in 1967 to 1970, *Science*, 168, 1582, 1970.
17. Gluekauf, E., in *Compendium of Meteorology*, Malone, T. F., Ed., The American Meteorological Society, Boston, 1951, 4.
18. Hughes, E., Paper presented at the American Chemical Society Fourth Mid-Atlantic Meeting, Washington, D.C., February 1969.
19. U.S. Standard Atmosphere, 1976, U.S. Government Printing Office, Washington, D.C., 1976, 3, 33.
20. Hyland, R. W., A correlation for the second interaction virial coefficients and the enhancement factors for moist air, *J. Res. Natl. Bur. Stand.* (U.S.), 79A, 551, 1975.
21. Wexler, A., *Meteorol. Abstr.*, 11, 262, 1970.
22. List, R. J., *Smithsonian Meteorological Tables*, 6th rev. ed., Smithsonian Institution, Washington, D.C., 1951, 333.
23. Harrison, L. P., in *Humidity and Moisture*, Vol. III, Wexler, A. and Wildhack, W. A., Eds., Reinhold, New York, 1964, 51.
24. Hyland, R. W., *J. Res. Natl. Bur. Stand.* (U.S.), 79A, 551, 1975.
25. Wexler, A. and Greenspan, L., Vapor pressure equation for water vapor in the range 0 to 100°C, *J. Res. Natl. Bur. Stand.* (U.S.), 75A, 213, 1971.
26. Besley, L. and Bottomley, G. A., Vapour pressure of normal and heavy water from 273.15 to 298.15, *J. Chem. Thermodyn.*, 5, 397, 1973.
27. Ku, H. S., Statistical concepts in metrology, in *Handbook of Industrial Metrology*, American Society of Tool and Manufacturing Engineers, Prentice-Hall, Englewood Cliffs, NJ, 1967, chap. 3.
28. Koch, W. F., Davis, R. S., and Bower, V. E., Direct determination of air density in a balance through artifacts characterized in an evacuated weighing chamber, *J. Res. Natl. Bur. Stand.* (U.S.), 83, 407, 1978.

29. Glaser, M., Schwartz, R., and Mecke, M., Experimental determination of air density using a 1 kg mass comparator in vacuum, *Metrologia*, 28, 45, 1991.

30. Schoonover, R. M., Measurement Science Conference, Anaheim, CA, January 2000.

31. Organisation Internationale de Metrologie Legale (OIML) International Recommendation R111, Weights of Classes E_1, E_2, F_1, F_2, M_1, M_2, M_3, OIML, 1994.

32. Organisation Internationale de Metrologie Legale (OIML) International Recommendation No. 33, Conventional Value of the Result of Weighing in Air, OIML, 1979.

33. List, R. J., *Smithsonian Meteorological Tables*, Smithsonian Institution, Washington, D.C., 1958, 266–267.

34. Davis, R. S., Equation for the determination of the density of moist air (1981/91), *Metrologia*, 29, 67, 1992.

35. Schoonover, R. M., A simple gravimetric method to determine barometer corrections, *J. Res. Natl. Bur. Stand.*, 85, 341, 1980.

36. Schoonover, R. M. and Taylor, J. E., Some recent developments at NBS in mass, Measurement, *IEEE Trans. Instrum. Meas.*, 35, 418, 1986.

37. Wise, J., Liquid-In-Glass Thermometry, NBS Monograph 150, U.S. Department of Commerce, 1986, 8–9.

38. Cross, J. L., Instructions, Triple Point of Water Cells, Jarrett Instrument Co., Wheaton, MD, Jan. 1964, 1–4.

39. Mangum, B. W. and Furukawa, G. T., Guidelines of International Temperature Scale of 1990 (ITS), NIST Technical Note 1265, U.S. Department of Commerce, 1990, 5.

40. Wexler, A., and Hasegawa, S., Relative humidity-temperature relationships of some saturated salt solutions in the temperature range of 0° to 50°C, *J. Res. Natl. Bur. Stand.* (U.S.), 53, 19, 1954.

41. Schoonover, R. M. and Jones, F. E., Parameters that can cause errors in mass determinations, *Cal. Lab.*, 5, 26, 1998.

42. Taylor, B. N. and Kuyatt, C. E., Guidelines for Evaluating and Expressing the Uncertainty of NIST Measurement Results, NIST Technical Note 129, 1994.

43. Jones, F. E. and Schoonover, R. M., A comparison of error propagations for mass and conventional mass, Measurement Science Conference 2000, Anaheim, CA, January 2000.

44. Pontius, P. E., *Science*, 190, 379, 1975.

45. Schoonover, R. M., Davis, R. S., Driver, R. G., and Bower, V. E., A practical test of the air density equation in standards laboratories at differing altitude, *J. Res. Natl. Bur. Stand.* (U.S.), 85, 27, 1980.

13

Density of Solid Objects

13.1 Development of a Density Scale Based on the Density of a Solid Object[1]

13.1.1 Introduction

The U.S. National Bureau of Standards (NBS) undertook programs to improve classical hydrostatic weighing[2] and to develop a density scale based on the density of a solid object.[3] This chapter discusses a continuation of that effort, which, in many instances, will eliminate the need for classical hydrostatic weighing and most of its inherent difficulties in determining the density of a solid. Additionally, the solid-object density standard is used to its full advantage.

In 1886, Thiesen[4] noted the benefits to be gained by completely immersing a balance in a liquid for the measurement of the density of an object. This idea lay forgotten as there was no practical way of pursuing the concept. With the development of the high-precision so-called "top-loading electronic balance," a form of Thiesen's idea has been successfully implemented.[5] The results of measurements presented here indicate that the density of standard kilograms could be assigned with a precision as high as 7 parts per million (ppm).

The work of Bowman and Schoonover[2] was primarily directed toward surmounting the difficulties encountered when a modern, single-pan analytical balance is used to weigh an object immersed in water to determine the density of the object to a very high accuracy.

The most notable difficulties involved cleanliness of the water, fabrication of the immersed portion of the balance suspension, degassing of the sample, and the awkwardness of weighing under the balance using a remote sample-loading device.

The uncertainty of 5 ppm or more that was usually assigned to the density of water was overcome with the development of the solid-object density scale[3] based on single-crystal silicon. The uncertainty assigned to the density of silicon crystals is 2 ppm or less. From this work came also experience with fluorocarbon liquid and its remarkable properties that are essential to the work discussed here.

It was speculated that a null-operated linear force balance, when immersed, would eliminate nearly all of the problems encountered in the density measurement. The success of the here discussed work has shown that opinion to be correct.

In the following is described a modification of a commercially available, servo-controlled, top-loading balance that permits operation, without loss of precision, while submerged in a fluid of special properties.

Next, data obtained by using the device to determine the volumes of various objects to an accuracy of better than 30 ppm in a fraction of the time that would have been required using conventional techniques are presented here.

Finally, industrial and scientific applications of the submerged balance are discussed.

13.1.2 Apparatus

The balance modified for this work was a Mettler Model PL12OO balance, the important specifications of which are:

Weighing range   0 to 1200 g
Reproducibillty   0.005 g
Linearity   ≤0.01 g

Significant mechanical and minor electronic modifications were introduced to the balance and its enclosure. These are described below along with descriptions of the thermostatically controlled bath and the fluorocarbon liquid used in the measurements.

13.1.2.1 Mechanical Modification of the Balance

The balance was supplied by the manufacturer in a modified form. That is, the measuring cell (electromagnetic force transducer) and the error-detecting electronics of the servo system were removed from the original chassis and remounted in a separate, identical chassis. Thus, the factory modification provided a measuring cell remote from the indicator unit. The latter contained most of the electronic circuitry and provided protection from hostile environments.

However, the remote measuring cell package in this configuration was excessively bulky for immersion in liquid and was replaced.

The weighing cell was removed from the chassis and remounted on an aluminum base plate of approximately 17-cm diameter. Three leveling screws were provided such that the center of gravity of the assembly was within the triangle that they defined.

Surrounding the measuring cell was an aluminum tube of the same diameter as the base plate and 12 cm in height with provision for passage of the electronic cable. Resting on the upper end of the tube was a cover plate of the same material with an opening for mounting the weighing pan on the measuring cell.

Buoyant forces on the immersed balance pan cause a zero-shift that could not be compensated for by the electronic circuitry of the balance, so additional tare weight was required. However, to facilitate reliable sample handling and to minimize corner load errors, the entire pan assembly was rebuilt.

The original weighing pan was disassembled. The spring element was then remounted beneath a square pan (5 × 5 cm) of sufficient mass to allow the instrument to zero and to indicate full scale.

The upper pan surface was divided into quarters by two shallow vee grooves at right angles to each other. This pan design easily accommodated samples in the shape of cylinders (on their side), spheres, and any object with a flat surface. A completely flat upper pan surface was avoided to reduce adhesion of a flat-bottomed weight to the pan.

Last, the cable connector at the weighing cell was removed and placed on the end of a 1-m length of 12-conductor ribbon cable; three unused conductors were removed from the ribbon cable to reduce its bulk. This modification provided a flexible cable for immersion in the fluid bath and kept to a minimum the path for heat leakage.

13.1.2.2 Electronic Modifications

The electronic elements submerged with the measuring cell were the magnetic force coil, the permanent magnet, and the error-sensing circuit of the servo controller. In addition, there was circuitry to maintain the magnetic flux of the stationary magnet constant as the ambient temperature changed.

The temperature signal for this circuitry was provided by a thermistor embedded in the stationary magnet. The heat-sink compound surrounding this thermistor was removed prior to immersion of the balance. The fluorocarbon fluid provided the needed thermal contact without running the risk of heat-sink compound floating about in the interior of the measuring cell.

When the measuring cell was submerged in the fluorocarbon and the balance was turned on, the pan began to oscillate and no reading could be obtained. To make the balance behave as desired, it was necessary to modify the analog servo electronics.

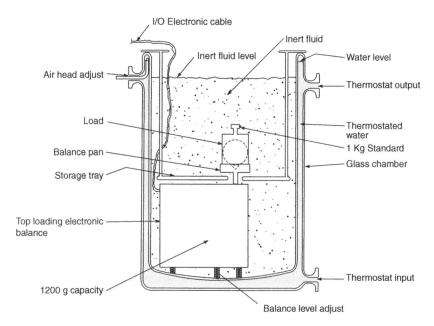

FIGURE 13.1 Liquid bath for complete immersion of measuring cell and samples.

Specifically, the 2.2 µF capacitor in the feedback loop of the analog amplifier was increased to 3.2 µF. (The input to this amplifier was proportional to the instantaneous position error of the balance. The output signal was a linear combination of the input, its derivative, and its integral.) After making this simple change, the balance performed as well in the fluid as it had previously performed in air.

13.1.2.3 Liquid Bath

A suitable liquid bath was designed for complete immersion of the measuring cell and a large number of samples in the fluid. A double-walled vessel was blown from borosilicate glass, as shown in Figure 13.1.

Several notable features of the fluid bath are summarized as follows:

1. Continuous inner bath surface for easy cleaning
2. Transparent structure for visual detection of air bubbles and dust prevention on the samples
3. Provision for adjustable air head to dampen pump pulsations produced by the thermostatic controller

Between the inner and outer walls of the bath, thermostatically controlled water was circulated at the rate of 7 l/min. A commercially available thermostated bath controller with a 200-W thermal capacity at 0°C seemed to be adequate for this work. Experience with this controller indicated a thermal stability of its internal reservoir of about 0.005°C/h. The temperature of the bath was set at room temperature (22°C).

Adequate sample manipulation was achievable by forceps, left in the bath at all times for thermal soaking. The forceps jaws were covered with soft plastic tubing to protect the samples.

13.1.2.4 Fluorocarbon Liquid

Clearly, the fluid in which an electronic measuring cell is submerged must have many special properties: it must be electrically insulating; it must be chemically inert; it must be optically transparent (in order for the servo optics to function properly); and it should not evaporate quickly.

These characteristics may be found in, for example, FC-75, a fluorinated fluid manufactured by the 3M Company. A comparison of some of the properties of FC-75 with those of water is given in Table 13.l. The values for FC-75 are supplied by the manufacturer. An additional noteworthy property of FC-75 is its immense appetite for gases; for example, the fluid is able to dissolve about 0.3 g of air per kilogram

TABLE 13.1   Comparison of Properties of FC-75 and Water

Property (at 25°C)	FC-75	Water
Density (g/cm³)	1.8	1.0
Coefficient of expansion (°C)⁻¹	1.6×10^{-3}	2.5×10^{-4}
Kinematic viscosity (cm²/s)	0.82×10^{-2}	0.89×10^{-2}
Vapor pressure (mm Hg)	30	24
Surface tension (dyne/cm)	15	72
Heat capacity (J/g-°C)	1.0	4.2
Thermal conductivity (W/cm-°C)	1.4×10^{-3}	6.1×10^{-3}

of fluid. This ability to dissolve atmospheric gases, combined with the low surface tension of the fluid, greatly inhibits bubble formation on immersed objects — one of the most serious problems in conventional high-precision hydrostatic weighing.

Finally, the fluid is 77% more dense than water at room temperature, thereby increasing the signal-to-noise ratio in comparison to a normal hydrostatic weighing in water.

The major disadvantages of this fluid as compared to water are its large coefficient of thermal expansion and its cost. However, use of FC-75 instead of water for conventional "hydrostatic" weighing has many advantages.

The density of FC-75 was not known accurately enough for the liquid to serve as a density standard. Instead, the fluid density was calibrated at the time of use by including a solid object of known mass and volume in the weighing scheme.

13.2   Principles of Use of the Submersible Balance

The use of the submersible balance is illustrated by finding the volume, V_A, of an object A when its mass, M_A, is known. Placing the object on the balance produces a reading, O_A, which is related to the other parameters through the equation:

$$O_A = K\left[M_A - \rho(t)V_A(t)\right],\tag{13.1}$$

where O_A is the difference in reading of the loaded and unloaded balance.

Here ρ is the density of the fluorocarbon and K is a constant scale factor that may be adjusted.

Both ρ and V_A are functions of the ambient temperature, t. In succeeding equations the functional dependence on temperature is not shown explicitly.

Normally, K is adjusted by the balance manufacturer or user so that the balance will read directly the mass of an object of density 8.0 g/cm³ in air of density 1.2×10^{-3} g/cm³, i.e., K = 1.000150. It was found convenient (although, of course, not essential) to readjust K to exact unity. Thus, K could be ignored in the succeeding equations.

Hence,

$$V_A = \left(M_A - O_A\right)\big/\rho.\tag{13.2}$$

The problem with using Eq. (13.2) is that the precision with which V can be measured far exceeds the accuracy with which ρ is known. Thus, for best results, one should also measure an object the mass and volume of which are known. Placing such an object on the balance essentially calibrates the density of the fluorocarbon at the time of weighing.

Let the known object be called S1. Then,

$$\rho = \left(M_{S1} - O_{S1}\right)\big/V_{S1},$$

and

$$V_A = \left[\left(M_A - O_A\right)/\left(M_{S1} - O_{S1}\right)\right] V_{S1}. \tag{13.3}$$

Further improvement in the results is realized by use of two objects of known mass, S1 and S2, the masses of which are near M_A and the volumes of which bracket V_A.

Hence,

$$V_A = V_{S1} + \left(V_{S2} - V_{S1}\right)\left[\left(O_{S1} - O_A\right) - \left(M_{S1} - M_A\right)\right]/\left[\left(O_{S1} - O_{S2}\right) - \left(M_{S1} - M_{S2}\right)\right]. \tag{13.4}$$

Note from the form of Eq. (13.4) that only differences in balance readings are required and, therefore, it is not necessary that the reading for the unloaded balance be on scale.

There are also applications, for example, in quality control, which only require the ratio of the volume of an object A to that of a working standard S1. Using the submersible balance, one can obtain the desired quantity:

$$\left(V_A/V_{S1}\right) = \left(M_A - O_A\right)/\left(M_{S1} - O_{S1}\right). \tag{13.5}$$

13.2.1   Measurements

The simplest measurement one can make using the submersed balance is to use Eq. (13.5) to determine the ratio of the volumes of two similar objects.

Two hermetically sealed, hollow, stainless steel weights each of which has a mass of 1 kg and a volume of about 337 cm^3 were chosen. Each weight was placed on the balance twice in a symmetric series, the sequence of measurements requiring less than 10 min for completion.

The computed ratio of volumes, 0.999156, is identical with that derived from a precise hydrostatic weighing. The agreement is somewhat fortuitous because round-off errors in reading the submerged balance would themselves limit the theoretical precision of the ratio measurement to 12 ppm.

One might expect to approach the theoretical limit, however, because the measurement consists merely of interchanging nearly identical objects and is insensitive to many of the limitations of a balance (e.g., nonlinearity).

The next test that was made of the submerged balance was to use Eq. (13.3) to find the volume of a 1-kg germanium crystal. The choice of a standard, S1, for use with Eq. (13.3) was designed to illustrate how a laboratory that lacks the means for precise hydrostatic weighing can obtain a suitable volume standard. The chosen standard was a commercially available precision ball bearing with a diameter of 6.35 cm. The mass of such a steel ball is approximately 1043 g. The bearing manufacturer was able to certify the sphericity and diameter of the ball to an accuracy that determines its volume at 20°C to better than 15 ppm.

A symmetric series of weighings was performed in which the steel ball was measured four times and a germanium crystal three times. The series of measurements required approximately 20 min to complete.

By using Eq. (13.3), the volume of the germanium crystal was computed to be 187.7193 cm^3, which is 24 ppm lower than that found by hydrostatic weighing. Round-off errors in the balance contributed an uncertainty of 25 ppm for this example.

The measurements suggested that the submerged balance was operating well. An additional experiment, consisting of the measurement on the submerged balance of a set of calibrated stainless steel weights of nearly equal density, showed the linearity of the balance to be unchanged by its modifications and immersion.

A more careful test of the balance was made using Eq. (13.4). The standards S1 and S2 were a kilogram of tantalum (volume 60 cm^3) and one of the two hollow weights described above (volume 337 cm^3, respectively).

TABLE 13.2 Summary of Results for Four Different Materials, [(measured volume/accepted volume) − 1], ppm

Run No.	Brass	Stainless	Germanium	Titanium
1	−5	−37	−25	−20
2	−28	−45	−22	−27
3	+7	−1	−15	−27
4	+17	−3	−18	−11
5	+18	−12	−1	−17
Ave. error	+2	−20	−16	−20
SD	19	20	9	7

The unknowns measured were all about 1 kg in mass: a brass weight (volume about 128 cm^3), the germanium crystal described above, and a titanium bar (volume 226 cm^3). All the objects were included in a symmetric series that required about 30 min to complete.

The six weights could all be kept below the surface of the fluorocarbon, on an immersed storage tray. The series was run five times over a period of several days. The results are summarized in Table 13.2.

It is noteworthy that measurements of the weights with large volume exhibit a lower standard deviation than do those of the higher-density weights. This is to be expected if the experimental scatter is due primarily to round-off errors in the balance. The remarkable feature is that the scatter is as low as is observed. The systematic errors may be due either to combined errors in the hydrostatic measurements of the standards and "unknowns" or to some unexplained source or sources of systematic error introduced by the experimental procedure.

13.2.2 Summary

Experience with the immersed balance warrants the following observations:

1. The linearity and reproducibility of the balance were unaffected.
2. The calibration factor, K, was also unchanged. A contamination of the fluorocarbon liquid by magnetic impurities may affect the balance.
3. The balance reading was stable even when the fluorocarbon liquid was gently agitated. Moving the forceps used in changing weights slowly through the liquid had no effect on the balance reading.

When weights were exchanged, the balance attained a steady reading within seconds of reloading the pan, although the liquid surface might still be in motion. Circulation of thermostated water through the outer chamber of the bath did not affect balance stability.

It was not certain why the electronic modification was necessary in order for the balance to operate acceptably when submerged. An hypothesis was that the initial motion of the weighing cell through the liquid generated turbulence, which, in turn, coupled to the cell. The servo electronics then responded to the changed force in such a way that instability resulted. A solution to the problem was, essentially, to increase the time constant of the integration response of the servo regulator. The fact that this tactic achieved the desired results does not, however, prove the hypothesis.

It is also worth noting that the electronic modification did not degrade the normal performance of the balance in air.

13.2.3 Discussion

Experience has shown that the immersed balance has two remarkable features that we believe can be used with outstanding advantage. First, although one might expect the immersed balance to be in a hostile environment, the opposite was found to be true. Second, the interchange of objects on the balance pan and the recovery of equilibrium were extremely fast compared with a classical hydrostatic weighing. These advantages can be utilized in two distinct, and what are believed to be important, applications.

First, there is scientific interest in density measurements, at the part per million level of accuracy, of solid samples having a mass in the range of l00 g to a few grams or less. Commercially available balances of the type used here have appeared with precision estimates of 1 or 2 ppm and ranges between 30 to 30,000 g. A similar treatment of these balances would conceivably result in density determinations with precision about the same as the uncertainty in the silicon density standard.

Perhaps a more interesting and important application of these instruments is in the realm of industrial metrology. Because the instrument is protected in a thermostated bath, it can perform well in areas that would ordinarily seem unsuitable. This feature, coupled with the extremely fast measurement cycle, makes possible measurements that were once considered to be too labor intensive or, in the case of mass production, too slow.

An example of such a measurement would be a production-line density determination of sintered turbine blades to assure their integrity. The standard for such a measurement could be a turbine of acceptable quality.

13.3 Determination of Density of Mass Standards; Requirement and Method[5]

13.3.1 Introduction

Beginning in 1965, Bowman, Schoonover, and others published a series of papers[1-3,7] that described the use of the mechanical one-pan two-knife balance in the high-precision determination of the density of a solid object. That work demonstrated that density measurements could be made with a precision better than 1 ppm.[3] In more recent years, the electronic force balance[8] has been perfected to a degree that it can replace most mechanical balances in both precision and capacity. Hence, the mechanical balance is rapidly disappearing from the scene.

In the following discussion, we present the need for accurate knowledge of the artifact densities in a mass measurement and explore the use of the electronic balance to measure the density of laboratory weights or of any solid object.

13.3.2 Requirements

The solution equation for the mass comparison between two artifacts is

$$X = \left[S\left(1 - \rho_a/\rho_s\right) - \delta\right]\big/\left(1 - \rho_a/\rho_x\right), \tag{13.6}$$

where S is the mass of the standard, ρ_s its density, ρ_x is the density of artifact of unknown mass X, ρ_a the density of air, δ the difference observed on the balance in mass units, and the remaining quantities relate to the unknown mass X. All terms have units that will result in mass X in grams.

A rigorous error analysis[9] reveals the need to measure the density of the mass artifacts to 1 part in 100,000 if the associated error is not to exceed 1 part in 1 billion. This requirement is compatible with the precision of the best mass comparators in use today.

13.3.3 Principles and Applications

There are several important gravimetric applications that can be performed on the electronic balance.[10] We have limited the discussion here to hydrostatic weighing principles, and, in particular, the determination of the density of a solid object is discussed in simple form and later extended to a more general form. Related topics of liquid density and glassware volume are briefly mentioned in Sections 13.4.11 and 13.4.12. Mass measurement is not necessary in the course of a density determination, but is addressed in the development of the general method.

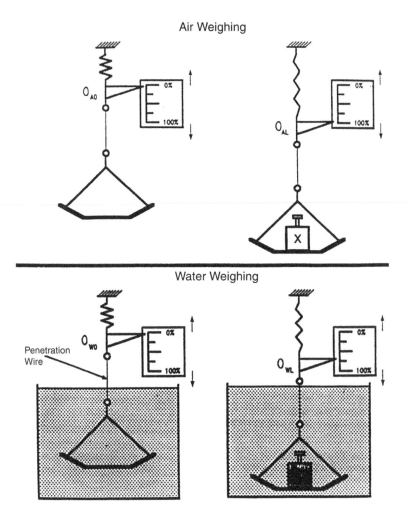

FIGURE 13.2 Illustration of detector observations required to weigh an object in air and then in water.

13.4   The Density of a Solid by Hydrostatic Weighing

It was shown in the 1967 work[2] that one does not need to use a mass standard to measure the density of a solid object by hydrostatic weighing. The only requirements are knowledge of the densities of air and water, and availability of a linear gravimetric force scale.

For completeness, we present the development of the equation required for a solid-object density determination when the object is suspended from a gravimetric force detector, first in air and then in water.

We have chosen a simple but hypothetically perfectly linear spring scale (force balance) as our detector. To make this concept clear we begin with a special case in which the air, the water, and the object are in mutual thermal equilibrium. Furthermore, the air density and the spring constant of the detector remain unchanged throughout the measurement, and the detector scale reads zero when the pan is empty both in air and in water.

Figure 13.2 illustrates the four detector observations required to weigh an object in air and then in water.

The following two equations are expressed in terms of the observables, the known and unknown components:

$$M_x g\left[1-\left(\rho_a/\rho_x\right)\right]=KO_{aL} \left(\text{air weighing}\right) \tag{13.7}$$

$$M_x g\left[1-\left(\rho_w/\rho_x\right)\right]=KO_{wL} \left(\text{water weighing}\right), \tag{13.8}$$

where M_x is the mass of object X, ρ_x is its density, ρ_a is the air density, ρ_w is the water density, K is proportional to the spring constant, g is the local acceleration due to gravity, and O_{aL} and O_{wL} are the detector observations when loaded in air and water, respectively.

Solving the above equations for the density of object X, ρ_x, one obtains the following equation:

$$\rho_x =\left(O_{wL}\rho_a -O_{aL}\rho_w\right)\big/\left(O_{wL} -O_{aL}\right). \tag{13.9}$$

The reader should note that the mass of X, M_x, does not appear in the above solution.

The difficulty associated with the use of Eq. (13.9) is the required thermal equilibrium and constant air density during the weighing cycles. Thermal equilibrium is nearly impossible to achieve in practice, and therefore it is desirable to have the water temperature cooler (by about 1°C) than the surrounding air; otherwise, water vapor condenses on the cooler balance mechanism causing a loss of measurement accuracy. However, equilibrium can be closely approximated in a stable laboratory environment.

Air density changes are related to changes in temperature, barometric pressure, relative humidity, and carbon dioxide (CO_2) content. Normal laboratory CO_2 variations and humidity excursions have a slight effect on air density and nature limits pressure variations to about 3%. Climatic control systems readily maintain air temperature within 0.5°C or better. Therefore, air density variations are limited to about 10% and may be much less on an hourly basis.

Although the equilibrium constraint is violated, the resultant errors may be acceptable to some users. To predict performance at the part per million level, a more general formulation is given later in the discussion.

A more detailed examination of these requirements is presented later. We repeat that a *mass standard* is totally unnecessary for this measurement.

13.4.1 The Force Detector

Although not perfect like the hypothetical spring-balance force detector described above, the performance of a modern electronic force balance is nearly perfect for many purposes. An overview of these instruments is given in Ref. 8 (see Chapter 10). A short summary of the principles of operation is given here. Detailed knowledge of the electronic circuits is unnecessary.

Figure 13.3 illustrates a representative mechanical structure of an electronic balance. When a downward force is applied to the balance pan (loaded with an object), it is opposed by a magnetic force generated by the interaction of two magnetic fields. One field is generated by a permanent magnet and the other by a controllable electromagnet. Usually, the magnetic force is applied through a multiplying lever. Sufficient magnetic force is generated to restore the mechanism (pan) to its unloaded position, or null point, relative to its structure as determined by a position sensor.

Obviously, the device is electromechanical, and we should expect errors both random and systematic to arise in the use of these instruments.

It is desirable in common weighing applications to tie the magnetic force to the unit of mass by calibration of the electronic circuit. The circuit is adjusted such that the algebraic sum of the gravitational and buoyant forces produces a balance indication approximately equal to the nominal value of the applied mass.

It is common practice for high precision balances to be supplied with a mass standard the density of which is about 8 g/cm^3 and with mass adjusted to a nominal value. This practice provides for a uniform response among balances to the given load at a given location. *This mass standard and the high degree of*

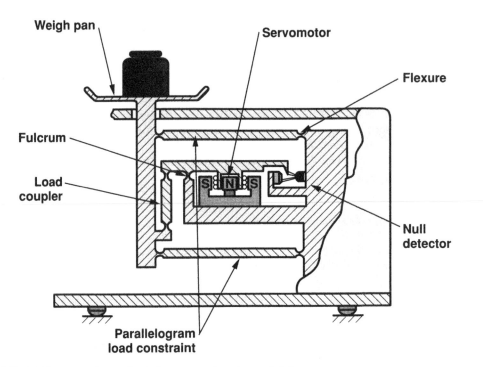

FIGURE 13.3 Representative mechanical structure of an electronic balance.

precision and linearity of the electronic force balance eliminates the need for a calibrated set of mass standards (weight set).

In pursuing the application of Eq. (13.9), it is unnecessary to quantify the mass in terms of any unit definition and we do not care about its density; it is merely a convenience to restore the spring constant if shifted from the initial value. However, when the calibration weight is tied to the mass unit, the electronic balance provides a convenient way to multiply and divide the mass unit within the capacity of the instrument.

13.4.2 Air Density

The *knowledge* of the density of air and water *embodies* the *information* that ties the above density measurement to the SI units. In the interest of obtaining the highest accuracy, the best available formulas are used in calculating these quantities. The air density equation for moist air used in this work is the CIPM 1981/91 recommendation.[11] This formula ties its predecessor, CIPM-81, to the International Temperature Scale of 1990 (ITS-90) and utilizes better estimates for the values of some of the constants and other parameters, in particular the more recent value of the universal gas constant. For brevity, we do not reproduce the formula here, but note that a 0.0004 mole fraction for CO_2 is assumed in our laboratory (see Chapter 12).

Uncertainties in the values of the parameters temperature, pressure, and relative humidity affect the uncertainty of the calculated air density. These parameters are measured at NIST with well-calibrated instruments with respective uncertainties of 0.01°C, 13 Pa (0.1 mmHg), and 2% RH. Based on these uncertainty estimates, an uncertainty of 0.0003 mg/cm³ (1 SD) has been assigned to the calculated air density values. It is this uncertainty estimate that is propagated in the analysis presented later.

13.4.3 Water Density

The work of Kell[12] was at this writing generally accepted as the best comprehensive treatment of water density (see Chapter 14). The Kell formula provides a value for the density of air-free water at 1 atm of pressure with an estimated uncertainty of 5 ppm. The formula assumes the use of the IPTS-1968 (t_{68}) temperature scale and temperatures measured in terms of the IPTS-1990 (t_{90}) must be converted to IPTS-1968. This is readily accomplished in the range between 20 and 30°C from the following approximate relationship:[13]

$$t_{90} - t_{68} = 0.006°C$$

Values of water density, ρ_{Kell}, obtained using Kell's formula may be adjusted for the effects of reaeration of the water after boiling, dissolved gases, and the compressibility of water using the correction terms developed by Bowman and Schoonover,[2] which give improved values:

$$\rho_w = \rho_{Kell}\left\{1 / 1 - C\left(B/760 + I/1033 - 1\right)\right\}$$
$$\left[1 - \left(2.11 - 0.053 t_w\right)\right]\left[1 - 1/\left(1 + D\right)\right] \times 10^{-6}, \tag{13.10}$$

where
ρ_w = water density
t_w = water temperature
D = number of days since boiling of the water
C = compressibility = 47.7 ppm/atm
I = immersed sample depth, cm
B = barometric pressure, in mmHg
ρ_{Kell} = Kell's water density

The originally published formula and units have been retained. The water temperature measurements here are estimated to be uncertain by 0.003°C (at $k = 3$) with a negligible effect on the calculation of water density.

13.4.4 A General Algorithm for Hydrostatic Weighing

The equilibrium conditions imposed by the use of Eq. (13.9) can be avoided with a more-detailed algorithm. The chief advantage is the achievement of improved estimates of measurement uncertainty. With a little extra effort, higher performance can be obtained from the balance. Both topics go hand-in-hand and are presented together here. However, before proceeding it is beneficial first to examine the effects of nonlinear balance response in a simple mass measurement.

In the above hydrostatic weighing derivation, the balance response is discussed in terms of a spring constant and the product of mass and local gravity is used to adjust the constant. In effect, the balance response is calibrated in terms of the gravitational force less the buoyant force exerted on the balance. This force is expressed by the following equation:

$$S\left[1 - \left(\rho_a/\rho_{st}\right)\right]g = KO_{1a}, \tag{13.11}$$

where S is the mass of a calibration weight and ρ_{St} its density at temperature t.

The force impressed on the balance by an object of unknown mass is

$$M_x\left[1 - \left(\rho_a/\rho_{xt}\right)\right]g = KO_{2a}, \tag{13.12}$$

where M_x is the mass of the unknown object and ρ_{xt} is its density at temperature t. The Os in these last two equations are balance observations.

Therefore,

$$S\left[1-\left(\rho_a/\rho_{st}\right)\right]/L = M_x\left[1-\left(\rho_a/\rho_{xt}\right)\right], \tag{13.13}$$

where L is the ratio of balance observations:

$$L = \left(O_{1a}-O'_{0a}\right)/\left(O_{2a}-O_{0a}\right). \tag{13.14}$$

The term, O'_{0a}, is the balance no-load indication during the calibration cycle and is set to zero, and O_{0a} is zero or near zero when the balance pan is empty during the weighing cycle.

Furthermore, the balance response when the calibration weight is engaged, $O_{1a} - O'_{0a}$, is redefined as O_C for this and all remaining applicable equations.

We can now express a solution for the unknown mass, M_x, in terms of the balance observations:

$$M_X = S\left[1-\left(\rho_a/\rho_{st}\right)\right]/\left\{\left[O_C/\left(O_{2a}-O_{0a}\right)\right]\left[1-\left(\rho_a/\rho_{xt}\right)\right]\right\}. \tag{13.15}$$

With a knowledge of the mass of the calibration weight and its density at the air temperature, t, from a simple weighing one can calculate the mass of an unknown object. Obviously the roles of S and X can be interchanged to perform an in situ *built-in mass calibration.*

In calibrating the balance, the manufacturer forces the no-load indication to be zero and, when the calibration mass is engaged, adjusts it to indicate its nominal value. The ideal balance response is, of course, a straight line connecting these points, and for some balances accuracy is preserved with extrapolation beyond these bounds. Usually, balances do not respond in the ideal manner and therefore any observation not at calibrated points may require correction for nonlinearity.

In the following discussion, it is assumed that the correction for nonlinearity has been applied to the balance observations for the unknown object during both the air and water weighings. The balance linearity is discussed in greater detail in Ref. 9 (see also Appendix C) and is considered briefly later in the discussion.

13.4.5 The General Hydrostatic Weighing Equations

It has been assumed that the air densities during the air and water weighings are different, and it follows that the temperatures, barometric pressure, and relative humidities involved may also be different. Temperature variations ensure that there will be two slightly different densities for the object during the measurement sequence. This can occur even with no lack of thermal equilibrium between the constituents during each of the weighings, because different temperatures may be encountered during air and water weighing.

However, as noted earlier, it is desirable to have the water temperature slightly cooler than that of the surrounding air during the water weighing cycle. Special precautions[2] must be taken to protect the measurement from the undesirable effects of this boundary condition.

The above simple weighing equations are now replaced with ones that permit variations in the ambient conditions that surround the objects during the weighings. Furthermore, the dependency of the balance calibration on air density is now taken into account. In addition, it is desirable to express the object density at a reference temperature that may be different from that of the measurement.

The expanded weighing equations are

For air weighing,

$$S\left\{\left[1-\left(Z_1\rho_{aa}/\rho_s\right)\right]/\left[O_C/\left(O_{2a}-O_{0a}\right)\right]\right\}g = M_x\left[1-\left(\rho_{aa}/\rho_{xtn}\right)\right]Xg, \tag{13.16}$$

For water weighing,

$$S\left\{\left[1-\left(\rho_{aw}/Z_2\rho_s\right)\right]/\left[O_C/\left(O_{2w}-O_{0w}\right)\right]\right\}g = M_x\left[1-\left(\rho_{aw}/\rho_{xtn}\right)\right]Yg. \tag{13.17}$$

Solving Eqs. (13.16) and (13.17) for the unknown density:

$$\rho_{Xtn} = \left\{\left[1-\left(\rho_{aa}/\rho_sZ_1\right)\rho_wY/\left[O_{Ca}/\left(O_{2a}-O_{0a}\right)\right]-\left[1-\left(\rho_{aw}/Z_2\rho_s\right)\right]\rho_{aa}X\right/\right.$$
$$\left[O_{Cw}/\left(O_{2w}-O_{0w}\right)\right]\right\}/\left\{\left[1-\left(\rho_{aa}/\rho_sZ_1\right)\right]/\left[O_{Ca}/\left(O_{2a}-O_{0a}\right)\right]\right. \tag{13.18}$$
$$\left.-\left[1-\left(\rho_{aw}/Z_2\rho_s\right)\right]/\left[O_{Cw}/\left(O_{2w}-O_{0w}\right)\right]\right\},$$

where

ρ_{xtn} = density of solid under test at 20°C
S = mass of calibration weight and ρ_s its density at 20°C
tn = 20°C

The reader will note that the density of the calibration weight now appears in the solution for ρ_{xtn} whereas its mass does not.

13.4.5.1 Air Weighing

Z_1 = $1/[1 + 3\alpha(t_{aa} - 20)]$
X = $1 + 3\beta(t_{aa} - 20)$
α = coefficient of linear thermal expansion of S
β = coefficient of linear thermal expansion of object undergoing test
t_{aa} = air temperature
ρ_{aa} = air density
O_{0a} = the balance "zero" reading (pan empty)
O_{Ca} = the balance indication with calibration weight engaged
O_{2a} = balance indication when loaded with the object of interest

Note: The balance "zero reading" may not actually be 0.

13.4.5.2 Water Weighing

Z_2 = $1/[1 + 3\alpha(t_{aw} - 20)]$
Y = $1 + 3\beta(t_{aw} - 20)$
t_w = water temperature
ρ_w = water density
t_{aw} = air temperature
ρ_{aw} = air density
O_{0w} = the balance "zero" reading (pan empty)
O_{Cw} = the balance indication with calibration weight engaged
O_{2w} = balance indication when loaded with the object of interest

The differences in elevation between the calibration weight, the air weighing pan, and the lower immersed pan require corrections for the gradient in the Earth's gravitational field.[2] This correction is approximately 300 µg/kg/m. In the case of the balance used here, the calibration weight and the air weighing pan are at the same elevation and the (lower) water weighing observations are adjusted to that elevation.

13.4.6 Linearity Test and Correction

The terms above, O_{2a} and O_{2w}, are assumed to have been corrected for any nonlinear balance response. The corrections were obtained by observing the balance response at 25, 75, and 100% of capacity. The balance is calibrated with a 100-g weight (50% of capacity) and the capacity of the balance is 200 g; therefore, an additional 100-g and 50-g weight is required for the test.

As noted earlier, the balance calibration forces a pan-empty indication of 0.0000 and the 100-g indication of 100.0000; these points are given and not observed. However, it is prudent to immediately check the calibration by reweighing the calibration weight or its replica. This provides assurance that an error-free calibration occurred. The procedure should be repeated as necessary as some balances require several cycles to obtain a stable calibration.

A detailed discussion of the balance linearity test and corrections is given in Ref. 11 and in Appendix C. The linearity corrections can be avoided by root-sum-squaring the balance nonlinearity with its precision when estimating the uncertainty of the density measurement.

13.4.7 Analysis

The method described by Ku[14] has been used to propagate errors in the functional relationship, $f(X_1, X_2, \ldots, X_n)$ of the uncorrelated variables $X_1, X_2, \ldots, X_n$. Tables 13.3 and 13.4 present for each variable its value, the estimated standard deviation, and an evaluation of the partial derivatives. At the bottom of each table is the estimated combined standard deviation for the function as given by the following relationship:

$$\left(SD_{\rho x}\right) = \left[\Sigma_{i=1}^{n}\left[\partial f/\partial X_i\right]^2\left(SD_i\right)^2\right]^{1/2}. \tag{13.19}$$

The measurement uncertainty for the density of 85-g silicon crystal is the root-sum-square of the estimated *effects* listed in the tables. The *effect* for each variable in the tables is calculated from the square root of both sides of the above equation. For the convenience of the reader, the partial derivatives are listed here, and they will be used to perform a sample calculation later.

The errors are propagated through the simple form of Eq. (13.9) rather than the more complex form of Eq. (13.18). The resulting calculated uncertainty from the following equations are nearly the same:

$$\left(\partial\rho_x/\partial O_{wL}\right) = -O_{aL}\left(\rho_a - \rho_w\right)\Big/\left(O_{wL} - O_{aL}\right)^2 \tag{13.20}$$

$$\left(\partial\rho_x/\partial O_{aL}\right) = O_{wL}\left(\rho_a - \rho_w\right)\Big/\left(O_{wL} - O_{aL}\right)^2 \tag{13.21}$$

$$\left(\partial\rho_x/\partial\rho_a\right) = O_{wL}\Big/\left(O_{wL} - O_{aL}\right) \tag{13.22}$$

$$\left(\partial\rho_x/\partial\rho_w\right) = -O_{aL}\Big/\left(O_{wL} - O_{aL}\right) \tag{13.23}$$

One important parameter in the error analysis is the balance reproducibilty as measured by the standard deviation. The balance used here, like many electronic balances, performed better when lightly loaded. Its standard deviation was found to be 49 µg from 0 to 100 g and 118 µg upward to 200 g. These standard deviations are combined in quadrature with the standard deviation of the linearity correction.

At the 200-g level of the air weighings, the combined standard deviation was 138 µg, whereas water weighings near 100 g were nearly free of nonlinearity and the standard deviation remained 49 µg.

TABLE 13.3 Error Propagation for a 200-g Silicon Crystal

Variable	Value	SD	Partial Deriv.	Effect
ρ_a	0.0012 g/cm³	0.0000003 g/cm³	−1.352942	−0.0000004 g/cm³
ρ_w	0.9974 g/cm³	0.0000017 g/cm³	2.352941	0.0000040 g/cm³
O_{aL} (bal. units)	200 g	0.00042/√6 g	−0.015856 cm⁻³	0.0000027 g/cm³
O_{wL} (bal. units)	85 g	0.000138/√6 g	0.027643 cm⁻³	0.0000016 g/cm³
			RSS =	0.0000051 g/cm³

The balance weighings were repeated six times, which is reflected in the error propagation in Table 13.3. It is noteworthy that the balance calibration reproducabilty (49 μg) is not improved by repeated calibration cycles and therefore the calibration is only performed once. This standard deviation cannot be obtained explicitly, but from the nature of digital circuits it is known to be less than ¹/₂ count, i.e., 50 μg, and the value (49 μg) determined by repeated weighings at the 100-g level was used here.

Furthermore, the balance indications are labeled as balance units. Clearly, if the mass of the built-in weight is a known mass, then the observations would be in SI units. It is only necessary to know the mass of the built-in weight if one wishes to make a mass measurement.

To recapitulate, to determine the uncertainty applicable to a density measurement made by the method described here, one only requires knowledge of the balance standard deviation and its lack of linearity, if any.

13.4.8 Balance Selection

Obviously, one needs a balance with a 200-g capacity for the silicon crystal density determination. If one examines the *effect* of the balance water weighing observation, O_{wL}, for Table 13.3, it is found to be the largest for all the variables except the density of water. The information regarding the density of water could not be improved, but a better-performing balance could improve the measurements.

However, the measurement error is approaching the limitation imposed by the water density and, if the improvement was not required, doing so would not be cost-effective.

Consider now a hypothetical measurement of the density of a 1-kg stainless steel mass standard. There are many balances available; the best (i.e., smallest standard deviation) cost, at this writing, about 24 times the least precise instrument. Tentatively, a balance that cost ¹/₁₂ as much as the most expensive one was selected.

Based on the manufacturer's specification, one can construct an error propagation table, Table 13.4, and see whether the cheaper balance will be satisfactory.

The pertinent specifications are

$$SD = 0.001 \text{ g}$$

$$\text{linearity error} = 0.003 \text{ g or less}$$

The goal here is to determine the density of the kilogram to 1 part in 100,000 (10 ppm).

TABLE 13.4 Error Propagation for a 1-Kilogram Mass Standard

Variable	Value	SD	Partial Deriv.	Effect
ρ_a	0.0012 g/cm³	0.0000003 g/cm³	−7.0	0.000002 g/cm³
ρ_w	0.9974 g/cm³	0.0000017 g/cm³	8.0	0.000014 g/cm³
O_{aL} (bal. units)	1000 g	0.001 g	−0.0639104	0.000064 g/cm³
O_{wL} (bal. units)	875 g	0.0032 g	−0.0557872	0.000178 g/cm³
			RSS =	0.00019 g/cm³

One can anticipate that the density of a kilogram of stainless steel, about g/cm³, could be measured with an error of 0.00019 g/cm³ or about 1 part in 40,000, and would not achieve our goal. If, on the other hand, one were able to correct the balance indications for the nonlinear response and record nine water weighing observations, one could achieve a standard deviation of about 0.001 g, comparable to that for the air weighing.

Because the balance span is calibrated at 1000 g, there is no nonlinearity correction to apply to the air weighing observations. With small additional effort one could achieve an error of 0.0000558 g/cm³ for the water weighing and an RSS of slightly better than 1 part in 100,000. It is then concluded that the selected balance will be adequate for the measurement goal provided the user applies the linearity correction and performs nine water weighing cycles.

The reader should note that both Tables 13.3 and 13.4 have the same standard deviations for the air density and water density. Although one could not improve these parameters very much, one could make them much worse, depending on how well the temperature, relative humidity, and barometric pressure are measured. See Ref. 9 and Chapter 12 for a similar treatment of these parameters.

13.4.9 Data Results

Six independent determinations of the density of a 200-g silicon crystal were made.[6] The crystal density is known to 2.25 ppm.[3] The data gathered using the electronic force balance and the method described here resulted in a standard deviation of 2 ppm with an offset of 2.4 ppm from the more accurate interferometric results described in Ref. 3.

13.4.10 Conclusions

It was believed that the investigations of the testing and use of the electronic balance support the contention that very respectable measurements at the parts per million level can be achieved. It is noteworthy that when the balance is used properly the need for well-calibrated laboratory weight sets is *eliminated*.

Most users would be well served by obtaining both a mass value and a density determination for the balance calibration weight from an appropriate calibration laboratory. The needs of many users would certainly be satisfied by accepting the value and tolerance assigned by a reputable balance maker.

Not all of the common uses of balances were covered. There are simple applications such as the tolerance testing of weights for regulatory use, i.e., weights and measures, to which these techniques can be adopted. It appears that many modern electronic balances perform so well that it might be beyond the ability of many laboratories to test and calibrate them adequately. This could well be a problem in the future where some applications, requiring certification, are limited by the ability of the certifying agency.

13.4.11 Appendix 1 — Liquid Density by Hydrostatic Weighing

An examination of Eq. (13.18) reveals the possibility of determining the water density, or the density of other liquids, with prior knowledge of the solid object density. Relabeling the terms ρ_w as ρ_L and ρ_x as ρ_{si} and rearranging terms, the density of an unknown liquid by hydrostatic weighing is

$$
\begin{aligned}
\rho_{\text{LBt}} = \Big\{ & \rho_{\text{Sitn}} \Big\{ \big[1-\left(\rho_{aa}/\rho_s Z_1\right) \big] \big/ \big[O_{Ca}/\left(O_{2a}-O_{0a}\right) \big] \\
& -\big[1-\left(\rho_{aL}/\rho_S Z_2\right) \big] \big/ \big[O_{CL}/\left(O_{2L}-O_{0L}\right) \big] \Big\} \\
& +\rho_{aa} X \Big\{ \big[1-\left(\rho_{aL}/\rho_S Z_2\right) \big] \big/ \big[O_{CL}/\left(O_{2L}-O_{0L}\right) \big] \Big\} \Big\} \Big/ \\
& Y \Big\{ \big[1-\left(\rho_{aa}/\rho_S Z_1\right) \big] \big/ \big[O_{Ca}/\left(O_{2a}-O_{0a}\right) \big] \Big\},
\end{aligned}
$$

(13.24)

where

    ρ_{LBt} = liquid density under test at the bath temperature, Bt
    ρ_{Sitn} = solid object density standard normalized at 20°C

13.4.12 Appendix 2 — Glassware Calibration

Flasks, burettes, and pipettes are usually calibrated by gravimetric means. A simplified equation is offered here for the calibration of general-use glassware. The capacity, C_t, is

$$C_t = S\left[1-\left(\rho_a/\rho_S\right)\right]\left[\left(O_F-O_E\right)/O_C\right]/\left(\rho_w-\rho_a\right),\tag{13.25}$$

where

    t = water temperature, in °C
    ρ_a = average air density of the weighing cycles
    ρ_w = water density
    ρ_S = calibration weight density at room temperature
    O_E = empty flask balance indication
    O_F = flask plus water balance indication

In the above equation, it is assumed that the balance has been zeroed before the empty and full weighings and the flask capacity is at the temperature of the water. The nominal value of the built-in calibration mass, S, is usually adjusted to be accurate within the least significant digit displayed by the balance and may not need additional calibration for this application.

It has been assumed that the weight is made from a material with a density near 8.0 g/cm^3 and has been adjusted to the "8.0 apparent mass scale" (see Chapter 15).

13.5 An Efficient Method for Measuring the Density (or Volume) of Similar Objects

13.5.1 Introduction

Schoonover and Nater[15] developed a method to determine the density of nearly identical objects. The method readily lends itself to automation and, because it utilizes a low-surface-tension fluorocarbon liquid in place of water, bubble formation is not a problem. The use of an artifact density standard eliminates the expensive air density instrumentation and water temperature measurement associated with the use of water as the density standard. A precision of 1 ppm was achieved for repeated volume determinations of a stainless steel 1-kg laboratory mass standard.

The method is directed to the density determination of OIML Class E$_1$ laboratory weights, but can be used *to measure the density of any similar objects*. The method takes advantage of the similarities in mass for a given nominal value and the restricted range in the material density from which the weights are constructed. The method is especially useful when the density measurement is to be automated using a high-precision electronic balance mass comparator in conjunction with an immersed weight handler.

This method has two other major advantages for the metrologist. As the method compares only volumes of objects with the same thermal coefficient of expansion, a high-accuracy temperature measurement is unnecessary and the associated expense is avoided. Additionally, the requisite skills required to develop the volume standard from first principles can be avoided by obtaining the calibration from a national laboratory. In essence, the method described here has the potential for placing a very high accuracy density (volume) measurement into a modest laboratory.

TABLE 13.5   Error Propagation for a Comparison Weighing of Two 1-kg
Weights of Densities Approximately 8.0 g/cm³

Variable	Value	SD	Partial Deriv.	Effect
m_S	1000 g	0.000022 g	1	22×10^{-6} g
ρ_S	8 g/cm³	0.00008 g/cm³	0.018753 cm³	1.5×10^{-6} g
ρ_X	8 g/cm³	0.00052 g/cm³	−0.150021 cm³	-78×10^{-6} g
δ	0.01 g	0.000020 g	−1.00015	-20×10^{-6} g
ρ_a	0.0012 g/cm³	0.0000009 g/cm³	−0.001251 cm³	-1.1×10^{-9} g
				RSS = 83×10^{-6} g

13.5.2 The Requirement

The OIML International Recommendation R111[16] defines the attributes of laboratory weights. The criteria that relate to the need for measuring the density for a class E_1 weight are as follows:

1. Maximum permissible error (MPE), the permitted departure from the nominal value of the mass
2. The uncertainty of a reported mass value must not exceed ⅓ of the MPE
3. The density domain for weights with mass above 100 g is 7.934 to 8.067 g/cm³

The MPE for an E_1 weight of nominal mass of 1 kg is 0.5 mg.

A density determination with an uncertainty of 0.033% is adequate to assure that the selected material is within the range. From examination of the simple weighing equation given below, we can estimate how well the density must be determined to meet requirement 2 above.

$$M_X = M_S\left\{\left[1-\left(\rho_a/\rho_S\right)\right]-\delta\right\}\big/\left[1-\left(\rho_a/\rho_x\right)\right], \qquad (13.26)$$

where M_X is the mass of weight X undergoing calibration and ρ_X is its density, M_S is the mass of the standard weight S and ρ_s is its density, ρ_a is the air density, and δ is the mass difference calculated from balance indications.

Table 13.5 is an error propagation for Eq. (13.26) and shows the values selected for the mass calibration of a 1-kg weight and the appropriate ancillary equipment. The partial derivatives are given in Ref. 9. Estimates for the values of M_S and ρ_s and their uncertainty may come from a national laboratory as in this section. Clearly, the root-sum-square (RSS) value from Table 13.5 (83×10^{-6} g) when multiplied by 2 meets the uncertainty requirement above. That is, 166×10^{-6} g $\leq$ (500×10^{-6} g ÷ 3). Therefore, the density of the unknown object must be determined to 0.0065% or better.

13.5.3 The Method

From OIML R111, we know that if density measurements are restricted to laboratory weights that meet the class E_1 requirements, the mass of a 1-kg weight will be equal to the nominal value within 0.00005%. The error in density that results from this mass error is 0.000004 g/cm³. If the mass errors were ten times larger, the method would still achieve an acceptable uncertainty. The weighing equation for a *density standard* suspended from an electronic balance and immersed in a liquid is

$$k\left(M_S-\rho_L V_S\right)=\delta_S, \qquad (13.27)$$

where k is the constant of proportionality $\cong 1.00015$ resulting from the balance internal calibration, M_s is the mass of the density standard, ρ_L is the density of the liquid, V_s is the volume of the density standard, and ρ_s is the difference in balance observations in mass units between the empty and loaded balance pan.

A similar equation can be written for another similar object, X:

$$k\left(M_X - \rho_L V_X\right) = \delta_X, \tag{13.28}$$

where M_X is the mass of the object whose unknown volume, V_X, is the volume to be determined of the object, and δ_X is the corresponding difference in balance indications in mass units. Because M_X is approximately equal to M_S, solving Eqs. (13.27) and (13.28) for V_X yields:

$$V_X = \left[\left(\delta_S - \delta_X\right)\big/ k\rho_L\right] + V_S. \tag{13.29}$$

We note that the density of object X is approximately equal to M_{nom}/V_X, where M_{nom} is the nominal mass value of the class E_1 weight undergoing the measurement.

Examination of Eq. (13.29) reveals that the balance merely detects a very small buoyant difference between the density standard and any similar object. However, knowledge of the liquid density is required. Schoonover and Nater chose not to use water, whose density is defined by a simple temperature measurement, and thereby avoided the difficulties associated with air bubbles clinging to the weights. Their choice of a fluorocarbon liquid (FC-75) avoids the bubble problem but does require a density determination.

13.5.4 The Measurement of Liquid Density

The balance used here and the automation apparatus present a special problem in determining ρ_L. It is more than inconvenient to disconnect the liquid weighing pan from the balance above. In addition, the balance is a mass comparator that utilizes a 100-g electronic weighing capacity in conjunction with built-in weights that can sum to 900 g, yielding a total weighing capacity of 1 kg. The apparatus is shown in Figure 13.4. The dials that manipulate these built-in weights are not automated and not easily reached by hand. The solution is to weigh the buoyant loss on the immersed density standard in terms of external standards placed on the balance pan in air. A few appropriate class E_1 weights will suffice for the measurement. The reader is referred to Figure 13.5.

Density Measurement Equipment

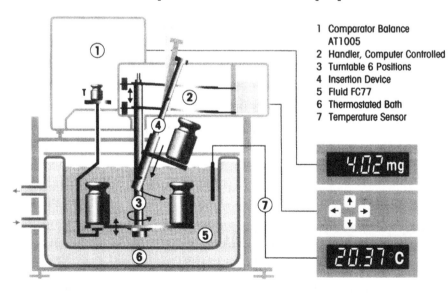

1 Comparator Balance AT1005
2 Handler, Computer Controlled
3 Turntable 6 Positions
4 Insertion Device
5 Fluid FC77
6 Thermostated Bath
7 Temperature Sensor

4.02 mg

20.37 °C

FIGURE 13.4 Apparatus for determining laboratory weight density.

Condition 1

δ1

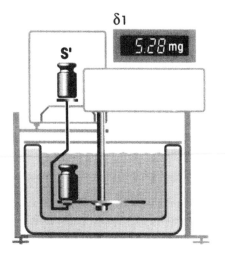

Condition 2

δ2

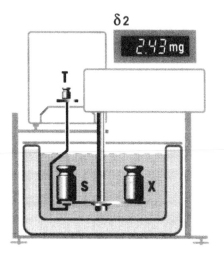

FIGURE 13.5 Conditions for determining liquid density.

M'_s and M_s are the masses of two class E_1 laboratory weights S' and S of the same nominal volume and can be used interchangeably in the following discussion. In the following discussion, weight S is the volume standard, i.e., the density standard. S' is placed on the balance air pan, and the balance dial weights are manipulated until a balance indication, δ_1, is achieved; see condition 1 of Figure 13.5. Then, S' is removed and S is placed on the immersed balance pan. The weight T of mass M_T is selected from a set of E_1 weights such that the balance indication δ_2, in mass units is within 10 g of δ_1; see condition 2 of Figure 13.5. Keeping δ_1 and δ_2 close to each other minimizes any nonlinearity errors. One can now write two force equations that describe both conditions of the balance that are solved for ρ_L:

$$F_1 = \left(M'_S - \rho_a V'_S \right) g = \left(\delta_1 / k \right) g, \tag{13.30}$$

$$F_2 = \left(M_S - \rho_L V_S \right) g + \left(M_T - \rho_a V_T \right) g = \left(\delta_2 / k \right) g. \tag{13.31}$$

In the above equations, V_T is the volume of mass T and is calculated from the nominal density 8.0 g/cm³; V_s and V_T are at the measurement temperature.

$$\rho_L = \left\{ \rho_a \left(V'_S - V_T \right) + M_T + \left[\left(\delta_1 - \delta_2 \right) / K \right] \right\} / V_S \tag{13.32}$$

13.5.5 Error Analysis

The error analysis is performed in the usual way following the ISO Expression of Uncertainty Guidelines.[17] Shown below are the partial derivatives for Eq. (13.29), the expression for V_X. For brevity, the partial derivatives for Eq. (13.29), the expression for ρ_L, and for the balance calibration factor k[18] (see Chapter 28) are not shown; none of them plays an important role as the analysis shows. For the last two terms, the uncertainty estimates were obtained in the same manner and the values are listed in the error budget of Table 13.6. The uncertainty contribution from these terms is dominated by the uncertainty in the volume standard (density standard), S, and are not significant here. In addition, covariant terms are ignored.

TABLE 13.6 Error Propagation for the Volume Determination of a 1-kg Weight of Density Approximately 8 g/cm³

Variable	Value	SD	Partial Deriv.	Effect
V_S	125 cm³	0.00125 cm³	1	0.00125 cm³
δ_s	0.4 g	0.000056 g	0.563 cm³/g	0.000032 cm³
δ_X	0.2 g	0.000056 g	−0.563 cm³/g	−0.000032 cm³
ρ_L	1.775 g/cm³	0.00003 g/cm³	−0.063 cm³/g	0.0000019 cm³
k	1.00015	0.000020	−0.113 cm³	0.0000023 cm³
				RSS = 0.00125 cm³

The partial derivatives are listed below:

$$\partial V_X / \partial V_S = 1$$

$$\partial V_X / \partial \rho_S = 1/k\rho_L$$

$$\partial V_X / \partial \delta_X = -1/k\rho_L$$

$$\partial V_X / \partial \rho_L = \left(\delta_X - \delta_S\right)/k\rho_L^2$$

$$\partial V_X / \partial k = \left(\delta_X - \delta_S\right)/k^2\rho_L$$

Examination of Table 13.6 reveals that the uncertainty contribution related to the balance performance was insignificant. The same insignificance applies to the liquid density measurement and to k. In fact, in practice the measurement is limited only by how well the volume standard can be calibrated by the national standards laboratories or any other laboratory.

13.5.6 Apparatus

The apparatus comprised a Mettler–Toledo AT 1005 mass comparator, a computer-controlled weight handler, a double-wall thermostated bath containing the fluorocarbon liquid, and a circulating-water thermostat.

The balance weighing house and weighing pan were removed; otherwise it is an off-the-shelf mass comparator described earlier. However, the built-in dial-operated weights act only as tare weight and once set for a given nominal value weight they remain unchanged. Only the electronic output constitutes the weighing observation.

The pan was modified to be a mere pedestal of about 1.2 cm in diameter. Doing so provided the clearance required for the central rotating axle of the weight handler and an increased weighing capacity. The increased capacity permitted the permanent installation of a substantial immersed weighing pan. The immersed pan is connected to the balance above with a suspension wire of sufficient torsional stiffness to prevent pan rotation. This rather rigid suspension is necessary to permit reliable automation of the measurement.

An off-the-shelf weight handler was modified to position six weights on the immersed weighing pan rather than the usual four. This required slight modification of the hardware and the computer program; otherwise, the principle and specifications of the weight handler, the thermostat, and the bath are well described in the literature.[2,7]

One noteworthy feature is the weight insertion tool that is built into the system. This feature allows an unskilled operator reliably and quickly to load and unload weights bubble-free.

Measurement production for the system is about 64 1-kg weights per 8-h day plus 16 check standard determinations.

TABLE 13.7 The Five Measurement Results for Liquid Density and the Volume of the 1-kg Check Standard of Density Approximately 8.0 g/cm^3

Sequence	Temp., °C	ρ_L, g/cm^3	Vol., cm^3, at 20°C
1	21.096	1.7757	125.6203
2	20.917	1.7760	125.6202
3	20.976	1.7759	125.6202
4	20.983	1.7759	125.6201
5	20.947	1.7760	125.6202

Mean = 125.6202 cm^3
SD = 0.000071 cm^3

13.5.7 Data

Efficient use of the system with the built-in quality check offered by a check standard requires two volume standards and four other similar objects. In this case, six OIML class E$_1$ 1-kg weights were used to demonstrate the method described here. NIST provided a density determination for the two weights used as the standard, S1, and check standard, S2; the others are labeled W1 through W4. The weighing sequence was S1, S2, W1, W2, W3, W4, S1. Well-known drift-elimination techniques were used to adjust the balance observations.

Five independent measurement sequences were performed with the weights removed from the fluid between runs and the measurements made on different days. At the beginning of each sequence, the liquid density was determined as described here. For brevity, we show only the data for the liquid density determination and the volume of the check standard. The results are presented in Table 13.7. The standard deviation in volume is better than 1 ppm. The volume assigned by NIST to S2 was 125.6208 cm^3 at 20°C with an uncertainty of 10 ppm.

13.5.8 Conclusion

It has been demonstrated that the method discussed here can achieve an uncertainty near 1 ppm. As a practical matter, 5 ppm is probably what is attainable for V_S by the national laboratories without exotic techniques and high cost. Experience with the apparatus has proved it to be highly reliable, efficient, and, particularly, cost-effective. This method could be quite useful to determine the density of objects other than mass standards, for example, the density of products mass-produced from sintered materials where the density is an indicator of the absence of defects. Schoonover and Nater have demonstrated that one could assign density values to fractions of the 1-kg density standard. That is, a 1-kg density standard could be ratioed for 500-g, 200-g, and 100-g weights.[19]

References

1. Bowman, H. A., Gallagher, W., and Schoonover, R. M., The Development of a Working Density Standard, Instrument Society of America, 20th Annual ISA Conference and Exhibit, Preprint No. 14.8-4.65, Los Angeles, CA, 1965.
2. Bowman, H. A. and Schoonover, R. M., with Appendix by Mildred W. Jones, Procedure for high precision density determinations by hydrostatic weighing, *J. Res. Natl. Bur. Stand.* (U.S.), 71C, 3, 1967.
3. Bowman, H. A., Schoonover, R. M., and Carroll, C. L., A density scale based on solid objects, *J. Res. Natl. Bur. Stand.* (U.S.), 78A, 13, 1974.
4. Thiesen, M., *Trav. et Mem. Bureau International des Poids et Mesures,* 5(2), 1886.

5. Schoonover, R. M. and Davis, R. S., Quick and accurate density determinations of laboratory weights, in *Proc. 8th Conf. IMEKO Technical Committee TC3 on Measurement of Force and Mass,* Krakow, Poland, Sept. 9–11, 1980; Paper in *Weighing Technology,* Druk, Zaklad Poligraficzny Wydawnictwa SIGMA, Warszawa, Poland, 1980, 1123-1127.

6. Schoonover, R. M., Hwang, M.-S., and Nater, R., The determination of density of mass standards; requirement and method, NISTIR 5378, February 1994.

7. Bowman, H. A., Schoonover, R. M., and Carroll, C. L., The utilization of solid objects as reference standards in density measurements, *Metrologia,* 10, 117, 1974.

8. Schoonover, R. M., A look at the electronic analytical balance, *Anal. Chem.,* 52, 973A, 1982.

9. Schoonover, R. M. and Jones, F. E., Examination of parameters that can cause errors in mass determination, presented at Measurement Science Conference, Pasadena, CA, 1994.

10. Schoonover, R. M. Taylor, J., Hwang, M.-S., and Smith, C., The electronic balance and some gravimetric applications (The density of solids and liquids, pycnometry and mass, in *ISA 1993 Proceedings of the 39th International Instrumentation Symposium,* 1993, 299.

11. Davis, R. S., Equation for the determination of the density of moist air (1981/91), *Metrologia,* 29, 67, 1992.

12. Kell, G. S., Density, thermal expansivity, and compressibility of liquid water from 0 to 150°C: Correlations and tables for atmospheric pressure and saturation reviewed and expressed on 1968 temperature scale, *J. Chem. Eng. Data,* 20, 97, 1975.

13. Magnum, B. W. and Furukawa, G. T., Guidlines for Realizing the International Temperature Scale of 1990 (ITS-90), NIST Technical Note 1265, 1990.

14. Ku, H. S., Statistical concepts in metrology, in *Handbook of Industrial Metrology,* American Society of Tool and Manufacturing Engineers, Prentice-Hall, Englewood Cliffs, NJ, 1967, chap. 3.

15. Schoonover, R. M. and Nater, R., An efficient method for measuring the density (or volume) of similar objects, presented at *Measurement Science Conference,* Anaheim, CA, Jan. 25–26, 1996.

16. OIML R 111, Edition 1994 (E), Organization Internationale de Metrologie Leagale, 11, Paris, France, 1994.

17. International Organization for Standardization, Guide to the Expression of Uncertainty in Measurement, International Organization for Standardization, Geneva, Switzerland, 1993.

18. Schoonover, R. M. and Jones, F. E., The use of the electronic balance for highly accurate direct mass measurements without the use of external mass standards, NISTIR 5423, National Institute for Standards and Technology, May 1994.

19. Schoonover, R. M. and Nater, R., A method for measuring the density (or volume) of laboratory weights, presented at Measurement Science Conference, Pasadena, CA, Jan. 23–24, 1997.

14

Calculation of the Density of Water

14.1 Introduction

A very convenient standard of density is pure air-free water of known isotopic content. Recent redeterminations of the density of water have been made by Patterson and Morris of the Australian CSIRO National Measurement Laboratory, by Watanabe of the National Research Laboratory of Metrology of Japan, and by Takenaka and Masui of the National Research Laboratory of Japan.

In the present chapter, corrections for change in density of water with air saturation, compressibility, and isotopic concentrations have been made to equations developed using the recent data. The corrected Patterson and Morris equation for calculation of water density is considered to supersede, or succeed, the work of Kell at the National Research Council of Canada.

The density of water is used as a reference in various areas. For example, water is used as the calibrating fluid in the calibration (gravimetric determination) of the volume of volumetric standards. The volume is calculated from the mass and density of the water.

14.2 Formulations of Wagenbreth and Blanke

Wagenbreth and Blanke[1] developed a formulation for the calculation of the density of water for *air-free* water. The Wagenbreth and Blanke formulation is a polynomial of fifth degree in temperature, °C on the 1968 International Practical Temperature Scale (IPTS-68).

Unless water is just freshly distilled, air will be absorbed and ultimately the water will be air-saturated. Wagenbreth and Blanke[1] calculated the correction for the difference between the density of air-free water and the density of air-saturated water.

14.3 Kell's Formulations

14.3.1 Density of Standard Mean Ocean Water

In 1977, Kell[2] of the National Research Council of Canada published an equation relating the density of Standard Mean Ocean Water, SMOW (ρ_{SMOW}), to temperature over the temperature range 0 to 150°C.

The concept of SMOW was introduced by Craig[3] in 1961 to provide a uniform standard for deuterium (D) and oxygen 18 (^{18}O) concentrations of natural waters. SMOW was first defined in terms of the U.S. National Bureau of Standards "reference sample 1."[4] Although there were several close approximations, no sample of SMOW existed. Therefore, there was no quantity of SMOW from which samples for density studies could be taken.

Kell's equation for the density of SMOW is[2]

$$\rho_{SMOW} = \left(999.8427 + 67.8782 \times 10^{-3}\,t + 103.1412 \times 10^{-6}\,t^3\right.$$

$$\left. + 15.95835 \times 10^{-9}\,t^5 + 636.8907 \times 10^{-15}\,t^7\right) \Big/$$

$$\left(1 + 9.090169 \times 10^{-6}\,t^2 + 1.4511976 \times 10^{-9}\,t^4\right.$$

$$\left. + 134.84863 \times 10^{-15}\,t^6 + 2.008615 \times 10^{-18}\,t^8\right).$$

(14.1)

Temperature t in this equation is on the International Practical Temperature scale of 1968 (IPTS-68).

14.3.2 Isothermal Compressibility of Water

Kell[5] also developed an equation for the isothermal compressibility of water, κ_T. The isothermal compressibility data used by Kell have been fitted[6] against temperature on the 1990 International Temperature Scale (ITS-90) over the temperature range 5 to 40°C. The resulting equation is

$$\kappa_T = 50.83101 \times 10^{-8} - 3.68293 \times 10^{-9}\,t$$

$$+ 7.263725 \times 10^{-11}\,t^2 - 6.597702 \times 10^{-13}\,t^3$$

$$+ 2.87767 \times 10^{-15}\,t^4,$$

(14.2)

where κ_T is the isothermal compressibility in $(kPa)^{-1}$.

At 20°C, the value of the isothermal compressibility of water is approximately 46.5 parts per million (ppm) per atmosphere. At locations where the atmospheric pressure is significantly different from 1 atm (101,325 Pa, 14.69595 psi, 760 mmHg), a correction for isothermal compressibility using Eq. (14.2) should be made. For example, at Boulder, CO (at an elevation of approximately 1400 m) in the United States the correction for isothermal compressibility is approximately –8 ppm at 20°C.

14.4 Conversion of IPTS-68 to ITS-90

Very simple equations relating ITS-90 temperature, t_{90}, to IPTS-68 temperature, t_{68}, have been used in the development of Eq. (14.2).[6]

In the temperature range 0 to 40°C the equation is

$$t_{90} = 0.0002 + 0.99975\,t_{68}.$$

(14.3)

In the temperature range 0 to 100°C the equation is

$$t_{90} = 0.0005 + 0.9997333\,t_{68}.$$

(14.4)

14.5 Redeterminations of Water Density

Density tables and the equations above rely heavily on the results of direct observations published by Thiesen et al.[7] in 1900 and by Chappuis[8] in 1907. Because of doubt concerning the accuracy of tables and equations based on the early work, remeasurement of the water density using several methods was undertaken by standards laboratories.[9]

TABLE 14.1 Densities of Air-Free SMOW
Calculated from Eq. (14.5), and Estimated SDs

Temperature, °C	Air-Free SMOW Density, kg/m³	SD, kg/m³
1	999.90125	0.0014
3	999.96594	0.00098
4	999.97358	0.00090
5	999.96537	0.00084
7	999.90319	0.00076
10	999.70166	0.00070
15	999.10168	0.00064
20	998.20569	0.00069
25	997.04593	0.00081
30	995.64801	0.00098
35	994.03222	0.0013
40	992.21489	0.0014

14.5.1 Measurements of Patterson and Morris

By weighing an evacuated hollow glass sphere of known volume in air-free water samples of measured isotopic content, Patterson and Morris[9] of the Australian CSIRO National Measurement Laboratory determined the absolute density of Standard Mean Ocean Water (SMOW) over the temperature range 1 to 40°C.

Measurements were made of the isotopic ratios $^{18}O^{16}O$ and D/H (deuterium to hydrogen) of the water used by Patterson and Morris.[9] Correction to convert the measured density of a water sample to the corresponding value for Vienna Standard Mean Ocean Water (V-SMOW)[10] was calculated.

Water Density Equation of Patterson and Morris

An equation, derived from the work of Patterson and Morris, that can be used to calculate the absolute density of air-free SMOW over the temperature range 1 to 40°C is

$$\rho = 999.97358 - \left[A\left(t - t_o\right) + B\left(t - t_o\right)^2 + C\left(t - t_o\right)^3 + D\left(t - t_o\right)^4 + E\left(t - t_o\right)^5 \right], \quad (14.5)$$

where
t_o = 3.9818°C
A = 7.0132 × 10⁻⁵ (kg/m³) (°C)⁻¹
B = 7.926295 × 10⁻³ (kg/m³) (°C)⁻²
C = −7.575477 × 10⁻⁵ (kg/m³) (°C)⁻³
D = 7.314701 × 10⁻⁷ (kg/m³) (°C)⁻⁴
E = −3.596363 × 10⁻⁹ (kg/m³) (°C)⁻⁵.

ρ is in kg/m³ and t (and t_o) is in °C on the ITS-90 temperature scale. The maximum density of V-SMOW was 999.97358 (±0.00089) kg/m³, the first term in Eq. (14.5); the temperature, t_o, at which water attains its maximum density was determined to be 3.9818°C.

In Table 14.1, the densities of air-free SMOW calculated from Eq. (14.5) and the estimates of standard deviation (SD) for temperatures between 1 and 40°C are tabulated.

14.5.2 Measurements of Watanabe

Watanabe[11] of the National Research Laboratory of Metrology of Japan measured the thermal dilatation (expansion) of pure water, obtained in terms of the ratio of the density at temperature t to maximum density, by detecting the change in buoyancy acting on a cylindrical hollow artifact made of fused quartz immersed in water. The range of temperature was 0 to 44°C.

In the series of measurements, the water bath temperature was stabilized at a value in the range 0 to 44°C and an observation was made of the "hydrostatic apparent weight" of the hollow cylinder using a balance. Repeated observations under the same conditions were averaged.

Watanabe Formulation for the Thermal Dilatation of Water

The formula for the thermal dilatation of water, free from air, developed for the temperature range 0 to 44°C by Watanabe is:

$$\rho/\rho_o = A + Bt + Ct^2 + Dt^3 + Et^4 + Ft^5 + Gt^6, \tag{14.6}$$

where

$A = 0.99986775$
$B = 6.78668754 \times 10^{-5}$
$C = -9.09099173 \times 10^{-6}$
$D = 1.02598151 \times 10^{-7}$
$E = -1.35029042 \times 10^{-9}$
$F = 1.32674392 \times 10^{-11}$
$G = -6.461418 \times 10^{-14}$.

Temperature t is on the ITS-90 temperature scale and the temperature of maximum density ρ_o is 3.9834°C. Watanabe gave no value for ρ_o.

14.5.3   Measurements of Takenaka and Masui

Takenaka and Masui of the National Research Laboratory of Japan measured the thermal expansion of purified tap water from which air was removed.[12]

Takenaka and Masui Equation for the Dilatation of Water

The equation developed by Takenaka and Masui is

$$\rho(t)/\rho_{max} = 1 - \left[(t-A)^2(t+B)(t+C)\right]/\left[D(t+E)(t+F)\right], \tag{14.7}$$

where $\rho(t)$ is the density of water (in kg/m³) at temperature t on the ITS-90 temperature scale in the range 0 to 85°C; and ρ_{max} is the maximum density of water, for which Takenaka and Masui used 999.9734 kg/m³.

The constants in Eq. (14.7) are

$A = 3.98152$
$B = 396.18534$
$C = 32.28853$
$D = 609628.6$
$E = 83.12333$
$F = 30.24455$.

An equation to be used to calculate water density $\rho(t)$ is obtained by multiplying the right-hand side of Eq. (14.7) by the maximum density, 999.9734 kg/m³.

14.5.4 Comparison of the Results for the Three Recent Formulations

Patterson and Morris compared the results of their calculations of dilatation, which they refer to as *d*, with those of Watanabe and Takenaka and Masui. The average fractional difference of Watanabe values from *d* is -0.509×10^{-6}, the SD is 0.41×10^{-6}. The average fractional difference of Takenaka and Masui values from *d* is -0.077×10^{-6}, the SD is 0.39×10^{-6}.

The Watanabe values and the Takenaka and Masui values are seen to be very close to the Patterson and Morris values. Several considerations lead to the choice of Eq. (14.5), derived from the work of Patterson and Morris, as the preferred expression of the relationship between the density of air-free SMOW and temperature, suitable for calculation of water density from measurements of temperature.

The Kell values, based on the much earlier work of Chappuis[8] and Thiesen et al.,[7] differ from Patterson and Morris derived values systematically and increase in difference monotonically with increasing temperature above 15°C. We consider the Patterson and Morris work to supersede, or succeed, the work of Kell.

14.6 Change in Density of Water with Air Saturation

Bignell[13] measured the change in the density of water with air saturation. He fitted the experimental points to develop the equation:

$$\Delta\rho\big(t\big) = -4.873 \times 10^{-3} + 1.708 \times 10^{-4} t - 3.108 \times 10^{-6} t^2, \tag{14.8}$$

where $\Delta\rho(t)$ is the change in water density in kg/m³, in the temperature range 0 to 20°C. The estimated total uncertainty of values calculated using Eq. (14.8) was 2×10^{-4} kg/m^{-3} at the 99% confidence level.

Bignell concluded in an earlier paper[14] that "there is probably not much need to extend the work to higher temperatures because the effect diminishes and the accuracy of density metrology at these temperatures would not warrant a more accurately known correction."

14.7 Density of Air-Saturated Water on ITS-90

The Bignell correction, Eq. (14.8), can be added to Eq. (14.5) to produce an equation to be used to calculate the density of air-saturated water in the temperature range 1 to 40°C on ITS-90.

The uncertainty in the density of air-saturated water for an uncertainty in temperature of 1°C is approximately 210 ppm or 0.21 kg/m³ at 20°C.

Table 14.2 is a tabulation of the values of the density of air-saturated water (SMOW).

TABLE 14.2   Densities of Air-Saturated SMOW

Temperature, °C	Air-Saturated SMOW Density, kg/m³
1	999.89654
3	999.96155
4	999.96934
5	999.96127
7	999.89756
10	999.69818
15	999.09867
20	998.20299
25	997.04338
30	995.64546
35	994.02952
40	992.21188

14.8   Compressibility-Corrected Water Density Equation

The density of air-free water, ρ_{wafc}, at an ambient pressure of P kPa is

$$\rho_{\text{wafc}} = \rho_{\text{waf}}\left[1 + \kappa_T\left(P - 101.325\right)\right], \tag{14.9}$$

where ρ_{waf} is calculated using Eq. (14.5) and κ_T is calculated using Eq. (14.2).
    The density of air-saturated water, ρ_{wasc}, at an ambient pressure of P kPa is

$$\rho_{\text{wasc}} = \rho_{\text{was}}\left[1 + \kappa_T\left(P - 101.325\right)\right], \tag{14.10}$$

where ρ_{was} is calculated by adding the Bignell correction, Eq. (14.8), to Eq. (14.5).

14.9   Effect of Isotopic Concentrations

If one wishes to calculate the absolute density of water accurately, the isotopic composition of the water must be taken into account. The isotopes of particular interest are ^{18}O, ^{16}O, D (deuterium), and H (hydrogen).
    The concentrations of the isotopes, [^{18}O] for example, are expressed as mole fraction or mol %. A mole is a quantity of a substance corresponding to the molecular (or atomic) weight of the substance. The mole fraction of a substance is the ratio of the number of moles of the substance to the number of moles of a mixture in which the substance is one of the constituents.
    The isotopic ratios [^{18}O]/[^{16}O], (R_{18}), and [D]/[H], (R_d), are measured for water samples, primarily by mass spectroscopy.
    For Vienna Standard Mean Ocean Water (V-SMOW), one defines:[9]

$$\delta_{18} = \left[R_{18}\Big/\left(2005.2 \times 10^{-6}\right)\right] - 1, \tag{14.11}$$

and

$$\delta_d = \left[R_d\Big/\left(155.76 \times 10^{-6}\right)\right] - 1, \tag{14.12}$$

where R_{18} and R_d are the ratios for the water sample.
    The difference between the density of a water sample and the density of V-SMOW can be calculated from[9]

$$\rho_{\text{sample}} - \rho_{\text{SMOW}} = \left(0.233\delta_{18} + 0.0166\delta_d\right) \text{kg}\Big/\text{m}^3. \tag{14.13}$$

    The isotopic ratios for laboratory water samples are affected by the history of the samples, the number of distillations, for example. The isotopic ratios for natural samples vary with location, climate, and other parameters.
    The range of published values of [^{18}O] and [^{16}O] is from 0.2084 to 0.1879 mol % (2084 to 1879 × 10^{-6}) and 99.7771 to 99.7539 mol %, respectively.
    The approximate range of values of [D] is from 0.0139 to 0.0152 mol % (139 to 152 × 10^{-6}).
    The isotopic ratios for nine of the Patterson and Morris[9] water samples ranged from R_{18} = 1982.4 to 1999.9 × 10^{-6} and R_d = 144.8 to 153.0 × 10^{-6}.
    The correction required to convert the measured density of a water sample to the corresponding value for V-SMOW is calculated by Eq. (14.13) to range from +0.00091 to +0.00382 kg/m^3 for the

Patterson and Morris samples. Compared to a water density of approximately 1000 kg/m³, the corrections correspond to approximately 0.91 to 3.8 ppm.

14.10 Estimation of Uncertainty in Water Density Due to Variation in Isotopic Concentrations

It is of interest to estimate the uncertainty in water density if isotopic concentrations were not measured. The work of Patterson and Morse[9] provided the data above, which can be used for such an estimate.

The water used by Patterson and Morse was generally doubly distilled natural water. If one assumed that the water of interest to metrologists were to have the isotopic concentrations of SMOW, the corrections of Patterson and Morse indicate for their set of nine samples the uncertainties at $k = 3$ of the corresponding values of water density, that is, uncertainties if isotopic concentrations were not measured. The ranges of these uncertainties are from 0.00091 to 0.00382 kg/m³, or 0.91 to 3.82 ppm.

14.11 Summary

Eq. (14.5) can be used to calculate the density of *air-free* water in the temperature range of 1 to 40°C in ITS-90 at an atmospheric pressure of 1 atm. Air-free water is understood to be water that has just been freshly distilled.

The density of *air-saturated* water in the same temperature range at 1 atm can be calculated by adding the right side of Eq. (14.8) to the right side of Eq. (14.5). Water can become air-saturated after sitting exposed to air for a relatively short period of time.

Eq. (14.9) can be used to calculate the density of air-free water, and Eq. (14.10) can be used to calculate the density of air-saturated water, in the temperature range 1 to 40°C at an atmospheric pressure of P kPa. The correction for compressibility is important when the sample is immersed in water for hydrostatic weighing.

To convert from SMOW density to the density of a water sample, Eq. (14.13) can be used.

References

1. Wagenbreth, H. and Blanke, W., Der Dichte des Wassers im Internationalen Einheitensystem und im der Internationalen Praktischen Temperaturkala von 1968, *PTB Mitt.*, 81, 412, 1971.
2. Kell, G. S., Effects of isotopic composition, temperature, pressure, and dissolved gases on the density of liquid water, *J. Phys. Chem. Ref. Data*, 6, 1109, 1977.
3. Craig, H., Standard for reporting concentrations of deuterium and oxygen-18 in natural waters, *Science*, 133, 1833, 1961.
4. Mohler, F. L., Isotopic abundance ratios reported for reference samples stocked by the National Bureau of Standards, U.S. NBS Technical Note 51, May 1960.
5. Kell, G. S., Density, thermal expansivity, and compressibility of liquid water from 0° to 150°C; correlations and tables for atmospheric pressure and saturation reviewed and expressed on 1968 Temperature Scale, *J. Chem. Eng. Data*, 20, 97, 1975.
6. Jones, F. E., *Techniques and Topics in Flow Measurement*, CRC Press, Boca Raton, FL, 1996, 28.
7. Thiesen, M., Scheel, K., and Disselhorst, H., Untersuchungen uber die themische Ausdehnung von festen und tropfbar flussigen Korpern, *Phys. Tech. Reichs. Wiss. Abh.*, 3, 1, 1900.
8. Chappuis, P., Dilatation de l'eau, *Trav. Mem. Bur. Int. Poids Mes.*, 13, D3-D40, 1907.
9. Patterson, J. B. and Morris, E. C., Measurement of absolute water density, 1°C to 40°C, *Metrologia*, 31, 277, 1994.
10. Marsh, K.N., Ed., *Recommended Reference Materials for the Realisation of Physicochemical Properties*, Blackwell Scientific, Oxford, 1987.

11. Watanabe, H., Thermal dilation of water between 0°C and 44°C, *Metrologia*, 28, 33, 1991.

12. Takenaka, M. and Masui, R., Measurement of the thermal expansion of pure water in the temperature range 0°C–85°C, *Metrologia*, 27, 165, 1990.

13. Bignell, N., The change in water density due to aeration in the range 0–8°C, *Metrologia*, 23, 207, 1986/87.

14. Bignell, N., The effect of density of dissolved air on the density of water, *Metrologia*, 19, 57, 1983.

15

Conventional Value of the Result of Weighing in Air

15.1 Introduction

Historically and at present, gravimetric measurements performed for governance incorporate an arbitrary normalization methodology. In the past, this normalization was given the name "apparent mass vs. brass" and currently the "conventional value of the result of weighing in air," also commonly referred to as "conventional mass" and "apparent mass." Discussed here is the concept of apparent mass or the conventional value of the result of weighing in air, the intent, use, benefits, and limitations.[1]

Weights and balances are known to have been in use for several thousand years. Their use prior to the scientific revolution of the 18th century was primarily for marketplace equity, and the variation caused by air buoyancy was of no concern. Since long ago, weights manufactured for commerce were made of brass and thus apparent mass vs. brass weighing standard has been in existence for some time. The British formalized such a scale[2] in 1856 and the United States in 1918.[2]

With the notion of mass as an inertial property of matter put forward by Newton,[3] the interest of Pascal[4] in measuring the variation of air pressure and Archimedes' principle,[5] the stage was set for the sanction of the metric system and the kilogram as the unit of mass in 1889. It follows that the present-day definition of the conventional value of the result of weighing in air would be based on the International Platinum Kilogram Standard of Mass (IPK).

The avoidance of making detailed buoyancy corrections to commercial weighings (legal metrology) is the primary impetus of the apparent mass system. In the past 100 years, the knowledge and ability to measure the parameters required to compute the density of air for the buoyancy correction have been available but not the computing power. Only since the advent of the inexpensive electronic calculator in the 1960s has it become commonplace, for those who desire, to work in mass.

The apparent mass system worked best if it were put on a formal basis. A definition for the basis in current use, the conventional value of the result of weighing in air, can be found in a document published by the International Organization of Legal Metrology (OIML), R33[6] and is given below. The discussion that follows should be helpful in dealing with the present-day promulgation of two systems of measurement, mass and apparent mass, that leads to the need for a third name, "true mass." The last name is given to mass measurements because if there is an apparent mass then there must be a true mass.

15.2 Conventional Value of Weighing in Air

In years past, weighing against brass weights had different meanings throughout the world. The weighings were performed for marketplace equity and the inertial property of matter was not known or quantified in terms of the IPK. It was recognized some time ago that a unified basis for the apparent mass vs. brass weighings was needed. Circular 3,[7] published by the U.S. National Bureau of Standards in 1918, provided a formal definition for the United States.

With the advent of stainless steel after World War II, the world community began to utilize stainless steel as a material from which to manufacture weights. By the mid-1960s, stainless steels had improved considerably and their use for weight fabrication began to displace the use of brass. Stainless steel is more attractive for use as mass standards because of its superior stability and durability as opposed to brass. Within a few years, brass was replaced by stainless steel of nominal density 8.0 g/cm^3 or 8000 kg/m^3.

Regardless of the basis for the apparent mass systems, since about 1903 they have been connected to the IPK. It follows that for a period of a few years stainless steel weights were calibrated and assigned values on the apparent mass vs. brass scale.

The difference in density between brass and stainless steel negated the convenience of the apparent mass system, and eventually the conventional value of the result of weighing in air came into general use. For some years this system was referred to as the "8.0 apparent mass scale"; today it is formally defined and called "conventional value of the result of weighing in air."

The definition given by the OIML in its document, International Recommendation No. 33,[6] is the following:

The conventional value of the result of weighing a body in air is equal to the mass of a standard, of conventionally chosen density, at a conventionally chosen temperature, which balances this body at this reference temperature in air of conventionally chosen density.

The conventionally chosen values of the physical constants in the defintion above are

Reference temperature: 20°C = t_0
Density of the standard of mass at 20°C: 8000 kg/m^3 = ρ_{ref}
Density of the air: 1.2 kg/m^3 = ρ_0

For clarity in this discussion, notation different from that used in International Recommendation No. 33 is used.

The relationship between m, the mass of the body, and m', the mass of the standard, is

$$m\left[1-\left(1.2/\rho_m\right)\right] = m'\left[1-\left(1.2/8000\right)\right], \tag{15.1}$$

where ρ_m is the density of the body at 20°C and the quantities in the square brackets are buoyancy correction factors.

By definition, the conventional value M of the result of weighing the body in air is

$$M = m',$$

or

$$M = m\left[1-\left(1.2/\rho_m\right)\right] \Big/ \left[1-\left(1.2/8000\right)\right]. \tag{15.2}$$

The OIML recommendation approximates M by

$$M = m\left[1 - 1.2\left(1/\rho_m - 1/8000\right)\right]. \tag{15.3}$$

15.3 Examples of Computation

For a body of mass m, and density ρ_m of 21,500 kg/m³, using Eq. (15.2),

$$M = m\left[1-\left(1.2/21500\right)\right]\Big/\left[1-\left(1.2/8000\right)\right]$$

$$= m\left(0.9999442/0.9998500\right),$$

$$M = 1.0000942\,m.$$

Using Eq. (15.3),

$$M = 1.0000942\,m.$$

Thus, the conventional value for the weighing of a body of density 21,500 kg/m³, the approximate density of platinum, is greater than m by 94.2 parts per million (ppm) or 94.2 mg/kg.

For a body of mass m, and density 2700 kg/m³, using Eq. (15.2),

$$M = m\left[1-\left(1.2/2700\right)\right]\Big/\left[1-\left(1.2/8000\right)\right],$$

$$M = 0.9997055\,m.$$

Using Eq. (15.3),

$$M = 0.9997056\,m.$$

Thus, the conventional value for the weighing of a body of density 2700 kg/m³, the approximate density of aluminum, is less than m by 294.5 ppm or 294.5 mg/kg using Eq. (15.2); using the Eq. (15.3) approximation, these quantities are 294.4 ppm and 294.4 mg/kg.

For a substance of mass m, and density 1000 kg/m³, using Eq. (15.2),

$$M = m\left[1-\left(1.2/1000\right)\right]\Big/\left[1-\left(1.2/8000\right)\right],$$

$$M = 0.9989498\,m.$$

Thus, the conventional value for the weighing of substance of mass m and density of 1000 kg/m³, the approximate density of water, is less than m by 1050.2 ppm or 1.0502 g/kg using Eq. (15.2); using the Eq. (15.3) approximation, these quantities become 1050.0 ppm or 1.0500 g/kg.

For the three examples above, the approximation equation, Eq. (15.3) yields results within 0.2 ppm of those yielded by the exact equation, Eq. (15.2).

For a variation in air density of less than or equal to 3%, the mass of standard, m', required to balance m of object or substance will vary. For an air density of 1.17 kg/m³ and density of m of 21,500 kg/m³,

$$m' = m\left(1-1.17/21500\right)\Big/\left(1-1.2/8000\right),$$

becomes $m' = 1.0000956\,m$. This represents a variation in m' of 95.6 ppm of m.

M is the conventional value of weighing in air and also the mass of material of conventionally chosen density and need not physically exist.

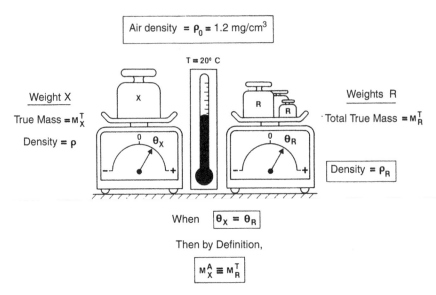

FIGURE 15.1 Illustration of the concept of the conventional value of weighing in air.

Eq. (15.2) is useful in tolerance testing of weights for legal metrology; buoyancy can be ignored. It is there but it is never examined. If the OIML R111 prescription is followed, the air density is within 10% of ρ_o and ρ_m is within the domain for the class of weight undergoing test. One does not then deal with the buoyant terms explicitly.

The conventional value of the result of weighing in air involves calculation of the outcome of an hypothetical air weighing performed at the specified arbitrary conditions and based on the mass of the reference material, M, and the mass of the body, m, of density ρ_m in air of density ρ_o.

For scientific metrology, buoyancy is accounted for in detail. One determines ρ_m and m and uncertainty estimates, and from Eq. (15.2) one calculates M for the legal metrologist. The determination of m is a measurement with an uncertainty estimate, and the tolerance test determines only whether the conventional value, M, is within the prescribed range. It is an approximation.

Figure 15.1, taken from Ref. 8 and slightly modified, clearly illustrates the concept of the conventional value of weighing in air.

Eq. (15.2) can be related to the unit of mass as embodied in the IPK with the following thought experiment. Conceptually, one can place the IPK on one pan of an equal-arm balance in vacuum and a weight constructed of the ideal material (8000 kg/m^3) on the other. One can adjust the mass of the weight of ideal material until it is in perfect balance with the IPK. The IPK can never really be placed in a vacuum for fear of changing its mass and thereby shifting the mass unit. Nevertheless, in principle, one can now allow air at the standard condition to enter the weighing chamber.

Furthermore, the IPK is replaced with a test weight of mass m and density ρ_m. Recall that remaining on the other balance pan is a weight adjusted to be equal to the IPK in vacuum and of density 8000 kg/m^3 at 20°C. The mass m is now adjusted until the equal-arm balance indicates that equal forces are imposed on each balance pan. One can now say that m has a conventional value of 1 kg as a result from weighing in air at the standard conditions; henceforth, for brevity one can simply refer to the weighing result as the conventional value of the result of weighing in air.

The basis for apparent mass or, in this instance, the conventional value, is arbitrary but provides a basis to adjust weights for use in governance made from material of standard density and used at the standard air density and temperature. The above definition (conventional value) does not refer to a conventional mass; it is commonplace in the measurement community to refer to the conventional value of the result of weighing in air as conventional mass, an undefined term.

15.4 Discussion

The conventional value definition specifies a reference material density but not the material itself. The OIML International Recommendation R111[9] for weight classes E_1, E_2, F_1, F_2, M_1, M_2, and M_3 likewise does not specify a material. However, R111 does specify many other attributes for weights, including the domain of material density. Taken together, these attributes are best met using stainless steel for the higher classes of weights. It is more economical to use other materials for the lower classes where the attribute restrictions are more permissive.

The ability of metal fabricators to make alloys the density of which is exactly 8000 kg/m³ is not perfect, nor is the standard air density achieved routinely. At some elevations it is never achieved. In reality, the conditions required by the definition of conventional mass are rarely attained. Therefore, one would expect some error in use.

Eq. (15.1) can be written with our notation to obtain an explicit expression of the conventional value of weighing in air, M:

$$M = m\left[\left(1-\left(\rho_o/\rho_t\right)\right)\right]\Big/\left[\left(1-\left(\rho_o/\rho_{\mathrm{ref}}\right)\right)\right], \tag{15.4}$$

where m is the mass of an object or weight, ρ_o is 1.2 kg/m³, ρ_t is the density of the object or weight at temperature t in °C, and ρ_{ref} is 8000 kg/m³.

Before proceeding, a brief discussion of mass (true mass) and its assignment by weighing is offered. Any object possessing the requisite attributes can be assigned a mass value in terms of the IPK.[10] The weighing equation for doing so follows:

$$S\left[1-\left(\rho_a/\rho_S\right)\right] - m\left[\left(1-\left(\rho_a/\rho_t\right)\right)\right] = \delta, \tag{15.5}$$

where S is the known mass of a standard weight, ρ_S is its density, ρ_a is the density of air during the weighing, and δ is the difference indicated by the balance in mass units.

The solution for the mass of the test weight, m, is

$$m = \left\{S\left[1-\left(\rho_a/\rho_S\right)\right] - \delta\right\}\Big/\left[1-\left(\rho_a/\rho_t\right)\right]. \tag{15.6}$$

Eq. (15.6) yields the mass (true mass), m, that can now be substituted into an equation of the form of Eq. (15.4) to compute the conventional value M for this object:

$$M = m\left[1-\left(1.2/\rho_t\right)\right]\Big/\left[1-1.2/8000\right]. \tag{15.7}$$

Example 15.1, below, demonstrates this computation.

Example 15.1

Suppose that the application of Eq. (15.6) to weighing data yields a mass of 1000.003600 g for m, with a corresponding density of 7810 kg/m³ at 20°C, a class E_2 kilogram weight. Substitution of these values into Eq. (15.4) yields:

$$M = 1000.003600 \times \left[\left(1-1.2/7810\right)\Big/\left(1-1.2/8000\right)\right]$$

$$M = 1000.000000 \text{ g}.$$

The conventional value of a weight can be expressed as the nominal value and its correction in milligrams as 1000 g + 0.000 mg.

There are occasions when one has knowledge of the conventional value, M, of a laboratory weight and desires the mass (true mass) value. The mass value is readily obtained by solving Eq. (15.7) for m, which yields:

$$m = M\left[1 - \left(1.2/8000\right)\right] \Big/ \left[1 - \left(1.2/\rho_t\right)\right]. \tag{15.8}$$

Example 15.2

The numerical values from Example 15.1 can be used to compute the mass of m from the conventional value, M:

$$m = 1000.000000 \times \left[\left(1 - 1.2/8000\right)\Big/\left(1 - 1.2/7810\right)\right]$$

$$m = 1000.003600 \text{ g}.$$

Having obtained the conventional value, M, for a weight, one is likely to use it at a nonstandard air density. The resulting error can be determined by constructing another conventional value, M_Q, for a nonstandard air density, ρ_a. From these definitions one can infer the forces imposed on the balance pan for the following four loads:

$$M\left[1 - \left(\rho_o/8000\right)\right]g = I_1 g,$$

$$m\left[1 - \left(\rho_o/\rho_t\right)\right]g = I_2 g,$$

$$m\left[1 - \left(\rho_a/\rho_t\right)\right]g = I_3 g,$$

$$M_Q\left[1 - \left(\rho_a/8000\right)\right]g = I_4 g,$$

where g is the local acceleration due to gravity and the I_i are balance indications in mass units.
The solution of these four force equations for M_Q is

$$M_Q = M\left[1 - \left(\rho_o/8000\right)\right]\left[\left(1 - \left(\rho_a/\rho_t\right)\right)\right]\Big/\left[1 - \left(\rho_o/\rho_t\right)\right]\left[1 - \left(\rho_a/8000\right)\right]. \tag{15.9}$$

Example 15.3

Eq. (15.9) is evaluated using values given in previous examples; additionally, one assumes an air density of 0.98 kg/m³ at 20°C for the nonstandard condition, an average air density for Denver, CO.

$$M_Q = 1000.000000\left[1 - \left(1.2/8000\right)\right]\left[1 - \left(0.98/7810\right)\right]\Big/\left[\left(1 - \left(1.2/7810\right)\right)\right]\left[\left(1 - \left(0.98/8000\right)\right)\right]$$

$$M_Q = 1000.000600 \text{ g}$$

$$M - M_Q = -0.0006 \text{ g}$$

Example 15.3 demonstrates the error of a nominally 1-kg test weight assigned a conventional value at a sea level air density of 1.2 kg/m³ and then transported to an elevation (approximately 1600 m) where the air density is 0.98 kg/m³. OIML Recommendation No. 33 provides a warning to users of the conventional value when the air density differs by 10% or more from 1.2 kg/m³. That is, 1.2 kg/m³ − 0.12 kg/m³ or 1.08 kg/m³. The average air density at Denver, CO (0.98 kg/m³) exceeds this 10% limit, which corresponds to an elevation of approximately 900 m.

OIML R111 specifies a hierarchy of weight classification designed to allow a weight of a higher class to act as a conventional value standard for a lower class weight. This is an efficient method for tolerance testing weights as the buoyancy correction detail is eliminated. We note that the above error of 0.6 mg is greater than $1/3$ the maximum permissible error (1.5 mg) for a class E_2 1-kg weight, the limit imposed by R111.

In Example 15.4 that follows, another thought experiment, one performs a calibration of a class F_1 1-kg weight using the class E_2 1-kg weight of Example 15.2 as the conventional value standard.

Example 15.4

The conventional value of the result of weighing in air, M, for the class E_2 weight is 1000.000000 g and the material density is 7810 kg/m^3. The class F_1 weight has an in-tolerance material density of 7390 kg/m^3. At sea level one asserts the weights are equal; that is, at an air density of 1.2 kg/m^3 the balance indicates zero difference. Therefore, the conventional result of weighing in air for the F_1 weight is 1000.000000 g.

The above calibration is repeated at Denver, CO at an air density of 0.98 kg/m^3. The above conventional value for the class F_1 kilogram can be adjusted for the Denver air density using Eq. (15.9), as was done for the class E_2 kilogram in Example 15.3. For the Class F_1 kilogram:

$$100.000000\left[1-\left(1.2/8000\right)\right]\left[1-\left(0.98/7390\right)\right]/\left[1-\left(1.2/7390\right)\right]\left[1-\left(0.98/8000\right)\right]=1000.002301\text{ g}.$$

At Denver, a balance would indicate a difference between these weights of

$$1000.000600-1000.002301=-0.001701\text{ g}.$$

Abiding by the specifications of R111, the measurement uncertainty for this weight should not exceed $1/3$ the maximum permissible error (5 mg ÷ 3). In Example 15.4, this has been exceeded with a type B uncertainty, not including the type A uncertainty that would be present in a real measurement.

15.5 Conclusions

The obvious conclusion that can be drawn from this look at the conventional value of a result of weighing in air is that for elevations between sea level and approximately 900 m one can perform tolerance testing of weights that meet the density specification of R111 (near 8000 kg/m^3) without detailed buoyancy correction. In addition, when a weight has a density of 8000 kg/m^3 at 20°C exactly, its mass (true mass) and conventional value are identical. The object density, ρ_t, is not restricted by the conventional value definition but by the imposition of R111.

The conventional result of weighing in air does provide a norm for adjusting scales and balances. Balances are adjusted to the conventional value of a test weight and when placed side by side will all give the same indication to a given load. This has not always been the case. For a period of time in the 1960s and 1970s there were balances in use adjusted to the older apparent mass versus brass scale (8.4) and those adjusted to the present conventional value (8.0). When these balances are sitting side by side at sea level and loaded with a 1-kg test weight, they will yield indications that are different by 7.1 mg.

Most of the world population resides at elevations between sea level and 1600 m, corresponding to air densities of 1.2 to 0.98 kg/m^3. If 1.09 kg/m^3 had been chosen as the reference air density in place of 1.2 kg/m^3, most of the measurements for governance in the United States would not violate the 10% air density rule. This is not the case with the present definition of the conventional value.

One may be tempted to apply corrections for elevation that exceed the 10% limit as we have shown that Eq. (15.9) does. However, the convenience of the conventional result of weighing in air would be lost, and one would need to obtain explicit densities of the weights. Obtaining the densities of weights is expensive and most users rely on weight manufacturers to guarantee only compliance

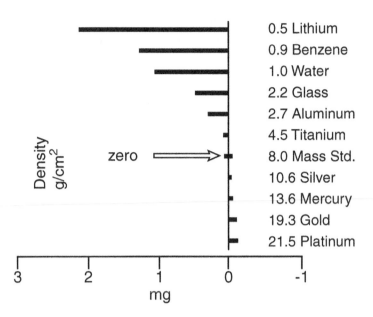

FIGURE 15.2 Bar graph illustrating buoyancy error.

with the ranges given in R111. Therefore, one cannot provide all the information required when using Eq. (15.9).

In the examples provided here using Eq. (15.9), the densities of the weights used were the lower extreme values for the appropriate ranges given in R111. One can calculate the errors associated with these extreme values but lacks the density knowledge to calculate an exact error of a real measurement. When specific knowledge of air density and the densities of the weights is available, one should work in mass and calculate the conventional value.

The difficulty that commonly arises when one has a choice between using mass (true mass) and the conventional value is knowing when not to use the conventional value for industrial and scientific measurements. The bar graph, Figure 15.2, is descriptive of this problem. The graph illustrates the buoyancy error that would exist, if buoyancy is ignored, when using a 1-g mass standard of density 8000 kg/m³ to determine 1 g of the other materials shown. Of course, when both objects have a density of 8000 kg/m³ the effect is indicated as zero, as it is for the conventional result of weighing in air.

For material denser than 8000 kg/m³, the error sign is opposite that of the less dense material. Although ignoring the buoyancy correction for gold and platinum results in a small mass error, its monetary value may be significant. On the other hand, the error in the mass of water is large, approximately 0.1%, and would result in comparable volume error when using the density of water to determine the capacity of a flask, a very important measurement. Since the the density of water is given in mass per unit volume (kg/m³), one would not use the conventional value, i.e., apparent mass, when making density or volume determinations.

The uncertainty of "the conventional result of weighing in air" assigned to a weight cannot be less than its mass uncertainty.[11] However, many users of the Conventional Value believe that the uncertainty associated with buoyancy is zero when it is not. Sometimes, mass measurements are performed using the methodology meant for tolerence testing. The users have no specific knowlege of the density of the weights and their uncertainty estimates do not include a component for bouyancy. As a consequence, their uncertainty estimates are too small and incorrect.

References

1. Schoonover, R. M., Conventional mass; the concept, intent, benefits, and limitations, in *Proceedings of Measurement Science Conference*, Anaheim, CA, January, 2001.
2. Pontius, P. E., Mass and Mass Values, NBS Monograph 133, U.S. Department of Commerce, Washington, D.C., June 1974.
3. Newton, I., Mathematical principles of natural philosophy, in *Britannica Great Books*, Vol. 34, William Benton, Chicago, 1952, 1.
4. Pascal, B., Scientific treatises, in *Brittanica Great Books*, Vol. 33, William Benton, Chicago, 1952, 355.
5. Archimedes, The works of Archimedes including the method, in *Brittanica Great Books*, Vol. 11, William Benton, Chicago, 1952, 403.
6. OIML International Recommendation No. 33, Conventional Value of Weighing in Air, International Bureau of Legal Metrology, Paris, 1979.
7. Circular of the Bureau of Standards No. 3, Design and Test of Standards of Mass, Washington, D.C., December 23, 1918.
8. Jaeger, K. B. and Davis, R. S., A Primer for Mass Metrology, NBS Special Publication 700-1, 1984, U.S. Department of Commerce, Washington, D.C.
9. OIML International Recommendation R111, Weights of Classes E_1, E_2, F_1, F_2, M_1, M_2, M_3, International Organization of Legal Metrology, Paris, 1994.
10. Schoonover, R. M. and Jones, F. E., Examination of parameters that can cause errors in mass determinations, *Cal. Lab.*, July/August, 26, 1998.
11. Jones, F. E. and Schoonover, R. M., A comparison of error propagations for mass and conventional mass, in *Proceedings of the Measurement Science Conference*, Anaheim, CA, January 20–21, 2000.

16

A Comparison of Error Propagations for Mass and Conventional Mass*

16.1   Conventional Value of the Result of Weighing in Air

The conventional value of the result of weighing, M, can be defined[2] as:

$$M = m\left[1 - \left(\rho_o/\rho_m\left(20°C\right)\right)\right]\Big/\left[1 - \left(\rho_o/\rho_{ref}\right)\right], \tag{16.1}$$

where m is the mass of an object, ρ_o = 1.2 kg/m^3 is the reference value of the density of air, ρ_{ref} = 8000 kg/m^3 is the reference value of the density of the object (for which the conventional value is to be determined), and $\rho_m(20°C)$ is the density of the object at the reference temperature 20°C.

16.2   Uncertainties in Mass Determinations

The following simple weighing equation applies to the determination of the mass, m, of a weight by comparison with a standard weight of mass S:

$$m = \left[S\left(1 - \rho_a/\rho_S\right) - \delta\right]\Big/\left(1 - \rho_a/\rho_m\right), \tag{16.2}$$

where ρ_a is the density of air in which the weighing is made, ρ_S is the density of the standard weight, δ is the mass difference calculated from balance indications, and ρ_m is the density of the weight of mass m.

16.3   Uncertainties in the Determination of m Due to Uncertainties in the Parameters in Eq. (16.2)

Eq. (16.2) expresses the relationship between m and various parameters. We undertake now to propagate the uncertainties in the various parameters using the method described by Ku[4] and also in the ISO Guide.[5]

---

*Chapter is based on Ref. 1.

According to Ku,

$$\left(\mathrm{SD}\right)^2 = \Sigma_i\left[\left(\partial m/\partial Y_i\right)^2\left(\mathrm{SD}_i\right)^2\right], \tag{16.3}$$

where SD is the estimate of standard deviation for m; $\partial m/\partial Y_1$ is the partial derivative of m with respect to the individual parameters Y_i, in this case in Eq. (16.2); and SD_i refers to the estimate of standard deviation for the individual parameters.

16.3.1 Balance Standard Deviation

The standard deviation, SD, of the balance is essentially equal to the standard deviation of the mass difference calculated from balance indications, δ. Therefore, in treatment later in this chapter of uncertainty trade-offs for δ and the density of the test weight, the trade-offs will be between the balance standard deviation and the standard deviation of the density of the test weight.

16.3.2 Application to R111

International Recommendation OIML R111, Weights of Classes E_1, E_2, F_1, F_2, M_1, M_2, M_3,[3] developed by International Organization of Legal Metrology (OIML) subcommittee TC 9/SC 3 "Weights," contains the principal characteristics and metrological requirements for weights that are used:

1. For the verification of weighing instruments
2. For the tolerance verification of weights of a lower class of accuracy
3. With weighing instruments

The recommendation applies to weights (of nominal mass from 1 mg to 50 kg) in classes of descending order of accuracy: E_1, E_2, F_1, F_2, M_1, M_2, and M_3.

E_1 weights are intended to ensure traceability between national mass standards (with values derived from the International Prototype of the kilogram) and weights of class E_2 and lower.

E_2 weights are intended to be used for the initial verification of weights of class F_1. E_2 weights can be used as E_1 weights if they comply with the requirements for surface roughness and magnetic susceptibility of class E_1 weights and if their calibration certificates give the appropriate data.

F_1 weights are intended to be used for the initial tolerance verification of weights of class F_2.

F_2 weights are intended to be used for the initial tolerance verification of weights of class M_1 and possibly M_2.

M_1 weights are intended to be used for the initial tolerance verification of weights of class M_2.

M_2 weights are intended to be used for the initial tolerance verification of weights of class M_3.

All tolerances are expressed in conventional value of weighing in air.

F_1 and E_2 weights are intended to be used with weighing instruments of accuracy class I. F_2 weights are intended to be used for important commercial transactions (e.g., gold and precious stones) on weighing instruments of accuracy class II. M_1 weights are intended to be used with weighing instruments of accuracy class II. M_2 weights are intended to be used in normal commercial transactions and on weighing instruments of accuracy class III. M_3 weights are intended to used on weighing instruments of accuracy classes III and IIII.

16.4 Comparisons of Weights

Returning to Eq. (16.2) and replacing m with E_1 and ρ_m with ρ_{E1},

$$E1 = \left[S\left(1-\rho_a/\rho_s\right)-\delta\right]/\left(1-\rho_a/\rho_{E1}\right). \tag{16.4}$$

The various partial derivatives $(\partial m/Y_i)^6$ are

$$\partial E1/\partial S = \left[1-\left(\rho_a/\rho_S\right)\right]\Big/\left[1-\left(\rho_a/\rho_{E1}\right)\right], \qquad (16.5)$$

$$\partial E1/\partial \rho_S = S\left(\rho_a/\rho_S^2\right)\Big/\left[1-\left(\rho_a/\rho_{E1}\right)\right], \qquad (16.6)$$

$$\partial E1/\partial \delta = -1\Big/\left[1-\left(\rho_a/\rho_{E1}\right)\right], \qquad (16.7)$$

$$\partial E1/\partial \rho_{E1} = -\left(\rho_a/\rho_{E1}\right)^2\left\{S\left[1-\left(\rho_a/\rho_S\right)\right]-\delta\right\}\Big/\left[1-\left(\rho_a/\rho_{E1}\right)\right]^2, \qquad (16.8)$$

$$\partial E1/\partial \rho_a = \left[\left(S-\delta\right)/\rho_{E1}-\left(S/\rho_S\right)\right]\Big/\left[1-\left(\rho_a/\rho_{E1}\right)\right]^2. \qquad (16.9)$$

16.4.1 Comparison of a Stainless Steel E_1 Weight with a Stainless Steel Standard of Mass S and Density 7.950 g/cm³

The class of the standard weight is not relevant; it must only have the characteristics of a good mass standard.

For class E_1 weights of nominal mass values greater than or equal to 100 g, the minimum and maximum limits for density are 7.934 and 8.067 g/cm³. The extremes of air density are taken to be 0.00091 and 0.0012 g/cm³.

The following are parameters to be used in calculations for comparisons of an E_1 weight of mass E1 with a standard of mass S:

S = 1000.0001 g
ρ_S = 7.950 g/cm³
δ = −0.00179 g, +0.00020 g
E1 = 1000.0002 g
ρ_{E1} = 8.067 g/cm³, 7.934 g/cm³
ρ_a = 0.00091 g/cm³, 0.0012 g/cm³ (corresponding to elevations of approximately 1400 m and sea level)

For ρ_{E1} of 8.067 g/cm³, ρ_a of 0.00091 g/cm³, and corresponding δ of −0.00179 g, an error propagation for Eq. (16.3) is given in Table 16.1. The root sum square (RSS) of the values in the last column is given as the combined standard uncertainty for $k = 1$.

TABLE 16.1 Error Propagation for Comparison of a Stainless Steel E_1 Weight of Density 8.067 g/cm³ with a Stainless Steel Standard

	$\partial E1/\partial Y_i$	SD	$(\partial E1/\partial Y_i)\cdot(SD)$, g
S	+0.999998	2.3×10^{-5} g	20.1×10^{-6}
δ	−1.000113	2.0×10^{-5} g	-20×10^{-6}
ρ_S	+0.014400 cm³	8.0×10^{-5} g/cm³	1.2×10^{-6}
ρ_{E1}	+0.013985 cm³	5.2×10^{-4} g/cm³	7.3×10^{-6}
ρ_a	−1.824545 cm³	9.0×10^{-7} g/cm³	-1.6×10^{-6}
			RSS = 31×10^{-6}

TABLE 16.2   Error Propagation for Comparison
of a Stainless Steel E_1 Weight of Density 7.934 g/cm³
with a Stainless Steel Standard

	$\partial E1/\partial Y_i$	SD	$(\partial E1/\partial Y_i) \cdot (SD)$, g
S	+1.000000	2.3×10^{-5} g	20.1×10^{-6}
δ	−1.000151	2.0×10^{-5} g	$−20 \times 10^{-6}$
ρ_S	+0.018989 cm³	8.0×10^{-5} g/cm³	1.5×10^{-6}
ρ_{E1}	−0.019066 cm³	5.2×10^{-4} g/cm³	$−9.9 \times 10^{-6}$
ρ_a	−1.824545 cm³	9.0×10^{-7} g/cm³	$−1.6 \times 10^{-6}$
			RSS = 31×10^{-6}

For ρ_{E1} of 7.934 g/cm³, ρ_a of 0.0012 g/cm³, and corresponding δ of +0.00020 g, an error propagation is given in Table 16.2. The RSS of the values in the last column is given as the combined standard uncertainty for $k = 1$.

16.4.2   Error Propagation for Conventional Value of Weighing in Air

The partial derivatives for the *conventional* value of weighing of the classes of weights in R111 are the products of the values of the quantity $[1 - \rho_o/\rho(20°C)]/[1 - \rho_o/\rho_{ref}]$ and the uncertainties for the individual parameters. For class E_1 weights, a value for the ratio ranges from 0.999999 to 1.000001; for the other classes of weights the ratio ranges very near 1.

Because only two digits are retained in the RSS, the error propagation for "true" mass and conventional mass are essentially equal.

16.4.3   Comparison of E_2 Weights with E_1 Weights

The following are parameters to be used in calculations for comparisons of an E_2 weight of mass E2 with an E_1 weight of mass E1:

E1 $= 1000.0002$ g
$\rho_{E1} = 8.067$ g/cm³
δ $= −0.0024$ g, +0.0056 g
E2 $= 999.9994$ g
$\rho_{E2} = 8.210$ g/cm³, 7.810 g/cm³
ρ_a $= 0.00091$ g/cm³, 0.0012 g/cm³

For ρ_{E2} of 8.210 g/cm³, ρ_a of 0.00091 g/cm³, and corresponding δ of −0.0024 g, an error propagation is given in Table 16.3.

For ρ_{E2} of 7.810 g/cm³, ρ_a of 0.0012 g/cm³, and corresponding δ of +0.0056 g, an error propagation is given in Table 16.4.

TABLE 16.3   Error Propagation for Comparison of a
Stainless Steel E_2 Weight of Density 8.210 g/cm³ with a
Stainless Steel E_1 Weight of Density 8.067 g/cm³

	$\partial E2/\partial Y_i$	SD	$(\partial E2/\partial Y_i) \cdot (SD)$, g
E1	+0.999998	83×10^{-6} g	83×10^{-6}
δ	−1.000111	50×10^{-6} g	50×10^{-6}
ρ_{E1}	+0.013985	8.0×10^{-5} g/cm³	1.1×10^{-6}
ρ_{E2}	−0.013502	5.2×10^{-4} g/cm³	$−7.0 \times 10^{-6}$
ρ_a	−2.159340	9.0×10^{-7} g/cm³	$−1.9 \times 10^{-6}$
			RSS = 97×10^{-6}

TABLE 16.4 Error Propagation for Comparison
of a Stainless Steel E_2 Weight of Density 7.810 g/cm^3
with a Stainless Steel E_1 Weight of Density 8.067 g/cm^3

	$\partial E2/\partial Y_i$	SD	$(\partial E2/\partial Y_i)\cdot$(SD), g
E1	+1.000005	83×10^{-6} g	83×10^{-6}
δ	−1.000154	50×10^{-6} g	50×10^{-6}
ρ_{E1}	+0.018443	8.0×10^{-5} g/cm^3	1.5×10^{-6}
ρ_{E2}	−0.019676	5.2×10^{-4} g/cm^3	-10×10^{-6}
ρ_a	+4.079685	9.0×10^{-7} g/cm^3	3.7×10^{-6}
			RSS = 97×10^{-6}

16.5 Maximum Permissible Errors on Verification

International Recommendation OIML R111[2] expresses the maximum errors permissible on initial and subsequent verification for each individual weight, related to conventional mass, by:

> For each weight, the expanded uncertainty U for k = 2 … of the conventional mass shall be equal to one-third of the maximum permissible error given in Table 1, except for class E_1 weights … [for which] U shall be *significantly less than* the maximum permissible error [emphasis added]."

Maximum permissible errors are the allowed maximum deviations from nominal conventional values.

From Table 1 of R111, the maximum permissible errors for class E_1 and class E_2 for a weight of nominal value 1 kg are 0.5 and 1.5 mg, respectively.

Note that the maximum permissible errors in service are left by R111 to the discretion of each state (OIML Member State).

Since the required expanded uncertainty for class E_1 weights is not specific, for our purposes we shall use the above-quoted specified limit for both class E_1 and class E_2 weights.

16.6 Uncertainty Trade-Offs

Using Tables 16.1 and 16.2, we shall now investigate trade-offs of two uncertainties, the SD of the mass difference calculated from balance indications (δ) and the SD of the density of test weights (ρ_{E1}).

Because the SD of the balance is essentially equal to the SD of δ, the trade-offs will actually be between the SD of the balance and the SD of the density of the test weight.

For class E_1 weights, one third of the maximum permissible error of 0.5 mg for a weight of nominal value 1 kg is 0.17 mg and the corresponding RSS uncertainty ($k = 1$) is 83 µg.

For the values of SD in Table 16.1, we can calculate a value of SD for ρ_{E1} that results in an RSS of 83 µg. That value is 5.6×10^{-3} g/cm^3.

Using the RSS value of 83 µg for E_1 weights and the values in Table 16.1 corresponding to S, ρ_S, and ρ_a, we calculate trade-offs for the SD for the balance (SD of δ) and SD for ρ_{E1}.

We now present an example using a SD of the balance of 50 µg, and the values of the partial derivatives in Table 16.1.

The values of the squares of the values in the third column of Table 16.1 for S, ρ_S, and ρ_a are 404, 1.4, and 2.56×10^{-12} g^2, for a total of 408×10^{-12} g^2.

If we use 50×10^{-6} g for the SD of the balance, the product in the last column in Table 16.1 is $50 \times (-1.000113) = -50.00565 \times 10^{-6}$ g, the square of which is 2500.6×10^{-12} g^2. This value summed with 408×10^{-12} g^2 is 2908.6×10^{-12} g^2.

The RSS value of 83 µg is 6889×10^{-12} g^2. The square root of the difference (6889 − 2908.6 = 3980.4) is 63.1×10^{-6} g. Dividing this value by the partial derivative of ρ_{E1}, +0.013985 cm^3, the required

TABLE 16.5   Trade-Off Uncertainty Values for E_1 Weights

Balance SD, µg	Req. SD of ρ_{E1}, g/cm^3	SD of ρ_{E1}, g/cm^3	Req. Bal. SD, µg
20	5.6×10^{-3}	1.0×10^{-4}	80.5
30	5.3×10^{-3}	3.0×10^{-4}	80.4
40	5.0×10^{-3}	5.0×10^{-4}	80.2
50	4.5×10^{-3}	1.0×10^{-3}	79
60	3.8×10^{-3}	3.0×10^{-3}	69
70	2.8×10^{-3}	5.0×10^{-3}	40
80	6.4×10^{-4}	5.58×10^{-3}	20

TABLE 16.6   Required Values of Balance SD for Various Values of SD of ρ_{E2}

SD of ρ_{E2}, g/cm^3	Req. Bal. SD, µg
1.0×10^{-4}	229.3
3.0×10^{-4}	229.2
5.0×10^{-4}	229.1
1.0×10^{-3}	228
3.0×10^{-3}	222
5.0×10^{-3}	207
1.0×10^{-2}	118

SD of ρ_{E1} is 4.5×10^{-3} g/cm^3. That is, for an uncertainty in the mass of the test weight equal to $\frac{1}{6}$ of the maximum permissible error of 0.5 mg and an SD of 50 µg for the balance, the SD of the density of the test weight (ρ_{E1}) must not exceed 4.5×10^{-3} g/cm^3.

For various values of SD of the balance, the required values of the SD of ρ_{E1} are listed in Table 16.5.

Rearranging the procedure and using the same parameters, required values of SD of the balance were calculated for various values of SD of ρ_{E1} and tabulated also in Table 16.5.

Using the values of the partial derivatives in Table 16.4 and the SD values: 83 µg for E_1; 8.0×10^{-5} g/cm^3 for ρ_{E1}; and 9.0×10^{-7} for ρ_a, the required SD for the balance, for various values of SD for ρ_{E2} have been calculated and tabulated in Table 16.6.

The required values of the SD of ρ_{E2} for values of balance SD in the range of the tabulated values can be estimated from Table 16.6.

16.7   Summary

Error propagations for mass for a standard weight compared with a weight of OIML R111 class E_1, and for a weight of OIML R111 class E_1 compared with a weight of OIML R111 class E_2 were developed. The combined standard uncertainties for mass and conventional mass were essentially equal.

Error propagations were used to study trade-offs of balance SD and the SD of the density of the test weight. These trade-offs can be effective in cost-saving in the procurement of balances and perhaps weights.

Tables taken from R111 are included in the Appendices to this book.

References

1. Jones, F. E., A comparison of error propagations for mass and conventional mass, presented at Measurement Science Conference 2000, Anaheim, CA, January, 2000.
2. OIML International Recommendation No. 33, Conventional Value of the Result of Weighing in Air, International Bureau of Legal Metrology, Paris, 1979.
3. OIML International Recommendation R111, Weights of Classes E_1, E_2, F_1, F_2, M_1, M_2, M_3, International Organization of Legal Metrology, Paris, 1994.
4. Ku, H. S., Statistical concepts in metrology, in *Handbook of Industrial Metrology*, American Society of Tool and Manufacturing Engineers, Prentice-Hall, Englewood Cliffs, NJ, 1967, chap. 3.
5. *Guide to the Expression of Uncertainty in Measurement*, International Organization for Standards, Geneva, Switzerland, 1992.
6. Schoonover, R. M. and Jones, F. E., Examination of parameters that can cause errors in mass determinations, *Cal. Lab.*, July/August, 26, 1998.

17

Examination of Parameters That Can Cause Error in Mass Determinations[1]

17.1 Introduction

Parameters that cause error in mass determinations are here examined in detail. Subjects covered are mass artifacts, mass standards, mass comparison, the fundamental mass relationship, weighing designs, uncertainties in the determination of the mass of an object, buoyancy, thermal equilibrium, atmospheric effects, cleaning of mass artifacts, magnetic effects, and instability of the International Prototype Kilogram (IPK).

17.2 Mass Comparison

The gravitational force exerted on a balance mechanism by a standard kilogram is compared to the gravitational force exerted by an artifact of mass and density nominally equal to those of the standard kilogram to determine the mass of the artifact. That is, the balance is a mass comparator.

If these forces are not equal, a second mass artifact (or a combination of artifacts), the mass of which (previously determined by an iterative process) is a small fraction of that of the standard kilogram, is required to calibrate the balance response in terms of the mass unit.

17.3 The Fundamental Mass Comparison Relationship

The fundamental relationship for the mass comparison of a standard of mass S and an object of mass X is expressed by the following equation:

$$\left\{ S\left[1-\left(\rho_a/\rho_S\right)\right] - X\left[1-\left(\rho_a/\rho_X\right)\right] \right\} g = \delta g, \tag{17.1}$$

ρ_a is the density of air, ρ_S is the density of the standard, ρ_x is the density of the object, g is the local acceleration due to gravity, and δ is the mass difference indicated by the balance. If the centers of gravity of the two weights are not in the same horizontal plane, there is a small correction due to gravitational gradient (Chapter 20).[2]

Solving Eq. (17.1) for X,

$$X = \left\{ S\left[\left(1-\left(\rho_a/\rho_S\right)\right)\right] - \delta \right\} \Big/ \left[1-\left(\rho_a/\rho_X\right)\right] \tag{17.2}$$

Conceptually, one could use the fundamental relationship to determine X from one δ observation. However, there are a number of weighing designs[3] (see Chapter 8) that allow a more precise determination of X and have the additional advantage of checking the consistency of the prototype kilograms.

NIST has the unique advantage of having *two* prototype kilograms. That being the case, the following weighing design can be employed:

K_{20}	K_4	X_1	X_2	Mass Diff.
1	−1	0	0	δ_1
1	0	−1	0	δ_2
1	0	0	−1	δ_3
0	1	−1	0	δ_4
0	1	0	−1	δ_5
0	0	1	−1	δ_6

where K_{20} and K_4 represent the prototype kilograms, X_1 and X_2 represent kilograms of unknown mass, and the δ are mass differences inferred from the balance observations.

The plus and minus 1 terms are used to indicate differences between masses; for example, in the first line the 1 under K_{20} and the −1 under K_4 mean that the difference in mass, $K_{20} - K_4$, is indicated by δ_1.

This weighing design is referred to as a "4–1's series," and in general is referred to as a "combinational" weighing design. δ_1 is the consistency check; i.e., the data are used to perform a statistical "*t*-test."

The solution equation for X_1 is the following:

$$X_1' = \left(-3\delta_2' - \delta_3' - 3\delta_4' - \delta_5' + 2\delta_6' + 4K'\right)\big/8$$

In the above equation, the primes indicate that buoyancy corrections have been applied to the various quantities. The restraint, K, is the sum of the two Pt-Ir kilograms. The estimate of the standard deviation resulting from the least-squares process is used in calculating the random (Type A) component of the uncertainty assigned to X_1 and X_2.

Subsequently, to protect the prototype kilograms from wear, X_1 and X_2 can be used as working standards at NIST. As a matter of practice, X_1 and X_2 are fabricated from stainless steel and not platinum-iridium, and therefore one can expect an increase in the uncertainty due to the uncertainty in the buoyancy, as discussed later.

17.4   Uncertainties in the Determination of X Due to Uncertainties in the Parameters in Eq. (17.2)

Eq. (17.2) gives the relationship between X and various parameters, Y_i. The uncertainties in the various parameters are propagated using the method described by Ku.[4]

According to Ku:

$$\left(\text{SD}\right)^2 = \Sigma_i \left(\partial X/\partial Y_i\right)^2 \left(\text{SD}_i\right)^2 \qquad (17.3)$$

where subscript i refers to the individual parameters, $\partial X/\partial Y_i$ is the partial derivative of X with respect to the individual parameter, and SD_i is the estimate of standard deviation for the individual parameter.

The various partial derivatives are:

$$\partial X/\partial S = \left[1 - \left(\rho_a/\rho_s\right)\right]\big/\left[1 - \left(\rho_a/\rho_x\right)\right] \approx 1 \qquad (17.4)$$

$$\partial X/\partial \rho_x = -\rho_a \left\{ S\left[1-\left(\rho_a/\rho_s\right)\right]-\delta \right\} \Big/ \rho_x^2 \left\{ \left[1-\left(\rho_a/\rho_x\right)\right]^2 \right\}$$
$$\approx -\left(S-\delta\right)\rho_a/\rho_x^2 \tag{17.5}$$

$$\partial X/\partial \rho_s = \rho_a S \Big/ \left\{ \rho_s^2 \left[1-\left(\rho_a/\rho_x\right)\right] \right\} \approx S\rho_a/\rho_s^2 \tag{17.6}$$

$$\partial X/\partial \rho_a = \left\{ S\left[1-\left(\rho_a/\rho_s\right)\right]-\delta \right\} \Big/ \rho_x \left[1-\left(\rho_a/\rho_x\right)\right]^2 \right\} - S \Big/ \left\{ \rho_s \left[1-\left(\rho_a/\rho_x\right)\right] \right\}$$
$$\approx S\left(1/\rho_x - 1/\rho_s\right) - \delta/\rho_x \tag{17.7}$$

$$\partial X/\partial \delta = -1 \Big/ \left[1-\left(\rho_a/\rho_x\right)\right] \tag{17.8}$$

Examination of the partial derivatives reveals the need for an uncertainty estimate for the air density, ρ_a.

At present, air density is calculated from the CIPM 81-91 formulation[5] and requires knowledge of air temperature (T), barometric pressure (P), relative humidity (RH), and CO_2 content, all measured in the weighing chamber. The partial derivatives of the air density equation with respect to the above parameters are

$$\partial \rho_a/\partial T = -\rho_a/T \tag{17.9}$$

$$\partial \rho_a/\partial P = \rho_a/P \tag{17.10}$$

$$\partial \rho_a/\partial\left(\text{RH}\right) = -0.0037960 e_s \left(\rho_a/P\right) \tag{17.11}$$

$$\partial \rho_a/\partial X_{CO2} = 12.011\left(\rho_a/M_d\right) \tag{17.12}$$

where X_{CO2} is the mole fraction of CO_2 in the air, M_d is the apparent molecular weight of dry air, and e_s is the saturation water vapor pressure.

In Table 17.1 are estimates of the uncertainties in air density that can be achieved for these parameters. The root sum square (RSS) uncertainty is carried forward as the air density uncertainty (Type A) (the coverage factor is 1). The Type B uncertainty arising from the constant parameters of the air density equation is insignificant.

TABLE 17.1 Estimates of Uncertainties in Air Density

Variable	Value	SD(y_i)	SD(ρ_a)$_i$, g/cm³
Temperature (T)	295 K	5 mK	0.020×10^{-6}
Pressure (P)	100,258 Pa	5.1 Pa	0.061×10^{-6}
Relative humidity (RH)	41%	1%	0.12×10^{-6}
CO_2 mole fraction	0.000440	0.000050	0.025×10^{-6}
			RSS = 0.14×10^{-6}

TABLE 17.2   Uncertainties in the Comparison of Pt-Ir Artifact with a U.S. National
Prototype Kilogram

Variable	Value	$SD(y_i)$	$\partial X/\partial y_i$	$[SD(y_i)^2 \cdot (\partial X/\partial y_i)^2]$
S	1000 g	2.3×10^{-6} g	1	5.3×10^{-12} g^2
ρ_S	21.5 g/cm^3	7.2×10^{-5} g/cm^3	-2.6×10^{-3} cm^3	3.5×10^{-17} g^2
ρ_X	21.5 g/cm^3	7.2×10^{-5} g/cm^3	-2.6×10^{-3} cm^3	3.5×10^{-17} g^2
δ'	0.01 g	1.0×10^{-6} g	-1	1.0×10^{-12} g^2
ρ_a	0.0012 g/cm^3	0.14×10^{-6} g/cm^3	4.7×10^{-4} cm^3	4.2×10^{-21} g^2
				RSS = 2.5×10^{-6} g

TABLE 17.3   Uncertainties in the Comparison of a Stainless Steel Artifact with a U.S.
National Prototype Kilogram

Variable	Value	$SD(y_i)$	$\partial X/\partial y_i$	$[SD(y_i)^2) \cdot (\partial X/\partial y_i)^2]$
S	1000 g	2.3×10^{-6} g	1	5.3×10^{-12} g^2
ρ_S	21.5 g/cm^3	7.2×10^{-5} g/cm^3	2.6×10^{-3} cm^3	3.5×10^{-14} g^2
ρ_X	8.0 g/cm^3	7.2×10^{-5} g/cm^3	-1.88×10^{-2} cm^3	1.8×10^{-12} g^2
δ'	0.01 g	1.0×10^{-6} g	-1	1.0×10^{-12} g^2
ρ_a	0.0012 g/cm^3	0.14×10^{-6} g/cm^3	78.5 cm^3	1.2×10^{-10} g^2
				RSS = 11×10^{-6} g

In Table 17.2, the uncertainties in the comparison of a Pt-Ir artifact with the U.S. National Prototype Kilogram are tabulated. From examination of Table 17.2, it can be seen that the two major uncertainties are the uncertainty on the National Prototype Kilogram provided by the Bureau International des Poids et Mesures (BIPM) (the coverage factor is 1), and the imprecision of the balance (from δ).

In Table 17.3, the uncertainties in the comparison of a stainless steel artifact with the U.S. National Prototype Kilogram are tabulated. All the uncertainties in Table 17.3 are of the order of 1 μg or higher, except for the uncertainty in the density of the National Prototype Kilogram, and have a very significant effect on the RSS uncertainty.

17.5   Buoyancy

When a mass comparison is made between a platinum-iridium standard and an artifact fabricated from a material of different density, the gravitational forces on the two bodies are opposed by buoyant forces the inequality of which must be taken into account.

Archimedes' principle[6] provides the necessary information to account for the buoyant forces. The gravitational force, $\mathbf{F}_1$, on an object,

$$\mathbf{F}_1 = Mg \qquad (17.13)$$

is *opposed* by a buoyant force, $\mathbf{F}_2$,

$$\mathbf{F}_2 = \rho_a V_M g, \qquad (17.14)$$

where M is the mass of the object, g is the local acceleration due to gravity, ρ_a is the air density, and V_M is the volume of the object.

The difference between the gravitational force and the buoyant force is most conveniently expressed as

$$\mathbf{F} = \mathbf{F}_1 - \mathbf{F}_2 = Mg\left[1 - \left(\rho_a/\rho_M\right)\right], \tag{17.15}$$

where ρ_M is the density of the object.

17.6 Thermal Equilibrium

Probably the limiting systematic error (Type B) remaining in the mass measurement is that due to the convective forces arising from the lack of thermal equilibrium between the mass artifact, the mass comparator, and the surrounding air.[7,8]

In practice, it is not possible to assure equality of temperatures of these three items; therefore, some convection will remain. It follows that the balance or comparator observation will always be biased by small convective forces and, therefore, the mass determination will have a systematic error on this account. Consequently, it is necessary to take precautions to assure thermal equilibrium as closely as possible.

Prior to the work of Schoonover and Keller,[7] these effects were usually ignored. To minimize the systematic error, safeguards, including the following, should be maintained:

1. Passive and active control of the thermal environment in and around the balance or comparator
2. Adequate thermal "soaking" of the artifacts and the comparator mechanism prior to mass measurement
3. Thermal sensors to assure that the safeguards are effective

17.7 Atmospheric Effects

The mass of mass standards, after specified cleaning, is affected by variation in atmospheric variables in an otherwise clean environment. The principal effect is due to variation in relative humidity and consequent variation in adsorption of water vapor (see Chapters 3 and 4).[9,10]

Schwartz[9] reported water vapor adsorption measurements on polished stainless steel mass standards. He found that the cleanliness of surfaces is the important factor that can influence the adsorption-induced mass changes. He concluded that sorption-induced mass changes of carefully cleaned stainless steel reference standards can be pratically neglected in the RH range of 30 to 60% at 20 to 22°C.

Schwartz appplied the BIPM cleaning procedures[11] and ultrasonic cleaning in ethanol.[12] At this writing, we are not aware of the existence of a similar detailed study of adsorption-induced mass changes of Pt-Ir mass standards. However, Kochsiek[10] found similar results for stainless steel plates and for Pt-Ir plates.

17.8 Magnetic Effects

In ultrahigh-accuracy mass determination, it is necessary to minimize the magnetic interaction between magnetic structures and the mass artifacts being compared. There may also be magnetic interaction between the mass artifacts and external magnetic fields; this interaction also must be minimized (see Chapter 20).

Davis[13] suggested the following strategies to reduce the force of interaction:

1. Minimize $\mu_R - 1$ by selecting materials with low magnetic susceptibility at low fields; μ_R is magnetic susceptibility.
2. Minimize the volume of magnetic material in the structure.
3. Maximize the distance between the mass artifact and the magnetic structures or magnetic fields.

Based on handbook values for platinum and iridium, the magnetic susceptibility for 90% Pt/10% Ir is 0.00027. For American Iron and Steel Institute (AISI) 316 stainless steel, the magnetic susceptibility is 0.003. A sample of AISI 304 has been found to have a value as high as 0.038; this alloy is slightly soluble in boiling water and in steam.

Using the above strategies, one should use standards manufactured from the AISI 316 alloy. Since the magnetic susceptibility is a limiting factor, items 2 and 3 merit careful consideration. OIML suggests that for weights of class E_1 the magnetic susceptibility of the metal or alloy should not exceed 0.01.[14]

17.9   Instability of IPK

There is evidence that the mass of Pt-Ir prototype mass standard artifacts has changed monotonically by about 50 µg, with respect to IPK, in the 100 years since the inception of the IPK as the standard of mass.[15] Because the IPK is one of the group of 40 original prototype kilograms, it is certainly conceivable that it is also changing in mass with time. We cannot know the magnitude of the change and the rate of change with time of the IPK.

Insofar as mass metrology as practiced assumes the value of the IPK to be invariant and that all other mass standards are referred to the initial mass value of the IPK, mass metrology can be practiced at the level of several parts in 10^9 billion (see Table 17.2) relative to the IPK. However, if there are experiments, for example, in which the *absolute mass* of an object must be known to better than 50 µg/kg the present system based on the IPK, which varies in value, is inadequate.

17.10   Cleaning

The mass of an artifact is dependent on the surface contamination present. Therefore, the mass and mass stability depend on the cleaning procedure prior to weighing. There are several cleaning procedures in use by BIPM and national standards laboratories (see Chapter 4).

The national prototype of the kilogram (platinum-iridium) as received by a recipient country from BIPM has been cleaned and washed by BIPM using the BIPM procedures.[11]

The mass assigned by BIPM is only applicable on a given date. Immediately following this date BIPM recommends adding 0.037 µg/day for a maximum period of 3 months.

The National Physical Laboratory of Teddington, England[15] found for its kilogram No. 18 a functional relationship between mass change and time, for BIPM data. The relationship is

$$M_t = M_o + 0.356097 \times t^{0.511678}, \qquad (17.16)$$

where M_t is the mass at time t after cleaning and washing, M_o is the mass at the time of cleaning and washing, and t is the elapsed time in days.

It has been suggested by Plassa[15] that the above equation calculates mass values to within a few micrograms for a period of up to 10 years following cleaning and washing, provided that the storage conditions can be carefully controlled.

The BIPM cleaning and washing procedures for platinum-iridium mass standard artifacts[11] involves solvent cleaning and steam washing. For cleaning, chamois leather is used that had been previously soaked for 48 h in a mixture of equal parts ethanol and ether after which the absorbed solvent is wrung out of the leather. This soaked chamois leather is rubbed over the entire surface of the mass standard artifact. In the steam-washing procedure, steam is directed to all parts of the surface of the artifact. National prototypes should be cleaned and washed using these BIPM procedures prior to use.

The BIPM practice for stainless steel mass standard artifacts omits the steam washing. However, national standards laboratories are free to clean stainless steel artifacts by whatever procedures they wish unless their stainless steel mass standards are calibrated by BIPM.

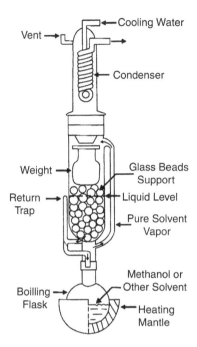

FIGURE 17.1 Sketch of a Soxhlet apparatus with a weight to be cleaned included.

During the development of the solid object density scale,[16–18] National Bureau of Standards (NBS) (now National Institute of Standards and Technology, NIST) studied the residue that remained on steel spheres after cleaning, and vapor degreasing was found to be the superior cleaning method. The method was used in preparation for very high precision mass measurement and diametric measurement by interferometry. Initially, inhibited 1,1,1-trichloroethane was used as a solvent for vapor degreasing. Ultimately, ethanol was used due to its availabilty; if a fume hood were available, methanol could be used. Figure 17.1 is a sketch of a Soxhlet apparatus with a weight to be cleaned included.

NIST studies have revealed that mass measurements of stainless steel kilogram artifacts of significantly different surface areas have comparable standard deviations. This finding indicates that vapor degreasing of the disparate surface areas does not contribute uncertainty to the measurements.

Numerous vapor degreasings over a period of a year did not result in different mass values. In the literature, it has been reported that alternative cleaning methods can change mass values and the variability of mass values. A newly manufactured mass standard artifact requires more rigorous and varied initial cleaning procedures to remove effects contributed by the manufacturing processes.

17.11   Conclusions

- From the above analysis, it is clear that the calibration of Pt-Ir kilograms is the simplest of the calibrations and does not require the highest accuracy in measurement of the parameters in the air density equation.

- Usually, the comparison of Pt-Ir kilograms is only made by BIPM. However, in the calibration of stainless steel kilograms, state-of-the-art measurements of the air density parameters are required to minimize the uncertainties on this account.

- If one is determining the mass of a silicon kilogram (by comparison with a stainless steel kilogram or with a Pt-Ir kilogram), such as would be done in an experiment to determine Avogadro's number, the state-of-the-art measurements are inadequate. They contribute almost all, 42 µg, of the uncertainty in the determination against stainless steel and 52 µg against Pt-Ir.

- For the Avogadro number experiment, the mass of the silicon kilogram should be determined in vacuum or by direct measurement of the air density[19] at the time of weighing if these relatively large uncertainties are to be avoided.

- Vacuum weighing of the IPK or the national prototypes would be an unacceptable practice because mass could be lost in vacuum.

- The Pt-Ir vs. stainless steel comparison requires near state-of-the-art measurements of the densities of the artifacts.

- The magnetic susceptibility of stainless steel should be checked to ensure that it is sufficiently low.

- The thermal history and thermal stability of the artifacts and their environment are crucial in the determinations of mass.

- The comparisons of artifacts the densities of which lie between 7.8 and 8.4 g/cm^3 require less rigor because the density difference is smaller than that between Pt-Ir and stainless steel.

17.12   Discussion

The 50 μg of instability, or possibly more, in the IPK is, at the present writing, not threatening to practical mass measurements, most of which never require accuracy better than 1 part per million (1 mg/kg).

As previously discussed, the drift has occurred over the course of 100 years and has only been detectable in recent times. The successful development of the NIST balance with a precision of better than 1 part per billion was a crucial step in highlighting the problem.

There is no reason to believe that the drift will not continue at the present rate, giving the metrology community time to find a time-invariant replacement for the IPK. However, as pointed out by work of the National Physical Laboratory of the United Kingdom, mercury amalgam on the surfaces of Pt-Ir artifacts may be another matter of concern (see Chapter 3).[20]

References

1. Schoonover, R. M. and Jones, F. E., Examination of parameters that can cause errors in mass determinations, NISTIR 5376, National Institute of Standards and Technology, February 1994.
2. Almer, H. E. and Swift, H. F., Gravitational configuration effect upon precision mass measurements, *Rev. Sci. Instrum.*, 46, 9, 1975.
3. Cameron, J. M., Croarkin, M. C., and Raybold, R. C., Design for the calibration of standards of mass, National Bureau of Standards (U.S.), Technical Note 952, 1977.
4. Ku, H. S., in *Handbook of Industrial Metrology*, American Society of Tool and Manufacturing Engineers, Prentice-Hall, Englewood Cliffs, NJ, 1967, 20.
5. Davis, R. S, Equation for the determination of the density of moist air (1981/91), *Metrologia*, 29, 67, 1992.
6. Quinn, T. J., The kilogram: the present state of our knowledge, *IEEE Trans. Instrum. Meas.*, 40, 81, 1991.
7. Schoonover, R. M. and Keller, J., National Bureau of Standards (U.S.), Special Publication 663, 39, 1983.
8. Schoonover, R. M. and Taylor, J. E., *IEEE Trans. Instrum. Meas.*, IM-35, 418, 1986.
9. Schwartz, R., Accurate measurement of adsorption layers on mass standards by weighing and ellipsometry in controlled environments, presented at 13th IMEKO Intl. Conf. on Force and Mass Measurement, Helsinki, 1993.
10. Kochsiek, M., Measurement of water adsorption layers on metal surfaces, *Metrologia*, 18, 153, 1982.

11. Girard, G., The washing and cleaning of kilogram prototypes at the BIPM, Bureau International des Poids et Mesures, Sèvres, France, 1990.

12. Schwartz, R., Untersuchung des Sorptionseinflusses bei massebestimmunger hoher Genauigkeit durch Ellipsometrie unter kontrollierten Umgebungsbedingungen, thesis, PTB-Bericht, 1993.

13. Davis, R. S., Using small, rare-earth magnets to study the susceptibility of feebly magnetic metals, *Am. J. Phys.*, 60, 365, 1992.

14. OIML International Recommendation RIII, Weights of classes E_1, E_2, F_1, F_2, M_1, M_2, M_3, International Organization of Legal Metrology, Paris, 1993.

15. Plassa, M., Working group "mass standards" report to CCM on activity 1991–1993, Comite Consultatif pour la Masse et les Grandeurs Apparentees, BIPM, Paris, June 1993.

16. Bowman, H. A., Schoonover, R. M., and Carroll, C. L., A density scale based on solid objects, *J. Res. Natl. Bur. Stand.* (U.S.), 76A, 13, 1974.

17. Bowman, H. A., Schoonover, R. M., and Carroll, C. L., The utilization of solid objects as reference standards in density measurements, *Metrologia*, 10, 117, 1974.

18. Bowman, H. A., Schoonover, R. M., and Carroll, C. L., Reevaluations of the densities of the four NBS silicon crystal standards, NBSIR 75-768, 1975.

19. Deslattes, R. D. et al., Determination of the Avogadro constant, *Phys. Rev. Lett.*, 33, 463, 1974.

20. Cumpson, P. J. and Seah, M. P., Stability of reference masses I: evidence of possible variations in mass of reference kilograms arising from mercury contamination, National Physical Laboratory, Teddington, U.K., January 1993.

18

Determination of the Mass of a Piston-Gauge Weight, Practical Uncertainty Limits

18.1 Introduction

The equilibrium pressure in a piston gauge depends on the gravitational force on the rotating parts and the area of the piston. The gravitational force is equal to the product of the local acceleration due to gravity and the masses of the rotating components.

Davis and Welch[1] focused on the problem of assigning mass values to piston-gauge weights of about 590-g nominal mass. The goal of the measurements was to accomplish an uncertainty in the mass calibration of the piston-gauge weights that would limit an uncertainty in the maximum presssure generated by the rotating assembly to no more than 1 ppm (1×10^{-6}).

Research under way at the time of the Davis and Welch work sought to achieve unprecedented levels of accuracy in the calibration of the selected piston gauges.

The gas thermometer mercury manometer[2] was used as the pressure standard. The error for this manometer was of the order of 1 part per million (ppm) in the pressure range from 10 to 130 kPa.

A stack of 590-g weights provides most of the rotating mass of a piston-gauge assembly. The mass uncertainy of such a stack of weights is the combined systematic uncertainties of the calibration of the individual weights and the random uncertainties of the individual weights.

Davis and Welch described in detail how the total uncertainty of their mass measurements was evaluated. The difficult-to-detect errors systematic to a given measurement technique amounted to 1 ppm. The National Bureau of Standards (NBS) (now National Institute of Standards and Technology, NIST) typically provided by contrast an uncertainty (Type A) of 0.04 ppm (at the estimate of one standard deviation) for calibrations of a single 500-g laboratory mass standard of the highest commercially available quality.

18.2 Assignment of Mass

An aggregate mass M_x is assigned to a piston-gauge weight plus the piston by comparison with a standard of mass M_s, under ambient conditions. M_x is determined using the following equation:

$$M_x = M_s \left(1 - \rho_a / \rho_s\right) / \left(1 - \rho_a / \rho_x\right) + \Delta, \tag{18.1}$$

where

ρ_a = density of the air within the balance case (about 1.2 mg/cm^3)

ρ_s = density of the standard (about 8 g/cm^3)

ρ_x = density of the piston-gauge weight (about 7.8 g/cm^3)

Δ = difference between balance readings corresponding to S and X (–2 mg < Δ < 2 mg), in mass units

The following were among the sources of error carefully examined (see Chapters 3, 4, 17, and 19):

1. Air-buoyancy corrections
2. Physically adsorbed surface moisture
3. Air convection within the weighing chamber

The mass of a 590-g piston-gauge weight was determined with a total uncertainty of 0.057 mg (0.1 ppm). It was concluded that significant improvement could not be realized with the conventional weighing techniques then available to most piston-gauge users.

References

1. Davis, R. S. and Welch, B. E., Practical uncertainty limits to the mass determination of a piston-gage weight, *J. Res. Natl. Bur. Stand.* (U.S.), 93, 565, 1988.
2. Guildner, L. A. et al., An accurate mercury manometer for the NBS Gas Thermometer, *Metrologia*, 6, 1, 1970.

19

Response of Apparent Mass to Thermal Gradients and Free Convective Currents

19.1 Thermal Gradients

19.1.1 Introduction

Following up on the work of Schoonover and Keller[1] on the necessity of thermal equilibrium between a mass artifact and the air surrounding the mass artifact in a balance chamber, Glaser[2] studied the change of apparent mass of 20-g stainless steel masses and tubes with lack of thermal equilibrium with ambient air. In 1946, Blade[3] performed a limited study of mass assignment errors caused by convection forces in measurements related to chemistry.

Schoonover and Keller[1] had concluded that to attain thermal equilibrium one simply needs to soak the weights in the balance chamber for 24 h in the case of 1-kg weights and 1 h for fractional gram weights. In addition to thermal soaking, the balance should be thermally shielded from the operator when robot handling is not used. Thermal quiescence should prevail.

The Glaser[2] experiment was performed with two 20-g masses and two 20-g tubes of nonmagnetic stainless steel with well-polished surfaces. One of the masses and one of the tubes were kept inside the balance chamber of a one-pan unequal-arm oscillating-beam balance. Before weighing, the other two artifacts were stored in a thermostat at a temperature differing from that of the balance.

The mass kept inside the balance chamber was used as a reference. The two artifacts stored in the thermostat were moved into the balance chamber before the start of a weighing series and were kept there until the end of the weighing series. A weighing series consisted of several identical weighing cycles.

A special mass support with a negligibly small effective surface for vertical air currents was used for measurements with different positive or negative initial temperature differences, $\Delta T = T_1 - T_2$, between the temperature of the thermostated artifacts, T_1, and temperature of the balance chamber, T_2. For comparison, a conventional balance pan having a large surface area (8.6 cm^2) was used at two separate temperature differences.

Schoonover and Taylor[4] improved the thermal conditions of weights and the weighing chamber, prior to weighing in the following ways. First, an aluminum soaking plate for masses undergoing calibration to be stored was placed beside the balance. The plate and the balance weighing chamber each contained a 100-Ω platinum resistance thermometer; these thermometers were part of a bridge and control circuit that forced the weights and the air inside the weighing chamber into thermal equilibrium. The long-term (days) temperature differences were less than 0.01°C.

Second, a body-heat simulator was placed in front of the balance to prevent operators from disturbing the equilibrium by their presence. The simulator, removed by the operator when the weighing operation began, was a nichrome heating element radiating in the infrared region.

All the experiments indicated the need for thermal equilibrium between the balance mechanism, the weights in use, and the air in the balance chamber. These conditions can be approached with use of the following:

1. Thermal shielding
2. Thermal soaking
3. Active soaking plates
4. Environmental controls
5. Robotic weight handling

19.1.2 Conclusions

If an artifact were not at thermal equilibrium, that is, the temperature of the artifact was not that of the balance chamber, the apparent mass of the artifact deviated from the value at thermal equilibrium.

The deviation was negative or positive depending, respectively, on whether the temperature of the artifact was higher or lower than that of the balance chamber.

It was concluded that buoyancy and sorption effects could not explain the observed deviations of apparent mass, and that a model based on free convective air currents and corresponding viscous forces gave satisfactory agreement between theory and experiment.

Viscous effects considered were those acting on the weighed artifacts and those acting on the balance pan, in those cases in which balance pans were used.

The required temperature homogeneity, artifact to balance chamber, was estimated by calculating the influences of free convective air currents and corresponding viscous forces.

The spurious forces could be prevented by keeping the artifact in the balance chamber before weighing, or by keeping it at the same temperature as the balance chamber, within limits.

ΔT ranged from −6.3 to +5.8°C and the corresponding initial differences of apparent mass ranged from +120 to −117 μg (in 20 g) for a mass artifact and from +256 to −234 μg for a tube, for the experiments.

19.2 Free Convective Currents

19.2.1 Introduction

Glaser and Do[5] followed up the work[2] on 20-g artifacts and tubes with an experimental investigation of the influence of free convective currents on the apparent mass of 1-kg mass standards.

19.2.2 Experimental

The balance used was a single-pan, electromagnetically-compensated mass comparator with a capacity of 1 kg, a standard deviation of less than 5 μg, and a resolution of 1 μg (Sartorius C 1000 S). The comparator was equipped with an automatic mass exchange device and a turntable with four positions. A separate unit controlled mass exchange and allowed specified programs of measurement.

Four stainless-steel mass standards with well-polished surfaces were used. Two identical standards were used for comparison. One served as a reference and was stored on a turntable inside the balance chamber at temperature T_1. Temperature T_1 was monitored by a thermocouple close to the turntable. The second mass standard was kept in a thermostat at temperature T_2.

The temperature difference $\Delta T = T_2 - T_1$ was taken as the initial value for the mass comparisons between the two standards. The mass comparisons lasted approximately 7 h.

19.2.3 Results and Discussion

It was conjectured that the change of air density with heat transfer caused convective air currents surrounding the heated or cooled (relative) surface of an artifact. Also, it was conjectured that the apparent mass of an artifact is changed by the corresponding convective force originating in the shear stress transmitted by the viscosity of the air and the momentum transfer of the flowing air.

Data were taken for the two 1-kg standards and for a 20-g mass artifact and a 20-g tube from the previous paper.[2]

ΔT ranged from −2.61 to +6.06°C for the 1-kg standard designated No. 1, and the corresponding range of initial differences of apparent mass was +399 to −1326 μg. For the 1-kg mass standard designated No. 2, ΔT ranged from −3.91 to +6.09°C, and the corresponding range of initial differences was +353 to −1190 μg.

For the 20-g mass standard, ΔT ranged from −6.3 to +5.8°C, and the corresponding range of initial differences was +120 to −117 μg. For the 20-g tube, ΔT ranged from −6.3 to +5.8°C, and the corresponding range of initial differences was +256 to −234 μg.

Simple convection models that combined free convection at a vertical wall with forced convection on horizontal surfaces were in "reasonable" agreement with the experimental results.

19.3   Temperature Differences and Change of Apparent Mass of Weights

In a paper published in 1999, Glaser[6] reported on measurements aimed at obtaining further information on the temperature accommodation and change in apparent mass of 1 and 50 kg weights; in particular,

a. the change of temperature of a heated or cooled weight as a function of time under different ambient conditions;

b. the change of apparent mass as a function of the temperature of a weight during a weighing series.

Two physical models were considered:

a. the change of temperature of a weight as a function of time;

b. the change of apparent mass of a weight as a function of temperature difference between the weight and ambient air.

The models were used to extrapolate the experimental results to predict, for weights of similar shape but other nominal values, the change of apparent mass and waiting times for given residual deviations and residual temperature differences.

References

1. Schoonover, R. M. and Keller, J., A surface-dependent thermal effect in mass calibration, *U.S. National Bureau of Standards* Special Publication 663, 39, 1983.
2. Glaser, M., Response of apparent mass to thermal gradients, *Metrologia*, 27, 95, 1990.
3. Blade, E., Differential temperature error in weighing, *Ind. Eng. Chem. Analyt. Ed*, 12, 330, 1943.
4. Schoonover, R. M. and Taylor, J. E., Some development at NBS in mass measurement, *IEEE Trans. Instrum. Meas.*, IM-35, 418, 1986.
5. Glaser, M. and Do, J. Y., Effect of free convection on the apparent mass of 1 kg mass standards, *Metrologia*, 30, 67, 1993.
6. Glaser, M., Change of the apparent mass of weights arising from temperature differences, *Metrologia*, 36, 183, 1999.

20

Magnetic Errors
in Mass Metrology

20.1 Introduction

In mass metrology, magnetic errors should be avoided. By magnetic error is meant an unsuspected vertical force **F** that is magnetic in origin.[1] If such a force exists, it would appear in a weighing as a mass of magnitude F/g, where g is the local acceleration due to gravity.

High-quality mass standards are assumed to be artifacts with an isotropic volume magnetic susceptibility, χ, the magnitude of which is much less than 1.[1] Also, artifact standards should have little or, ideally, no permanent magnetization.[1]

Magnetic susceptibility is the ratio of the intensity of magnetization produced in a substance (or body in the present case) to the intensity of the magnetic field to which it is exposed.

20.2 Magnetic Force

To a good approximation, the magnetic force (unwanted) can be given by

$$\mathbf{F} = \left(-\mu_o/2\right)\left(\partial/\partial z\right)\left[\int \chi' \mathbf{H} \cdot \mathbf{H}\, dV\right] - \mu_o\left(\partial/\partial z\right)\left[\int \mathbf{M} \cdot \mathbf{H}\, dV\right], \tag{20.1}$$

where χ' is the effective volume magnetic susceptibility of the standard, defined as $\chi - \chi_A$, where χ_A is the volume suceptibility of air (3.6×10^{-7}); **M** is the permanent magnetization of the standard (defined as the magnetic moment per unit volume in zero field); **H** is the local magnetic field strength: and the z-axis is parallel to g. The parameter μ_o is the vacuum permeability, identically equal to $4\pi \times 10^{-7}$ N/A^2 (newton per square ampere).

The integrals are taken over the volume of the standard. χ' is assumed to be a scalar. χ_A is sufficiently small to be neglected in most examples given by Davis.[1]

20.3 Application of a Magnetic Force Equation

A simplified force equation, successor to Eq. (20.1), has been applied to problems of mass metrology by Gould[2] and Kochsiek.[3]

Gould[2] concluded that an alloy used for mass standards should be chosen both for its low magnetic susceptibility and for resistance to permanent magnetization upon exposure to high fields. He found that for stainless steels the alloy with the lowest susceptibility was also the most difficult to magnetize. He recommended an alloy with $\chi \approx 0.003$ when measured in a unified field strength of 16 kA/m. The permanent magnetization was less than 1 A/m after exposure to a "suitably large" uniform field.

Kochsiek[3] argued that, once demagnetized, the normal use of mass standards should not subject them to fields great enough to remagnetize them. He made a strong recommendation against selecting inferior stainless steel alloys, known to be easily magnetized. He argued that secondary 1-kg mass standards used by national laboratories should have a volume susceptibility below 0.003.

We see, then, that Gould and Kochsiek, focusing on different aspects of the problem, arrived at nearly the same guidelines for selecting stainless steel alloys suitable for the the highest quality of mass standards.

The Organization Internationale de Metrologie Legale (OIML) recommended that Class E_1 and E_2 mass standards have volume susceptibilities less than 0.01 and 0.03, respectively.[4]

Davis described in detail the calibration and operation of a susceptometer developed at the Bureau International des Poids et Mesures (BIPM), which was optimized for the determination of the magnetic properties of 1-kg mass standards.[5]

In practice it might be necessary to weigh objects of materials that are magnetic or are susceptible to magnetic fields. Some weighing strategies are: (1) weighing below the balance pan in a shielded space, (2) weighing using shielded containers, and (3) using a balance that does not emit a field and is manufactured from low magnetic susceptibility material. One might test a top-loading balance for magnetic coupling with the object of interest by performing the weighing first with a magnetically inert spacer below the object and then on top of it. There is no magnetic effect if the balance indication is constant.

References

1. Davis, R. S., Determining the magnetic properties of 1 kg mass standards, *J. Res. Natl. Inst. Stand. Technol.*, 100, 209, 1995.
2. Gould, F. A., Tests on highly non-magnetic stainless steels for use in the construction of weights, *J. Sci. Instrum.*, 23, 123, 1946.
3. Kochsiek, M., Anforderungen an Massenormale und Gewichtstucke fur hochste Genauigkeitsanspruche, *Wagem Dosieren*, 9, 4, 1978.
4. Organization International de Metrologie Legale, International Recommendation on Weights of Classes E_1, E_2, F_1, F_2, M_1, M_2, M_3, International Recommendation No. R111, OIML, Paris, 1994.
5. Davis, R. S., New method to measure magnetic susceptibility, *Meas. Sci. Technol.*, 4, 141, 1993.

21

Effect of Gravitational Configuration of Weights on Precision of Mass Measurements

21.1 Introduction

Mass measurements or mass comparisons consist of making comparisons between the gravitational attraction forces exerted on the standard and unknown weights by the Earth. Almer and Swift[1] have shown that an effect coupling the gravitational force exerted on a weight to the height of the center of gravity of the weight is significant when weighings of the highest precision are attempted, and that this effect will cause significant systematic errors in mass measurements if it is not accounted for.

The effect, the "gravitational configuration effect," arises because, for weights of nominally equal mass, the distance of the center of gravity above the base of each weight (and hence the effective distances from the center of the Earth) depends on the size and shape of the weight. Since the acceleration due to gravity decreases with the inverse of the square of elevation, the magnitude of the gravitational force on weights of equal mass but of different size and shape can be different.

21.2 Magnitude of the Gravitational Configuration Effect

Almer and Swift[1] have shown that a simple geometric approach can be taken to calculate the gravitational configuration effect for two cases:

1. Two nominal kilogram weights constructed of the same material with centers of gravity spaced 1 cm apart vertically
2. A National Prototype Kilogram Standard and a stainless steel working standard

So long as the elevation of a point z (center of gravity of a weight in this context) above mean sea level is small compared to r_e, the radius of the Earth (6.378140×10^6 m), the value of the acceleration due to gravity at height z, g_z, is[2]

$$g_z \approx g_0\left[1 - 3.14 \times 10^{-7} z\right], \tag{21.1}$$

where g_0 is the acceleration due to gravity at sea level and z is in meters.

For the first case, the separation between the centers of gravity was 1 cm (0.01 m) for the two weights each of nominal mass 1 kg, as stated above.

As a consequence of Eq. (21.1), the correction to be applied to account for the separation is 3.14 μg (3.14 parts in 10^9). This correction can be added to the observed mass of the weight at the higher elevation.

For the second case, the National Prototype Kilogram Standard is a right-circular cylinder of platinum-iridium alloy of height (39 mm) approximately equal to its diameter. Stainless steel kilogram working standards of about the same proportions are 54.6 mm high. The centers of gravity of stainless steel working standard kilograms are 7.8 mm (0.0078 m) higher above their bases than is the center of gravity of the National Prototype Kilogram Standard.

The gravitational configuration effect correction for the direct comparison of the National Prototype Standard and stainless steel working standard kilograms, as consequence of Eq. (21.1), is 2.45 μg (2.45 parts in 10^9).

The gravitational configuration effect is also important when performing hydrostatic weighing to determine the density of solids and liquids.[3] Usually, a solid object is suspended below a balance where the standard resides. The usual practice is to adjust the object to the standard above. The adjustment is approximately –314 μg/kg/m; see also Chapter 13.

21.3   Significance of the Gravitational Configuration Correction

The precision of the best kilogram balance weighing is of the order of 1 μg (1 part in 10^9) or less. It is clear then that gravitational configuration corrections of 2.45 and 3.14 parts in 10^9 are very significant for weighing on the best balance.

For weighing in air at this level of precision, however, if one does not know the air density correction well enough, the gravitational configuration correction is much less significant than it is for weighing in vacuum.

References

1. Almer, H. E. and Swift, H. F., The gravitational configuration effect upon precision mass measurements, *Rev. Sci. Instrum.*, 49, 1174, 1975.
2. Hess, S. L., *Introduction to Theoretical Meteorology*, Holt, Rinehart and Winston, New York, 1959, 8.
3. Bowman, H. A., Schoonover, R. M., and Carroll, C. L., A density scale based on solid objects, *J. Res. Natl. Bur. Stand. (U.S.)*, 78A, 13, 1974.

22

Between-Time
Component of Error
in Mass Measurements

22.1 Introduction

It is occasionally observed that when sets of measurements are made on successive days the results differ. The standard deviation for measurements made on a single day can be referred to as the "within-series"[1] standard deviation; the standard deviation for measurements made on different occasions is designated s_T. If s_T is significantly greater than the within-series standard deviation, the difference is attributed to a between-time standard deviation, s_b.

We now digress to Eisenhart's definitive paper[2] on the precision and accuracy of instrument calibration systems. Eisenhart defined and discussed "repetition" of a measurement:

> Repetitions [by the same *measurement process*] will undoubtedly be carried out in the same place, i.e., in the same laboratory, because if it is to be the same measurement process, the very same apparatus must be used. But a 'repetition' cannot be carried out at the same *time*. How great a lapse of time should be allowed, nay *required*, between 'repetitions'?"[2]

Eisenhart observed that an answer to this question was given by Student:[3]

> After considerable experience I have not encountered any determination which is not influenced by the date on which it is made; from this it follows that a number of determinations of the same thing on the same day are likely to lie more closely together than if the repetitions had been made on different days.

> It also follows that if the probable error is calculated from a number of observations made close together in point of time, much of the secular error will be left out and for general use the probable error will be too small.

> Where then the materials are sufficiently stable it is well to run a number of determinations on the same material through any series of routine determinations which have to be made, spreading them over the whole period.

Experiments to determine the precision and systematic error of a measurement process must be based on an appropriate random sampling of the range of circumstances commonly met in practice.[2]

> The experiments must be capable of being considered a *random* sample of the population to which the conclusions are to be applied.[3]

A quotation from Airy[4] is also relevant here.

When successive series of observations are made, day after day, of the same measurable quantity, which is either invariable … or admits of being reduced by calculation to an invariable quantity…; and when every known instrumental correction has been applied …; still it will sometimes be found that the result obtained on one day differs from the result obtained on another day by a larger quantity than could have been anticipated. The idea then presents itself, that possibly there has been on some one day, or on every day, some cause, special to the day, which has produced a *Constant Error* in the measures of that day.

The existence of a daily constant error … ought not to be lightly assumed. When observations are made on only two or three days, and the number of observations on each day is not extremely great, the mere fact, of accordance on each day and discordance from day to day, is not sufficient to prove a constant error. The existence of an accordance analogous to a "round of luck" in ordinary changes is sufficiently probable…. More extensive experience, however, may give greater confidence to the assumption of constant errors … first, it ought, in general to be established that there is possibility of error, constant on one day but varying from day to day.

To judge the accuracy of physical or chemical determinations and to "provide an overall check on procedure, on the stability of reference standards, and to guard against mistakes, it is common practice in many calibration procedures, to utilize two or more reference standards as part of the regular calibration procedures."[2] For example, in the calibration of laboratory standards of mass at National Institute of Standards and Technology (NIST), known standard weights are calibrated side-by-side with the unknown weights (see Chapter 8).

Measurements obtained in routine remeasurement of the differences between pairs of mass standards constitute realistic repetitions of the calibration procedure as do measurements for a single check standard.

A procedure used to investigate the possible existence of a between-time standard deviation or between-time component of uncertainty and to estimate its magnitude is presented in Reference 1. That procedure will be generally followed here.

To estimate these standard deviations, sets of measurements are made on several, m, occasions (or days); each set of measurements consisting of the same number, p, of measurements.

The standard deviations from the m occasions or days are pooled to obtain an estimate, s_w, of the within-group standard deviation based on $(p - 1)m$ degrees of freedom. The estimate of the standard deviation of the mean of any of the p measurements groups is $s_w/(p)^{1/2}$.

The square of the pooled standard deviation for the standard deviations for the measurements on the separate days is given by:

$$\left(\text{Pooled SD}\right)^2 = \Sigma n_i \, SD_i^2 / \Sigma n_i = \bar{s}_w^2 \qquad (22.1)$$

where SD_i is the standard deviations for the separate days and n_i is the number of degrees of freedom for the SD_i, the number of observations less 1, $(p - 1)$.

The pooled SD is designated $\bar{s}_w$. The estimate of the standard deviation of the mean of any of the p measurements groups is $\bar{s}_w/p^{1/2}$. $\bar{s}_w$ is an estimate of the within-group standard deviation based on $(p - 1)m$ degrees of freedom, where m is the number of days on which measurements are made.

For the set of means for the m days, an estimate of the standard deviation, s_T, is computed based on $(m - 1)$ degrees of freedom.

If s_T is significantly greater than $\bar{s}_w/(p)^{1/2}$, the difference is attributed to a between-time standard deviation, s_b, defined as:

$$S_b = \left(s_T^2 - \bar{s}_w^2/p\right)^{1/2}. \qquad (22.2)$$

In the present chapter, the F test is used to infer whether s_T is significantly different from $\bar{s}_w/p^{1/2}$. The F test is a measure of the ratio of two variances, $(SD)^2$, in the present context the ratio of s_T^2 to $\bar{s}_w^2/p$:

$$F = s_T^2 \Big/ \left[\bar{s}_w^2 \Big/ p \right]. \tag{22.3}$$

F provides critical values that will rarely be exceeded if the squares of the two SD values are estimates of the same variance, that is, if the two SD values are estimates of the same SD.

Various calculations are illustrated for data for one of the four groups of measurements made in the present study.

22.2   Experimental

Laboratories are frequently encountered where the environment or techniques may induce weighing error. A good example is when two different observers are asked to assign state-of-the-art mass values to a 1-kg mass standard. Each observer alternates weighing until a large number of mass assignments have been collected. Subsequent data reduction and analysis reveal two distinct groups of mass values differing by 100 μg and each group is clearly associated with one of the operators. This situation came about because one operator stored both the standard and object of unknown mass in the balance 24 hours prior to weighing, while the other stored the unknown elsewhere and placed it in the balance immediately prior to weighing.

These unsuspecting operators were not aware of the problems related to the lack of thermal equilibrium in weighing. However, had they examined the data early on for a between-time component and tested the means they would have been known that a measurement problem existed. In this case, no time was available to repeat the measurements and, therefore, measurement uncertainty had to be increased by an embarrassing amount.

The difficulties described above occur when laboratory procedures are not well defined, or adhered to, and the measurement process is not well understood. These circumstances coupled with multiple operators can result in unrealistic uncertainty estimates when left uncorrected.

What follows are several small groups of weighings made under differing conditions in the same laboratory by the same operator. They reflect actual operating techniques found in mass laboratories.

A 500-g stainless steel weight has been compared by the method of double substitution weighing to a summation 500-g weight comprised of a 200-g weight and a 300-g weight. That is, the mass difference 500g − Σ500g was measured six times. All weights were fabricated from the same alloy and had the same density; therefore, a buoyancy correction was not needed. The weighings were performed on a 1-pan-2-knife-edge balance with a precision of approximately 50 μg. The weighings for each group are a set of six double substitutions and were continued for 6 days.

For Group 1, all weights resided on an aluminum soaking plate *adjacent to the balance*. The plate was intended to be thermally regulated to pace any change in the balance's interior temperature. Additionally, a human body heat simulator of approximately 40 W was placed in front of the balance when the operator was not present. However, the thermally regulated soaking plate at times would be out of phase with the balance's interior temperature because the laboratory itself cycled too rapidly for the plate regulator. The effect of this sometimes out-of-phase condition may have made the results for attempts at thermal equilibrium between the weights and balance interior worse than if the soaking plate were unregulated.

The Group 2 measurements were conducted with the weights stored adjacent to the balance directly *on the concrete pier* with only thin paper under the weights for surface protection. Body heat regulator was not used during this group of measurements. These conditions reflect the normal weighing procedures for many laboratories.

The Group 3 measurements proceeded with the 500-g weight stored *inside the balance* and the Σ500-g weight *outside the balance* on a piece of foam insulation wrapped in aluminum foil. The body heat simulation was omitted and sunlight was allowed to enter the laboratory by removing an insulated

Table 22.1 Means and Standard Deviations
for the Data of Group 3

Day	$\bar{x}$, g	SD, g
1	0.000483	0.0000516
2	0.00056	0.0000802
3	0.00058	0.0000678
4	0.00045	0.0000797
5	0.00053	0.0000719
6	0.000596	0.0000858

$$(\text{Pooled SD})^2 = (5/30)[(0.0000516)^2 + (0.0000802)^2$$
$$+ (0.0000678)^2 + (0.0000797)^2$$
$$+ (0.0000719)^2 + (0.0000858)^2]$$
$$= (3.257478 \times 10^{-8})/6 = 5.429130 \times^{-9} = \bar{s}_w^2$$
$$\bar{s}_w = 0.00007368 \text{ g}$$
$$\bar{s}_w/(6)^{1/2} = 0.00003005 \text{ g} = 30.05 \text{ μg}$$
Mean of the six means = 0.0005332 g
SD of the six means = 0.0000571 g = s_T
$$s_T^2 = 3.261 \times 10^{-9} \text{ g}^2$$
$$[\bar{s}_w/(6)^{1/2}]^2 = 9.0283 \times 10^{-10} \text{ g}^2$$

$$F = (3.261 \times 10^{-9})/(9.028 \times 10^{-10})^{1/2}$$
$$F = 3.6120$$

window plug. These conditions simulate laboratories with very poor weighing conditions, and the weights were stored on a wood table adjacent to the balance.

For Group 4, all weights were stored *inside the balance* and the body heat simulator was present when the operator was not. The body heat simulator was also adjusted to a slightly higher output. As a consequence, the balance's interior temperature remained nearly constant throughout the weighings. These last conditions are the best one can achieve without robotic weight handling. Unfortunately, the interior of many balances, as in the one used here, are not large enough to store all of the weights of a decade, i.e., 500, 200, 200, 100, Σ100, and the check 100 g inside the balance along with two starting kilogram standards. This requirement is ideal for the combinational weighings encountered with the use of weighing designs.

We now illustrate the calculations using the data for Group 3. The mean of the measurements, $\bar{x}$, (500 g − Σ500 g), and the SD for the six measurements for each day are listed in Table 22.1.

The critical value of F for 5 (for s_T) and 30 [for $\bar{s}_w/(6)^{1/2}$] degrees of freedom at 0.05 probability (found in many statistics texts) is 2.5336. Therefore, F exceeds the critical value and the two SDs can be considered to be different.

One can calculate s_b, the between-time component of uncertainty:

$$s_b = \left(3.261 \times 10^{-8} - 9.028 \times 10^{-10}\right)^{1/2}$$

$$s_b = 0.00004856 \text{ g} = 48.56 \text{ μg}$$

s_b in this case is shown by the F test to be significant.

The F test results for the other 3 groups are given below:

Group 1 4.4862679
Group 2 1.3901337
Group 4 0.7373270

For Groups 1 and 3 evidence exists for a between-time component of uncertainty; for Groups 2 and 4 it does not.

22.3 Discussion

The four groups of measurements generated values of $\bar{x}$ and SD which were used to test the hypothesis that the within-series and overall SDs were not different.

If one looked at these two parameters only visually, it might not be apparent that a between-times component of uncertainty does or does not exist.

The varied conditions under which the measurements were made generated varied values of F, which for two groups indicated that between-time components of uncertainty existed. The conditions for Groups 2 and 4 were superior to those for Groups 1 and 3, and produced better precision.

For those measurements with between-time components the relevant SD is s_T, which is larger than the SD for the within-series results. The within-series SD, $\bar{s}_w$, underestimates the uncertainty; s_T is, therefore, the relevant SD.

For those measurements for which between-time components do not appear to exist, the within-series pooled SD, $\bar{s}_w$, is the relevant uncertainty.

In practice, after s_b has been determined from measurements made on a series of successive days, s_b is root sum squared with the within-series SD to calculate the relevant uncertainty for future measurements.

Periodically, measurements over time should be examined to investigate whether significant between-time components of uncertainty exist. This procedure can reveal the possible sources of systematic errors and suggest the desirability of carefully examining and, if possible, improving the measurement system.

The strength of a laboratory's standard operating procedure (SOP) being followed by every operator in the laboratory is highlighted by the example described here. Although rigorous application of a measurement SOP may yield consistent results, unrevealed bias not accounted for in the estimated uncertainty might still remain.

References

1. Taylor, J. K. and Opperman, H. V., Handbook for the Quality Assurance of Metrological Measurements, NBS Handbook 145, 1986, pp. 8.3, 8.4.
2. Eisenhart, C., Realistic evaluation of the precision and accuracy of instrument calibration systems, *J. Res. NBS*, 67C, 161, 1963.
3. Student, 1917, 415 (referred to in reference 2).
4. Airy, G. B., *On the Algebraical and Numerical Theory of Errors of Observations and the Combination of Observations*, Macmillan, Cambridge, 1861.

23

Laboratory Standard Operating Procedure and Weighing Practices

23.1 Introduction

The smaller the uncertainty issued by a mass calibration laboratory, the greater the attention that should be given to routine laboratory maintenance and certain statistical surveillance. The areas of interest are as follows:

1. Laboratory environmental control
2. Air density instrumentation
3. Balances
4. Mass standards
5. Weight cleaning
6. Statistical surveillance
7. Routine bookkeeping

Precise weighing conditions are sensitive to the Earth's vibration, air currents, air pressure change, temperature and relative humidity variation, contamination, instability of the mass standards, weight cleaning technique, and the operators. Presumably, a well-designed laboratory initially has an appropriate operating procedure. With time, all things change and without vigilance and timely corrective action estimates of uncertainty may become untrue. The following review is not a recommended standard operating procedure (SOP) but serves to illustrate the key elements for the most rigorous measurements. What is required in a particular laboratory may be less.

23.2 Environmental Controls and Instrumentation

Weighing results are influenced most by the lack of thermal equilibrium and the buoyancy correction, i.e., the measurement of the parameters of air temperature, air pressure, and relative humidity. The environmental controls must always remain consistent with the operating limits of a laboratory for these parameters. Of course, air pressure is usually not controlled but unwanted rapid variation in pressure may be caused by a structure change to a facility. This might be a new external doorway, elevator shaft, or air duct nearby. The other parameters are usually monitored continuously and compared to the laboratory history and any departure corrected immediately. Air filters are usually replaced on a fixed calendar schedule.

Similarly, laboratory background vibration in balance piers may change. Here, experience is helpful in determining what is acceptable and what is not. For example, a balance performs well as determined by standard deviation, and with the addition of some new apparatus does not. If a new radio frequency transmitter is affecting balance circuitry and therefore performance, its use should be banned. Vibration problems might not be so obvious and operators can teach themselves about vibration by observing a pool of mercury next to the balance. Images viewed by light reflected from the surface should be steady. A safer alternative is a 25-cm-diameter beaker filled with water. The operator views the laboratory by looking at the image reflected from the underside of the water–air surface (meniscus). The image should be perfectly still, but usually is not and some vibration can be tolerated. With experience, one can judge what amount of vibration is permissible.

The calibration techniques for temperature, pressure, and humidity sensors is discussed in Chapter 12. The techniques are also valid for checking calibrations performed elsewhere. Laboratories should decide what is an acceptable check interval for each of these instruments. Clearly, a mercury-in-glass thermometer does not need the attention that is required to ascertain the correctness of a thermistor thermometer. If the mercury moves freely in the capillary on the day the thermometer was manufactured, it will probably do so for the life of the thermometer unless the glass is broken. Dimensional changes will occur in the thermometer bulb with time; many years will go by before this is detectable. The only concerns with a mercury-in-glass thermometer are depth of immersion, placement, and whether the uncertainty is adequate for the measurement. Therefore, an SOP for a mercury-in-glass thermometer would be a search for broken glass upon use and a periodic inspection of placement and immersion. Should the mercury column separate, retire the thermometer. Electronic thermometers can be intercompared with others, if available, on a weekly or monthly basis. Otherwise, the techniques discussed in Chapter 12 can be used. Of course, records of all instrument checks should be maintained.

The mercury barometer is similar to the mercury thermometer. If it is not broken and if it is used properly, it is probably functioning. The electronic pressure sensors can be treated like the thermistor thermometer and compared with others on a routine basis. Otherwise, the weighing technique of Chapter 12 can be used on a routine basis.

Some types of humidity sensors are rendered inaccurate with exposure to certain chemicals, so caution while cleaning around the balance must be exercised. The thin-film type of detector usually will change with time, and dew-point meters require periodic mirror cleaning and thermistor checks. Relative humidity instruments may require a more frequent test interval than the other sensors. The dew-point gauge relies on first principles, whereas the other sensors require calibration. Recalibration above saturated salt solutions is just as convenient as intercomparing instruments and should be the first choice. This technique is described in Chapter 12. With the experience of use, one is able to best judge what the most efficient test interval should be for each type of relative humidity instrument.

23.3 Balances

Balances need cleaning with use. The balance itself may function well with a dirty pan, but the contamination is spread to objects being weighed. However, a dirty balance mechanism may affect its standard deviation and linearity. A balance can be very nonlinear and have a very good standard deviation. Besides an obvious outright failure, it is quite possible for a balance standard deviation or linearity to deteriorate slowly with time. Apart from routine weighing chamber and pan cleaning, one should not routinely disassemble the balance for cleaning unless indicated by a failed function. Extensive disassembly will often leave the balance with a larger standard deviation than prior to the cleaning. The successful disassembly and cleaning of a balance is very dependent on one's skill.

The best criterion to judge balance performance is standard deviation, which can be checked with an *F*-test (see Chapters 6 and 8). Mass comparators are least likely to be affected by small changes in linearity but this too can be checked by the use of a check standard and the *t*-test (see Chapters 6 and 8). However, a change in the check standard could be coupled to a change in the local mass standard. If a linearity problem is suspected, it can readily be checked independently (see Chapter 28). A mass calibration laboratory doing more than simple tolerance testing should consider techniques that estimate the standard deviation and calibrate a check standard with each use (see weighing designs, Chapter 8).

23.4 Mass Standards

Mass standards require periodic cleaning whether used or not, unless very effective storage techniques are employed. Failure to clean mass standards can go undetected even when check standards are used unless the check standards are likewise cleaned. Cleaning, in and of itself, can cause significant mass change to a laboratory weight. The most convenient (reliable) cleaning method is by vapor degreasing or by the Soxhlet apparatus (a variant); see Chapter 4. For rigorous mass calibrations, no other cleaning method is recommended. Wiping a 1-kg stainless steel weight with a cloth soaked with ethyl alcohol can impart mass variations of 100 μg.

It is also wise to place an antiscuff material under mass standards and periodically replace the material. Chemical filter paper is ideal for this protection and should be changed weekly or any time it is obviously contaminated with a foreign substance.

The t-test performed on the measured difference between two like standards or on check standards is the best indicator of stability. Unfortunately, two identical weights may wear away at the same rate with equal use, which is the case with the starting kilograms in the example of Chapter 8. The only action that can be taken in this case is to send these standards to higher authority for recalibration on a periodic basis. The period depends directly on use unless an accident occurs. If check standards are not used in a measurement process, then all standards must be recalibrated on a periodic basis. Stainless steel mass standards that have little or no use may be stored for many years and then cleaned just prior to use.

23.5 Weight Cleaning

As mentioned above, there are only two recommended methods for cleaning one-piece steel mass standards. These are by vapor degreasing and by the Soxhlet apparatus. Both methods can be used with methyl or ethyl alcohol and each requires surveillance. Some people object to methanol for reasons of personal safety, but it is a synthesized chemical, somewhat purer than ethanol and not very toxic in this application. With use, the solvents will become contaminated, the equipment will require cleaning, and the solvent will need replacement. The life cycle of either method will be extended if the weights are precleaned with a spray or in a bath of the same clean solvent. Most laboratories calibrate a customer's weight only once before shipping and the check standard is the usual overall check on the process. This, of course, assumes that the check standard undergoes periodic cleaning itself. The best check on the performance of these methods of cleaning is to observe the hot weight as it is removed from either apparatus. On the upper flat surfaces will be pools of rapidly evaporating solvent. This solvent should evaporate completely. If a stain appears, the apparatus should be cleaned and the solvent replaced.

23.6 Weighing

We recall the simple weighing equation:

$$Xg = \left\{ \left[S\left(1 - \rho_a/\rho_s\right) + \delta \right] \Big/ \left(1 - \rho_a/\rho_x\right) \right\} g,$$

where X is the mass of a test weight, g is the local acceleration due to gravity, S is the mass of a standard weight, ρ_a is the density of air, ρ_X is the density of the test weight, ρ_S is the density of the standard weight, and δ is the mass difference indicated by the balance.

We note the absence of all forces in the equation except for the gravitational force and the buoyant force. In fact, the equation assumes that a state of thermal equilibrium exists at the time of weighing.

Under the best circumstances, there is always some lack of equilibrium during the weighing of the standard weight and the test weight, and, therefore, the value obtained for the mass of an object will be uncertain by the error in the standard and errors due to whatever additional forces are present and unaccounted for when the above equation is used to obtain the mass value X.

The usual approach taken in weighing is to suppress unwanted forces if it is known that they may become significant. Because most weighings are made with an operator sitting at a balance, heat transmitted from the operator assures that thermal equilibrium cannot exist between the objects being weighed and the surrounding air (see Chapter 19). Additionally, the balance itself may be a source of unwanted forces due to self-heating, electric and magnetic fields, loading errors, etc. In the following are some simple interventions that can be taken for some common weighing problems.

23.7 Balance Problems

23.7.1 Balance Support

Balances must be rigidly supported. Many users are inclined to use soft vibration isolators to support the balance but doing so may allow the balance mechanism to shift with respect to the Earth's gravitational field as the balance is used. A simple test for this is to load the balance and observe the indication as a significant mass is moved from one location to another on the weighing table. The balance indication should remain constant. Similarly, motion of the operator at the weighing position should not affect the balance indication. Often concrete floors are not sufficiently rigid and couple motion of the operator to the balance causing harmful support motion. Ideally, the balance support is decoupled from the surrounding floor. Massive blocks of concrete or granite when used as balance support attenuate micromotion (vibration) that might otherwise degrade balance performance.

Nonrigid wooden structures, wooden subfloors for example, should be avoided. Supports with low resonant frequencies may also be harmful to balance performance. Simply exciting a support with a tap of the hand should not affect the balance. These disturbances often can be viewed with a large beaker of water placed next to the balance as previously mentioned. Ideally, images reflected from under the water surface will be stationary but a certain amount of motion is usually present and is not harmful. The inexperienced metrologist can quickly learn the acceptable limits of vibration using this simple technique.

23.7.2 Loading Errors

Some balances require the operator to place the object being weighed on the balance pan in such a way that the center of gravity of the object coincides with the center of the pan; doing so prevents "off-center" loading error. When this is not readily accomplished, a load-centering pan may be obtained or below-the-pan weighing by suspension wire will provide centering.

23.7.3 Electronic Forces

Balances may be affected by forces associated with electric and magnetic fields. Balances are fabricated from materials that have low magnetic susceptibility and that are electrically conductive, to avoid these forces. Glass used as windows may be coated to dissipate electric charge. Operators may install α-particle emitters to prevent electric charging. Nevertheless, it may be necessary to weigh materials such as glass, plastics, and anodized aluminum, all of which may sustain an electrical charge. The modern electronic balance contains strong magnets and coils producing magnetic and electric fields that may not always be completely shielded. There are many simple steps that can be taken to avoid these forces. Prominent ones are discussed here.

Piers used to support balances should not contain materials that may be magnetic, such as reinforcement steel. Electrical connections should be well grounded. Proximity to large electric motors and large objects of ferrous metal (milling machines, etc.) should be avoided. Relative humidity should be maintained above 25% to prevent the formation of electric charge on the balance and the object being weighed.

When materials are weighed that are susceptible to electrostatic charge, they can be weighed wrapped in aluminum foil that is otherwise kept grounded. Charge can be neutralized by bathing objects with pure methanol. During recovery from the methanol evaporation, the object is stored in a grounded Gaussian cage, a simple metal container. Handling forceps should be made of electrically conductive materials and connected to an electrical ground during use.

If it is necessary to weigh magnetized objects, one can weigh below the balance pan using a suspension wire. The area surrounding the suspended object can be shielded using an enclosure made of pure aluminum. One can test for magnetic force interaction between the balance and the object being weighed by first weighing on the balance pan and then using a suspension wire. If unaffected, the result will be the same except for a very small gravitational gradient correction of 300 μg/kg/m (see Chapter 13).

23.7.4 Convection

Convection arises from the lack of thermal equilibrium between the balance, the object being weighed, and the surrounding air. Convection and the resulting drag force on the balance are minimized by controlling the temperature of the weighing environment. A windowless interior weighing room is desirable. Ideally, a constant temperature is desirable. However, a slowly rising temperature is preferable to one that is falling or cyclic. Falling temperature disturbs the natural thermal gradient, resulting in air motion in the weighing chamber. Beyond environmental control, there are several simple measures that can be taken to avoid errors caused by convection currents in the weighing chamber.

In weighing rooms where high accuracy is desired, the environmental control system should be adjusted and maintained for minimum variation in temperature and humidity. The lighting system may be adjusted for minimum heat load, but overhead lighting does provide a stable thermal gradient in the room. Lighting should be in the in-use condition at all times and never turned off.

Balance self-heating is present in most modern electronic balances. To the extent that the manufacturer is unable to isolate the weighing chamber from the heat sources (servo motors and circuit board components) the chamber becomes an oven. This becomes a problem when objects to be weighed cannot be stored inside the weighing chamber for thermal soaking prior to weighing. Remedial actions are to relocate the offending circuit boards or to preheat the object to be weighed to the same temperature as the weighing chamber.[1]

The operator's body heat will disturb any equilibrium that exists in the balance weighing chamber in a few minutes. This operator influence can be eliminated by a robotic loading device, the balance can be operated from a distance, or the operator can be thermally shielded with an infrared reflecting drape and insulated gloves. The latter works well for critical nonroutine measurements. For some nonroutine measurements the operator can approach the balance and make a measurement with great speed and leave the area before his or her influence is felt. Fortunately, convection forces will cancel each other if the standard and the object being weighed are at the same temperature, have the same thermal conductivity, and the same surface-to-mass ratio, that is, are twins. Therefore, one can take advantage of the latter simply by thermally soaking the weights.

In many routine measurements such as laboratory weight calibration, the weights have a nearly minimum surface-to-mass ratio because they are cylinders of height equal to diameter. A sphere has the minimum ratio but is not practical in this instance. Laboratory weights are fabricated from stainless steel, which has a poor thermal conductivity but is nearly the same for every weight. Therefore, if one thermally soaks the weights in the weighing chamber sufficiently long before weighing, an equilibrium condition is approached. If the chamber is not large enough, an aluminum soaking plate adjacent to the balance can be used for weight storage. Any remaining convection force is nearly identical for like objects and is effectively canceled.

For some mass measurements where the object being weighed does not have the minimized surface-to-mass ratio, it is better to designate a look-alike standard, the mass of which can be determined under very idealized conditions. Then, in-use advantage can be taken of the cancellation property described above. An example of this standard might be a glass flask used to examine the mass of all glass flasks in a production process. Convection forces will be nearly canceled as will be the buoyant forces. The glass flask mass standard can be further enhanced by coating it with a nearly invisible layer of tin oxide that will make its surface electrically conductive and provide some immunity to electric charging.

Of course, when air temperature is changing rapidly the object being weighed will thermally lag the air and the buoyant force will not be appropriately accounted for.

23.7.5 Unnatural Pressure Variations

Nature provides a constantly changing air pressure that is accepted in weighing except for a few sophisticated measurements where pressure is controlled. This natural variation becomes too severe for weighing with the passage of low-pressure weather fronts associated with hurricanes. Otherwise, pressure changes of less than 3 mmHg/h can be tolerated. Unnatural rapid pressure excursions will cause an incorrect accounting for the buoyant force that may be significant to some measurements. These can occur when a building is pressurized by the environmental control system and door openings cause a sudden change in pressure. Rapid pressure changes may also occur near an elevator shaft when the elevator is in motion. Obviously, door openings can be limited during weighing operations and balance proximity to elevator shafts avoided.

23.7.6 External Air Motion

Air motion external to the balance is almost always present and may be coupled to the environmental control systems, the placement of doors, walls, and windows. When air currents are detrimental to balance performance, one should always look to reducing airflow if possible. Baffles and deflectors can also be installed at vent sites and near balances. If necessary, external enclosures can be placed around a balance. Often, it is unnecessary to enclose the face opposite the operator. Sometimes all that is necessary is to seal the balance case by sealing its seams with tape and improving existing balance door seals.

23.8 Statistical Surveillance

Data from the routine statistical tests, t and F, are collected over a long period of time. It is assumed, of course, that these tests are acceptable and, if failed, that the failure was determined to be from a misreading or an erroneous balance load. These data can be posted to control charts and also periodically subjected to a rigorous statistical analysis. Trends may become apparent in the values of check standard and balance standard deviations. On the basis of this analysis, adjustments may be made to the so-called accepted values or corrective action might be taken. Abrupt and repeated failure of these tests requires immediate interdiction by laboratory personnel.

23.9 Routine Bookkeeping

Most organizations have internal record-keeping requirements that go beyond quality assurance. This may be for legal liability protection or auditing purposes. These *pro forma* requirements may not be important to quality measurements but may be a strict internal requirement and must be given, at least, the minimum attention.

References

1. Schoonover, R. M. and Taylor, J. E., Some recent developments at NBS in mass measurements, *IEEE Trans. Instrum. Meas.*, IM-35, 10873, 1986.

24

Control Charts

24.1 Introduction

In this chapter, control charts[1] are developed and used to demonstrate attainment of *statistical control* of a *mass calibration process*. Extensive use is made of Ref. 1.

An appropriate check standard weight is weighed at established intervals and the measurement results are plotted on a chart. Usually, check standards are employed with every use of a weighing design (see Chapter 8). The measured values are represented on the ordinate (*Y*-axis) and the sequence of the measurements is represented on the abscissa (*X*-axis). A horizontal line is drawn representing the mean of the measured values.

Based on statistical considerations, control limits are indicated on the chart within which results of measurements are expected to be randomly distributed. The mass calibration system is considered to be in *statistical control* if subsequent measurements fall within the control limits. The use of control charts can also detect long-term drift in a measurement process.

The statistical information on which the control limits are based can be used to calculate confidence (uncertainty) limits for measurements made while the measurement system is demonstrated to be in statistical control.

24.2 Procedure

24.2.1 System Monitored

The system monitored is considered to consist of

1. The balance used
2. The standard operating procedure (SOP)
3. The laboratory environment
4. The operator
5. Any other sources that could contribute to the variance or bias of the measurement data

Any of the above that can be considered to be constant or negligible contributors to the variance may be consolidated and monitored by a single control chart. Any that cannot be so considered may require individual control charts.

For equal-arm balances, the precision is a function of the load and, in principle, a separate control chart is needed for every load tested. Control charts utilizing mass reference standards at two or three load ranges, appropriately spaced within the range of the test, are satisfactory.

24.2.2 Check Standards

1. Check standards used in high-precision calibration measurement should be stable, have the same characteristics as the objects under test, and should be dedicated to this use.
2. Check standards for lower-order calibrations should simulate the primary standards of the laboratory to the extent feasible; they should be calibrated with an accuracy equal to or better than the potentiality of the process monitored.

3. To prevent damage or deterioration of the check standards, they should be cared for as primary standards and not used as other than check standards except as starting standards when weighing designs are used in abbreviated weight sets (see Chapter 8).

4. Check standards never require recalibration as long as the process is in control and in constant use. However, long-term system drift may occur due to wear of standards and check standards.

24.2.2.1 Recommended Check Standards for Typical Test Situations

Balance	Range of Measurement	Check Standard(s)
Microbalance 20 or 30 g capacity	1 mg to 20 or 30 g	1 mg, 10 mg, 100 mg, 1 g, 10 g
100 or 160 g capacity	20 g to 100 g	100 g
1 kg capacity	100 g to 1 kg	1000 g
3 kg capacity	1 kg to 3 kg	3 kg
30 kg capacity	3 kg to 30 kg	30 kg
Large capacity balance	50 kg to 1000 kg	100 kg, 200 kg, 500 kg, 1000 kg

24.2.2.2 Establishing Control Chart Parameters

The control chart parameters consist of the following:

1. The central line, the best estimate of the mean of measurements of the check standard
2. Control and warning limits that represent probabilistic limits for the distribution of results around the central line

The parameters are evaluated on the basis of a reasonable number of initial measurements and are updated as additional measurement data are accumulated.

To establish the control chart parameters:

1. Make at least 7 and preferably at least 12 independent measurements of the check standard under the same conditions used to make routine measurements. No two measurements should be made on the same day, to ensure that the long-term standard deviation can be estimated to the extent feasible (see Chapter 22).
2. Calculate the mean, $\bar{X}$ and the estimate of the standard deviation, SD_X, in the conventional manner.
3. Establish the control chart parameters as follows:

$$
\begin{aligned}
\text{Central line} &= \bar{X} \\
\text{Upper control limit} &= \bar{X} + 2SD_X \\
\text{Lower control limit} &= \bar{X} - 2SD_X
\end{aligned}
$$

24.2.2.3 Upgrading Control Chart Parameters

When a significant amount of additional data is available or when the previously determined parameters are no longer pertinent due to changes in the system, control chart parameters should be upgraded.

Ordinarily, upgrading is merited when the amount of new data is equal to that already used to establish the parameters in use, or when at least 7 additional data points, and never more than 28 points, have been accumulated.

Upgrade the control chart parameters as follows:

1. Calculate $\bar{X}$ and SD_X for the new set of data and examine for significant differences from the former value using the t-test and the F-test, respectively (see Chapter 6).
2. If either of the new values is significantly different from the former value, the reason for the difference should be determined if possible and a decision made whether corrective actions are required.
3. If no corrective actions are required, new values of the parameters should be established using the most recent data.
4. If no significant differences are found, all data should be pooled and new values of the control chart parameters should be calculated based on all existing data. See Ku[2] for discussion of pooling.

24.2.3 Frequency of Measurement

1. The check standard should be measured every time the process is used to generate a calibration.
2. Whenever a long period of inactivity has occurred, it is good practice to make a series of calibration measurements to demonstrate that the process remains in a state of statistical control.

Control charts should be updated periodically. It is sufficient to note the *F*-test and *t*-test values for each calibration as pass–fail information. If the *F*-test and the *t*-test are failed with repeated measurements, one must look at balance performance or at the stability of the standard or the check standard.

Changes due to accumulation of dirt or wear usually occur very slowly and may be detected over long periods of time by the control chart.

24.3   Types of Control Charts

Several kinds of control charts will be found to be useful.

The simplest of the control charts is based on the repetitive measurement of a stable test object and either the results of a single measurement (*X* chart) or the means of several measurements ($\bar{X}$ chart) are plotted with respect to sequence or time of measurement.

The results should be randomly distributed about the mean ($\bar{X}$) in the case of an *X* chart and about the mean of means ($\bar{\bar{X}}$) in the case of an $\bar{X}$ chart when the measurement system is in a state of statistical control. The results should lie within defined limits, based on statistical considerations.

An $\bar{X}$ chart is preferable to an *X* chart because a change in performance will be indicated more conclusively by mean values than by individual values. This advantage must be weighed against the increased effort required to maintain the $\bar{X}$ chart; an $\bar{X}$ chart based on the average of two measurements is a good compromise, when possible.

When a property-value control chart (*X* chart or $\bar{X}$ chart) and a precision chart (SD chart, for example) are maintained in parallel, diagnosis of out-of-control situations, as due to imprecision or bias, and the identification of assignable causes for the out-of-contol situations are facilitated.

24.3.1   *X* Control Chart

To generate an *X* chart, single measurements are made of a stable test object, at least a few times preferably on different days (if a measurement system is to be maintained in a state of statistical control over a period of time).

The results are plotted sequentially and the measurement process is considered to be in control when the results are randomly distributed within limits as defined below.

24.3.2   Initial Control Limits, *X* Chart

24.3.2.1 Central Line, $\bar{X}$

Measure the test object on at least 12 occasions (recommended) but no more frequently than daily. The initial central line is the mean of the *n* measurements, X_i, for $i = 1, 2, \ldots, n$.

$$\text{Central line, } \bar{X} = \Sigma X_i / n$$

Calculate SD_X, an estimate of the long-term standard deviation of *X* in the usual manner.
Calculate the upper and lower control limits as:

$$\text{UCL} = \bar{X} + 2\text{SD}_x$$

$$LCL = \bar{X} - 2\text{SD}_x$$

If the system is in a state of statistical control, approximately 95% of the plotted points should fall within the control limits (LCL and UCL).

24.3.3 $\bar{X}$ Control Chart

Measure a test object in replicate, periodically; these duplicate measurements should be made at least once on each test day or at least monthly, whichever is more frequent. The means of the measurements, $\bar{X}$, are plotted sequentially.

When the plotted points are randomly distributed within the control limits, the system is judged to be in a state of statistical control.

24.3.4 Initial Control Limits, $\bar{X}$ Chart

24.3.4.1 Central Line, $\bar{\bar{X}}$

Measure the test object on an least 12 occasions (recommended) but no more frequently than daily. The initial central line is the mean of n duplicate measurements, $\bar{X}$, for i = 1, 2, ..., n.

$$\text{Central line, } \bar{\bar{X}} = \Sigma \bar{X}_i / n$$

Calculate $SD_{\bar{x}}$, an estimate of the long-term standard deviation of $\bar{X}$ in the usual manner.

If the long-term standard deviation exceeds the short-term standard deviation by more than a factor of 2, the quality control should be improved to decrease the long-term standard deviation to more acceptable values.

24.3.4.2   Control Limits

$$\text{UCL} = \bar{\bar{X}} + 2SD_{\bar{x}}$$
$$\text{LCL} = \bar{\bar{X}} - 2SD_{\bar{x}}$$

If the system is in a state of statistical control, approximately 95% of the plotted points should fall within the control limits.

24.4 Updating Control Charts

The control limits may be updated after additional control data are accumulated (at least as much as originally used). To see whether the second set of data for $\bar{X}$ or $\bar{\bar{X}}$ is significantly different from the first set, a *t*-test is performed.

If the second set is not significantly different, all data may be combined to obtain a new and more robust estimate of $\bar{X}$ or $\bar{\bar{X}}$. If the second set is significantly different, only the second set should be used in revising the control chart. Investigation should determine the cause of the offset.

The standard deviation, SD, for the second set of determinations should be compared with the first estimate using the *F*-test to determine whether to pool it with the first estimate or to use it separately in setting new control limits.

A smaller value of SD may result from improvement of the precision as a result of a learning experience, for example, or a better-performing balance is obtained.

A larger value of SD could be due to to an original poor estimate of the standard deviation of the measurement process, or to an increase in imprecision resulting from an assignable cause, for example, balance deterioration.

In either case, the reason for the smaller or larger value should be ascertained.

All of the values of R, the difference between the largest and smallest values (measured values of X), may be combined to obtain an updated estimate of $\bar{R}$, from which updated control limits can be computed, if the values of R show no systematic trends and if $\bar{R}$ has not changed significantly.

Judgment of the significance of apparent changes in $\bar{R}$ can be made by computing the corresponding values of SD and then conducting an F-test.

24.5 Interpretation of Control Chart Tests

If the measurement system is in a state of statistical control, plotted points should be randomly distributed within the control limits.

If repeated measurements lie outside the control limits, repeat the test. If repeated test measurements fall outside the control limits, corrective action must be taken.

In well-controlled processes, failures of F-tests and t-tests are usually due to placement of incorrect weights on the balance pan. To evaluate the uncertainty of measurements, control charts may be used in some cases.

When an appropriate control chart is maintained, an X chart or an $\bar{X}$ chart may be used to evaluate bias and to document the standard deviation of the measurement process. Then, the values for SD on which control limits are based may be used in calculations of confidence limits for measurement values. Control charts may reveal periodic or seasonal shifts that are manifestations of between-time components (see Chapter 22).

References

1. Taylor, J. K. and Opperman, H. V., Handbook for the Quality Assurance of Metrological Measurements, SOP No. 9, Recommended Standard Operations Procedure for Control Charts for Calibration of Mass Standards, NBS Handbook 145, November 1986.
2. Ku, H. S., Statistical concepts in metrology, in *Handbook of Industrial Metrology*, American Society of Tool and Manufacturing Engineers, Prentice-Hall, Englewood Cliffs, NJ, 1967, chap. 3.

25

Tolerance Testing
of Mass Standards

25.1 Introduction

In this chapter, SOP (standard operating procedure) No. 8, Recommended Standard Operations Procedure for Tolerance Testing of Mass Standards by Modified Substitution, of National Bureau of Standards Handbook 145, "Handbook for the Quality Assurance of Metrological Measurements,"[1] is followed closely.

The SOP describes procedures to be followed for determining whether or not mass standards are within the tolerances specified for a particular class of weights as set forth in OIML R111 or similar specifications. It is suggested that these procedures be used with conventional results of weighing in air rather than mass (see Chapter 15).

25.2 Prerequisites

1. Mass standards must be available with calibration certificates traceable to a national standards laboratory.
2. The balance used must be in good operating condition.
3. The operator must be experienced in precision weighing techniques.

25.3 Methodology

25.3.1 Scope, Precision, Accuracy

The tolerance testing method is applicable to all mass tolerance testing. If the uncertainty of the mass measurement is no more than 0.1 of the permissible tolerance of the mass standard tested, the precision of the tolerance determination will not be a factor.

25.3.2 Summary

A modified substitution procedure is used to compare the mass to be tested with a calibrated reference mass standard.

The reference standard is placed on a balance (a single-pan, an equal-arm, or a full-electronic balance may be used) to obtain a convenient reference point and a sensitivity test is conducted.

The error (departure from nominal value) of the weight tested is determined by comparing the balance reading for it to the balance reading for the reference standard.

A weight is considered to be within tolerance when its error does not exceed the tolerance established for the particular class of weight.

25.4 Apparatus/Equipment

1. Single-pan, equal-arm, or full-electronic balance with sufficient capacity for the load tested, and with readability equal to or less than 0.1 of the acceptable tolerance tested.
2. Mass standards calibrated with an accuracy of 0.1 or less than the tolerance tested. The calibration must be traceable to a national standards laboratory.
3. Calibrated sensitivity weights.
4. Counterweights, *T*, uncalibrated, of approximately the same mass as the standard weights (for Option C, below).

25.5 Procedure — Option A, Use of Single-Pan Mechanical Balance

1. Select a reference standard of the same nominal value as the weight under test. Place the standard on the balance pan. Adjust the optical scale reading to approximately midscale and record the reading.
2. Add a sensitivity weight equal in mass to approximately 0.25 of full scale reading, and record the reading.
3. Calculate the value of a scale division. If it is within ±2% of the nominal value (usual case), the nominal value of a division can be used for tolerance testing.
4. Remove the sensitivity weight and adjust the optical scale to account for the corrected value of the standard used.

Example: If the nominal range of the optical scale is 100 mg and the reference standard has a correction of –2.5 mg, the optical scale is adjusted to read 47.5 mg when the standard is on the pan. Under this condition, the reading of 50.0 mg represents the nominal mass of the standard.

5. Remove the standard.
6. Place weight to be tested on the balance pan and read the optical scale. The error in the weight is the amount by which the indication deviates from the midscale reading. If the weight indication is more than the midscale value, the weight is heavy by the indicated difference; if the indication is less than the midscale value, the weight is light. Record the error.
7. After several (no more than ten) weights have been tested, put the standard on the balance pan and record the reading. The difference between this indication and the previous one for the standard indicates a balance drift. This drift will ordinarily be very small. If the drift ever exceeds 25% of the tolerance applicable to the weights under test or affects a measurement result to the extent that a weight may be out of tolerance, the measurement should be repeated and more frequent checks of the standard should be made.
8. Readjust the optical scale at any time that a significant difference is observed when rechecking a standard.

25.6 Procedure — Option B, Use of Full-Electronic Balance

1. Select a reference standard of the same nominal value as the weight under test, and place the standard on the pan. If the standard is light, add small calibrated weights with the standard, equivalent to the correction for the standard. Record the reading. If the weight is heavy, do nothing at this point but follow instructions of item 5 below. Zero the balance so errors can be read directly from the balance indications.
2. Add a calibrated sensitivity weight, sw, of mass greater than or equal to two times the tolerance; and record the reading. Verify whether the nominal scale division is within ±2% of the actual value. In this case, the nominal value of the scale division may be used.
3. Remove the sensitivity weight and zero the balance.

4. Remove all weights from the balance pan.
5. Place the weight to be tested on the balance pan. If (and only if) the standard used is heavy, add small calibrated weights equal to the correction required for the standard and carry these along with every weight tested. Record the balance reading, which indicates, directly, the error of the weight tested. If the reading is positive, the weight is heavy by the indicated amount; if the reading is negative, the weight is light by that amount.
6. After several (no more than ten) weights have been tested, recheck the zero as in item 3 above and record the reading. The difference between this indication and the previous one for the standard indicates a balance drift. This drift will normally be small. If the drift should ever exceed 25% of the tolerance applicable to the weight under test or affect a measurement result to the extent that the weight may be out of tolerance, the measurement should be repeated and more frequent checks of the standard should be made.
7. Readjust the zero at any time that a significant difference is observed when rechecking a standard.

25.7   Procedure — Option C, Use of Equal-Arm Balance

1. Place a reference standard of the same nominal value as the weight under test on the left balance pan, together with small calibrated weights equal in mass to the correction required for the standard if it is light. If the standard is heavy, do nothing further at this point but follow instructions in item 4 below. Add sufficient counterweights to the right balance pan to obtain a sum of turning points of approximately midscale value. If necessary, number the graduated scale such that adding a weight to the left pan will increase the balance reading (see Chapter 5). Record the rest point as O_1.
2. Add an appropriate calibrated sensitivity weight to the left balance pan and record the rest point as O_2. Calculate the sensitivity:

$$\text{Sensitivity} = \text{AM}_{sw}\big/\big(O_2 - O_1\big),$$

where AM_{sw} is the apparent mass of the sensitivity weight.
3. Remove all weights from the left balance pan.
4. Place the weight to be tested on the left balance pan. If the standard used in item 1 above was heavy, add small correction weights, equivalent to the correction required for the standard, to the left balance pan. Add small calibrated weights as required to the left balance pan or the right balance pan to obtain an approximate balance and record the rest point as O_3.
5. Calculate the error of the weight tested, as follows:
   If the added weight, AW, is placed on the left balance pan,

$$\text{ERROR} = \big(O_3 - O_1\big)\big[\text{AM}_{sw}\big/\big(O_2 - O_1\big)\big] - \text{AW}$$

   If the added weight, AW, is placed on the right balance pan,

$$\text{ERROR} = \big(O_3 - O_1\big)\big[\text{AM}_{sw}\big/\big(O_2 - O_1\big)\big] + \text{AW}$$

6. After several weights (no more than ten) have been tested, recheck rest point O_1, as described in item 1 above. Only a small difference should be observed; if this difference is significant, use a new value for O_1 in subsequent measurements. If this change should ever exceed 25% of the applicable tolerance or affect a measurement result to the extent that a weight may be out of tolerance, the measurement should be repeated and more checks of the standard should be made.

25.8   Tolerance Evaluation

1. Compare the error in the weight tested with the tolerance for the class of weights to which it belongs. If the error is numerically smaller than the tolerance, the weight is considered to be within tolerance. If the error is larger than permissible, the weight is considered to be outside of tolerance and appropriate action should be taken. It is recommended that weights with errors within 10% of the tolerance limit be adjusted.

Reference

1. Taylor, J. K. and Oppermann, H. V., Handbook for the Quality Assurance of Metrological Measurements, NBS Handbook 145, SOP No. 8, 1986.

26

Surveillance Testing

26.1 Introduction

Almer and Keller[1] published an exhaustive presentation of surveillance test procedures. Later, Jaeger and Davis[2] included some of the Almer and Keller material in their "A Primer for Mass Metrology." The present chapter includes material from these two earlier publications.

Surveillance testing looks for signs that one or more members of a weight set may have changed value since the latest calibration. The basic idea is to ensure the self-consistency of the weight set.

For example, the 100-g weight can be checked against the sum of 20-g, 30-g, and 50-g weights to see whether the difference is within expected limits.

It is also advantageous, when possible, to compare one weight of the set (usually the largest) against an independent standard, S. This comparison establishes whether the entire weight set has undergone a change, even though there may still be self-consistency within the set.

The basic motivations for surveillance testing are as follows:

1. To verify the values of newly calibrated weights
2. To establish the stability of a new weight set
3. To determine whether an accident (such as being dropped on the floor) has changed the value of a weight or weights involved

It may not be necessary to make buoyancy corrections in surveillance testing, since weights within a set are fabricated from the same alloy, and the surveillance test is carried out at an air density not too different from the calibration air densities. The magnitude of appropriate buoyancy corrections should be compared to the surveillance limts to determine whether it is worthwhile to make the buoyancy corrections.

26.2 Types of Surveillance Tests

Two distinct types of surveillance test are referred to as Type I and Type II tests. Type I is presented here in detail. The reader is referred to the references for detailed presentation of Type II tests. The Type II technique is similar to Type I technique except that a 3–1 weighing design is used (see Chapter 8).

26.3 Type I Test[2]

The object of the Type I surveillance test is to intercompare all weights in a set using a minimum number of steps. It is preferable that one standard weight, designated here S, be a member of the set. S should have a mass as large as that of the member of the set with the largest mass, or as large as convenient.

We illustrate here surveillance testing using a Type I surveillance test.

For the comparison measurement:

1. Start out with the largest weight and compare it with a summation of weights next in magnitude such that the sum is equivalent to the largest weight.
2. A weight from the first summation is compared with a lower summation.
3. Continue with the process until all the weights in the weight set have been used.

If a standard S has been included in the set, S should be compared first with the weight of mass nominally equal to that of S.

Example

The set consists of a standard weight S of *nominal* mass 100 g and weights of *nominal* mass 100, 50, 30, 20, 10, 5, 3, 2, 1, 0.5, 0.3, 0.2, 0.1, 0.05, 0.03, 0.02, 0.01, 0.005, 0.003, 0.002, and 0.001 g, which are referred to here as M_1 to M_{21}, respectively.

For the measurement sequence, first the standard weight S is compared with M_1 of *nominal* mass 100 g. Then, one works down to M_{21} of nominal mass 1 mg (0.001 g) in a minimum number of steps to include all of the weights.

The designation M_i refers to the nominal value of the ith weight and M_{iT} refers to the true mass of the ith weight. The symbol $'$ refers to a summation of weights in the weight set and δ is measured mass difference.

1st Measurement:

$$M_{1T} - S = \delta_1, \text{ where } M_{1T} = 100 \text{ g, } S = 100 \text{ g, nominally}$$

2nd Measurement:

$$M_{1T} - M_{2'T} = \delta_2, \text{ where } M_{2'T} = \left(M_2 + M_3 + M_4\right)$$
$$= \left(50 + 30 + 20\right) \text{ g}$$
$$= 100 \text{ g}$$

In words, for illustration, the second measurement is of the difference in true mass of the weight M_1 and the true mass of the summation of the weights M_2, M_3, and M_4 of nomimal masses 50, 30, and 20 g. This difference is designated δ_2. The sum of the nominal masses of the three weights is designated M_2'.

3rd Measurement:

$$M_{4T} - M_{3'T} = \delta_3, \text{ where } M_{3'T} = \left(M_5 + M_6 + M_7 + M_8\right)$$
$$= \left(10 + 5 + 3 + 2\right) \text{ g}$$
$$= 20 \text{ g}$$

4th Measurement:

$$M_{8T} - M_{4'T} = \delta_4, \text{ where } M_{4'T} = \left(M_9 + M_{10} + M_{11} + M_{12}\right)$$
$$= \left(1 + 0.5 + 0.3 + 0.2\right) \text{ g}$$
$$= 2 \text{ g}$$

5th Measurement:

$$M_{12T} - M_{5'T} = \delta_5, \text{ where } M_{5'T} = \left(M_{13} + M_{14} + M_{15} + M_{16}\right)$$

$$= \left(0.1 + 0.05 + 0.03 + 0.02\right) g$$

$$= 0.2 \text{ g}$$

6th Measurement:

$$M_{16T} - M_{6'T} = \delta_6, \text{ where } M_{6'T} = \left(M_{17} + M_{18} + M_{19} + M_{20}\right)$$

$$= \left(0.01 + 0.005 + 0.003 + 0.002\right) g$$

$$= 0.02 \text{ g}$$

7th Measurement:

$$M_{19T} - M_{7'T} = \delta_7, \text{ where } M_{7'T} = \left(M_{20} + M_{21}\right)$$

$$= \left(0.002 + 0.001\right) g$$

$$= 0.003 \text{ g}$$

We note that there is truncation in the seventh measurement as there remain only two weights.

The mass differences, δ_1 through δ_7, can now be compared with the differences calculated using the previously known or accepted values of the masses of the weights. These new differences should be plotted and compared chronologically with similar differences from previous sets of measurements.

The weight set is then monitored using predetermined uncertainty limits. A similar surveillance test can be devised for the 5,2,2,1,Σ1 weight sequence (see Chapter 8).

26.4   Surveillance Limits

After any necessary buoyancy corrections have been made, the measured mass differences, δ_i, will not in general be exactly equal to those values, δ_{ic}, calculated from the mass values given on the calibration certificate for the weight set.

To judge whether the differences, $\delta_{ic} - \delta_i$, could be due to combinations of calibration uncertainties for the weight set and the random uncertainty of the measurements that is associated with the balance used to make the surveillance measurements, surveillance limits associated with each δ_i must be established.

We now designate the surveillance limit (SL), to be

$$\left(\text{SL}\right)\left(\text{at } k = 2, \text{ see Chapter 7}\right) = 2\left[U^2 + \left(\text{SD}\right)^2\right]^{1/2},$$

where U is the systematic uncertainty as determined from the calibration report, and SD is the limit to random uncertainty. The SD used is the estimate of the standard deviation of the balance being used for the surveillance measurements; this uncertainty is known from many measurements made previously on the balance.

One can expect that any value of δ_i should fall within $\delta_{ic} \pm \text{(SL)}$.

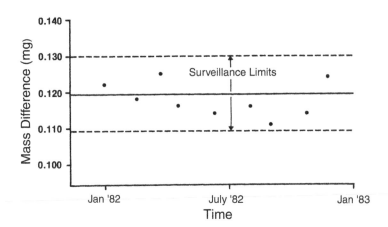

FIGURE 26.1 Example of a surveillance chart.

For surveillance purposes, U can be approximated by the root-sum-square of the individually reported calibration uncertainties:

$$U = \left(\Sigma U_i^2 \right)^{1/2}.$$

The value of 3 SD gives a 99.7% confidence interval for the random uncertainty.

26.5   Surveillance Charts

After surveillance limits have been established, surveillance charts can be constructed; one chart for each δ is preferable. Figure 26.1 is an example of a surveillance chart given by Jaeger and Davis.[2]

Values of δ_4 are plotted against time. The solid horizontal line is the value of δ calculated from the most recent calibration report for the weight set. The dashed horizontal lines are the surveillance limits.

In this example, all the values of δ_4 determined throughout the time period fall within the surveillance limits. It can be assumed that the weights represented by δ_4 have not changed significantly throughout the time period.

For each δ_i, a surveillance chart would be similarly constructed and studied to detect values, if any, that fall outside the surveillance limits. If any fall outside, one would deduce which of the weights in the weight set had changed significantly between successive surveillance measurement sequences.

26.6   Identification of Weights Whose Mass Has Changed

If a measurement, δ_i, has changed between surveillance measurement sequences, it is necessary to determine which individual weights have changed in mass. The example of Jaeger and Davis[2] is used to illustrate the determination procedure.

For the surveillance measurement sequences above, a case in which the value of δ_4 was outside its surveillance limits but the value of δ_3 was inside its surveillance limits is investigated. It is known, therefore, that there has been some change (or changes) in mass in the subset of weights $M_4 = 1$ g + 0.5 g + 0.3 g + 0.2 g. These four nominal masses correspond to the weights M_9 through M_{12}, respectively.

Three measurements are now made:

1st Measurement:

$$M_{9T} - M_{4T''} = \delta_{4''}, \text{ where } M_{4T''} = M_{10} + M_{11} + M_{12}$$

2nd Measurement:

$$M_{10T} - M_{5T''} = \delta_{5''}, \text{ where } M_{5T''} = M_{11} + M_{12}$$

3rd Measurement:

$$M_{11T} - M_{6T''} = \delta_{6''}, \text{ where } M_{6T''} = M_{11} + M_{13}$$

The surveillance limits for $\delta_{4''}$, $\delta_{5''}$, and $\delta_{6''}$ are then calculated and the surveillance plots will be examined.

1. If $\delta_{4''}$ lies outside the surveillance limits but $\delta_{5''}$ and $\delta_{6''}$ do not, it is probable that the mass of the M_9 weight (1 g) has changed since its last calibration.
2. If $\delta_{4''}$ and $\delta_{5''}$ lie outside the surveillance limits by opposite amounts and $\delta_{6''}$ is inside the limits, it is probable that the mass of the M_{10} weight (0.5 g) has changed since its last calibration.
3. If $\delta_{4''}$ and $\delta_{5''}$ lie outside the surveillance limits by the same amount and $\delta_{6''}$ is outside the limits by the same amount and in the opposite direction, it is probable that the mass of the M_{11} weight (0.3 g) has changed since its last calibration.
4. If $\delta_{4''}$, $\delta_{5''}$, and $\delta_{6''}$ all lie outside the surveillance limits by about the same amount in the same direction, it is probable that the mass of the M_{12} (0.2 g) has changed since its last calibration.

If none of above four conditions above is met, it is probable that the mass of more than one of weights in the weight set has changed. Ref. 1 presents a more thorough analysis of surveillance and should be consulted in this case and for additional information.

References

1. Almer, H. E. and Keller, J., Surveillance Test Procedures, NBSIR 76-999, May 1977.
2. Jaeger, K. B. and Davis, R. S., A Primer for Mass Metrology, NBS Special Publication 700-1, November 1984.

27

The Mass Unit Disseminated to Surrogate Laboratories Using the NIST Portable Mass Calibration Package

27.1 Introduction

The surrogate laboratories project began[1] with the premise that a National Institute of Standards and Technology (NIST)-certified calibration could be performed by the user in the user's laboratory. With this goal in mind, a very informal low-budget project was undertaken to expose the technical difficulties that lay in the way.

Here, the highlights of earlier work[2,3] are briefly touched on. Data from the third package are presented, with discussion, that it is believed adequately demonstrates the successful achievement of the initial goal.

Most importantly, the chapter conclusion discusses what are believed to be the benefits and weaknesses of the original premise and comments are made on other possible uses of the method.

Initially, the project looked primarily at the mass assignment to 1-kg artifacts in the participant laboratories. The results of earlier demonstrations[4] indicated that this would not be a problem for ordinary one-piece laboratory weights made of stainless steel with densities of about 8 g/cm³.

However, these demonstrations expose difficulties when large buoyant forces are encountered, as in the comparison of kilograms with large volume differences, i.e., aluminum, stainless steel, and tantalum, as well as problems related to surface-dependent thermal effects and thermal inertia.

Other work undertaken by NIST[5] and since verified by others[6] suggests that the lack of thermal equilibrium during the measurements can present a serious obstacle in this undertaking, not necessarily with the comparison of identical single artifacts but for the aggregates that occur in combinational weighing. Unfortunately, surface-dependent thermal effects are not readily separable from the large buoyant effects mentioned above. After completing the initial round-robin, it was obvious that some of the participants needed assistance with data reduction and nearly all lacked adequate accuracy in the pressure measurement, without which one cannot make reliable buoyancy corrections.

In addition, taking into account the effects of temperature and humidity would require additional support. In essence, it was decided to supply the participants with a computer, software, thermometer, barometer, humidity sensor, mass standards, documentation, and other items. The participant was required to provide a laboratory, the mass comparator, and a skilled operator. It was believed that this was a realistic approach to overcome apparent inadequacies in the equipment and staff experience of the participant laboratories.

With support for the project very limited, the California State Laboratory was chosen as the only participant, besides NIST, in the third and last phase. The choice was based on the technical support and the certainty that a strong astute critical review of the project would be forthcoming from California.

Early on, it was realized that this approach to calibration would also be useful in a rigorous examination of mass laboratories, and for the training of their staff. These last two issues are addressed later in this chapter.

27.2   Review

The first round-robin involved a normal one-piece laboratory kilogram weight and special 1-kg artifacts with exaggerated surfaces and volumes. Additionally, the package supplied a description of the required measurements and an appropriate data-reduction algorithm.

The results of these measurements were disappointing, especially when judged in terms of experience with a mass measurement assurance program (MAP), which for the most part has been very good.

Two very striking characteristics became evident from the experience with the first package; i.e., not all laboratories were qualified from a staff and equipment point of view, and others lacked adequate environmental control to perform measurement at the 0.025 part per million (ppm) level of precision.

High levels of precision can be achieved in very poor laboratories, but systematic errors cause large offsets in the results.

In summary, the inability to determine air density accurately, very poor laboratory temperature control, and low volume, high velocity airflow environmental control systems accounted for most of the non-operator difficulties.

Items of interest addressed to improve the second mass package were automatic computer data acquisition, on-site computer data reduction, accurate instrumentation for measuring the air density parameters of temperature, barometric pressure, and relative humidity, and active temperature control of the artifacts.

Changes in the artifacts included the addition of a right circular cylinder with a high aspect ratio (height/diameter) for examining surface-dependent thermal effects.

In general, the second circulation of the package was successful but did not completely avoid the difficulties encountered in poor thermal environments. That is, comparing like geometries is inherently more accurate, in a normal laboratory environment, than comparing the unlike geometries such as those encountered in weighing weight summations.

The other noteworthy handicaps encountered were inadequate maintenance of the State kilograms and the absence of precise knowledge of their density.

The lack of periodic ties to NIST by some of the participants made calibrations in terms of the State standards pointless. This problem was solved in the third package by including two 1-kg mass standards with an uncertainty on the sum of about 0.06 ppm, i.e., 60 µg.

The importance of a small starting uncertainty at the 1-kg level is discussed in greater detail later in this chapter. We alert the reader at this time that the above uncertainty is relative to the U.S. national prototype kilograms (K_{20}, K_4) and does not include any drift in the defining artifact maintained at the Bureau International des Poids et Mesures (BIPM), the so-called grand mass.

This aspect of the uncertainty is likewise discussed later.

27.3   The Third Package

After completion of the second package, it appeared certain that a complete calibration of a weight set from 1 kg to 1 mg could easily be accomplished with the NIST mass calibration software, a pair of starting 1-kg standards, and appropriate check standards.

It was decided to limit the demonstration to NIST and the State of California. The calibration of the starting kilograms and especially the check standards is very time-consuming and entails high cost.

Additional labor was expended by including a laptop computer and writing a menu-driven version of the NIST mass code. A menu-driven code is not only unnecessary but also very inefficient for skilled operators to use and entailed additional and unnecessary labor.

In summary, the State of California was provided with everything required for an *in situ* calibration of its primary mass standard in the range mentioned earlier except for the balances and the skilled operator. California not only provided the balances but a good laboratory environment for their use.

Changes to the package for the third round were a better computer, more-generalized software, added standards, and the elimination of the active weight "soaking" (temperature) plate and its replacement with a passive plate.

Additionally, density measurements were performed on the California kilograms.

27.4 Hardware and Software

In hardware content, the package remained unchanged in function other than the discontinued use of the active soaking plate with the temperature controller. The plate was used without the active electronic circuitry.

Other hardware changes did not change the measurement function but improved the operational reliability. The most substantial change was in the computer and software.

Any computer that permits the operation of the software and the automatic data acquisition is adequate, and it would not be useful to describe it here. However, much effort was expended in developing a documented menu-driven version of the so-called mass code software.

At that juncture, it was felt that it would require less time to use a menu structure than to teach others the details of the software. This decision was made before the project was limited to just NIST and the State of California.

The software itself was developed to limit the options of the statistical analysis program (mass code) to only the geometric progression of the state standards, $5,3,2,1,\Sigma1$. This was accomplished with the menu-controlled software and an editor.

The operator is directed through the data collection and analysis, including the report of calibration, and detailed knowledge of the underlying software was not a requirement. Sufficient information for setting up the hardware and initiating the software control of the measurement process was provided in written documentation.

The program gave the operator flexibility in halting and continuing the measurements at junctures not detrimental to the process. Without sufficient computer skills and special knowlege, the operator would not be able to deviate from measurement design. The software did not prevent deviation for those skilled in the measurement and computers, but no information was provided in the documentation on how to proceed.

27.5 The Measurements

The measurement sequence was three complete calibrations performed in Sacramento, CA and three identical calibrations performed at NIST-Gaithersburg (MD) using the same starting kilograms at both locations. These measurements were followed with check measurements of the starting kilograms to ensure a secure tie to the definition of the mass unit.

The set of weights, including the starting standards T_1 and T_2, the check standards, and the California set are completely described by the following progression:

$$1 \text{ kg } T_1, \ 1 \text{ kg } T_2, \ 1 \text{ kg } S_1, \ 1 \text{ kg } S_2, \ \Sigma1 \text{ kg}, \ \Sigma100 \text{ g}, \ 100 \text{ g}, \ \text{chk } 100 \text{ g},$$

$$\Sigma10 \text{ g}, \ 10 \text{ g}, \ \text{chk } 10 \text{ g}, \ \Sigma1 \text{ g}, \ 1 \text{ g}, \ \text{chk } 1 \text{ g}, \ \Sigma100 \text{ mg}, \ 100 \text{ mg},$$

$$\text{chk } 100 \text{ mg}, \ \Sigma10 \text{ mg}, \ 10 \text{ mg}, \ \text{chk } 10 \text{ mg}, \ 1_1 \text{ mg}, \ 1_2 \text{ mg}, \ \text{chk } 1 \text{ mg}.$$

In this notation S_1 and S_2 are the California state kilograms, the prefix "chk" indicates a check standard supplied by NIST and sum (Σ) indicates a weight summation with a 5, 3, 2 progression. That is, there is a summation of weights for each decade shown, beginning with the sum, 500 g + 300 g + 200 g, and ending with the sum, 5 mg + 3 mg + 2 mg.

Therefore, a complete set calibration is accomplished using one series known as "five 1's" and repeating a second series, referred to as a "5,3,2,1,1,1," six times for a total of seven series.

The series designs are as follows:

Five 1's Weights					5,3,2,1,1,1 Weights					
1_1,	1_2,	1_3,	1_4,	sum 1	5,	3,	2,	1,	chk 1,	sum 1
+	–				+	–	–	+		–
+		–			+	–	–		+	–
+			–		+	–	–	–		+
+				–	+	–	–			
	+	–			+	–	–	–		–
	+		–			+	–	+	–		–
	+			–		+	–	–	+		–
		+	–			+	–	–	–		+
		+		–		+	–	–			–
			+	–			+		–		–

The plus and minus signs in the above designs indicate the force differences to be measured, usually by a balance. Each line of the design gives the weight combinations to be compared; i.e., all of the + signs are weights of one load and all of the – signs are weights of the other load.

There are various schemes for making the comparisons and in this particular case four balances of different capacities are used in an effort to minimize the balance error contribution to the uncertainty assigned by the process.

The reader who is unfamiliar with the method will find many fine hours of entertainment by reviewing the literature. However, Chapter 8 provides the reader with an introduction to the use of weighing designs.

One complete calibration of the weight set requires 76 different weighings and about 7 h of time in front of the balances. The air density data collection was automatic as were some of the balance observations.

Thus, after the last balance observation was taken, the computer provided, via hard copy and magnetic record, a complete data analysis as part of the report of calibration. The time required for a complete calibration is initially somewhat longer as many pieces of data must be entered into the computer.

Additional time is allotted after moving the weights between the balances to attain thermal equilibrium.

The process cannot be shortened without great risk of increasing the magnitude of systematic errors. As mentioned earlier, three complete calibrations of the weight set were performed by each participant, and therefore one can expect to take statistical advantage of the multiple values in assigning the final uncertainties.

The data were accumulated in a deliberate fashion, i.e., by completing a full set calibration before beginning another. This format permits all the random effects that one may not perceive to influence the measurement process. These effects, if any, are then more apt to be observable, i.e., between-time components.

Table 27.1 presents the mass assignments, in grams, to each weight in the set for all six calibrations. The calibration data are for the California mass standards from 1 kg to 1 mg. Three complete calibrations were performed at each location, all based on the same pair of starting kilogram standards.

27.6 Data

First, there is nothing remarkable about the data given in Table 27.1. The results were expected based upon the many years of experience the NIST Mass Group had with round-robins since the early 1960s.

Table 27.1 Calibration Data for the California Mass Standards from 1 kg to 1 mg; Three Complete Calibrations Performed at Each Location, All Based on Same Pair of Starting Kilogram Standards

Nom.	NIST 1	NIST 2	NIST 3	CAL 1	CAL 2	CAL 3
1 kg$_1$	1000.00820474	0.00827267	0.00818025	0.00830024	0.00828930	0.00826765
1 kg$_2$	1000.00720165	0.00727038	0.00716522	0.00723753	0.00718703	0.00720483
500 g	500.00227748	0.00231827	0.00226679	0.00229987	0.00228465	0.00228174
300 g	300.00150764	0.00154964	0.00150565	0.00156125	0.00154915	0.00155505
200 g	200.00088066	0.00089292	0.00086855	0.00090348	0.00091583	0.00090483
100 g	99.99976777	0.99975361	0.99979796	0.99979664	0.99975488	0.99976121
50 g	50.00103108	0.00104375	0.00102987	0.00105582	0.00104387	0.00102251
30 g	30.00117527	0.00117647	0.00116796	0.00117343	0.00117539	0.00117178
20 g	20.00010863	0.00013458	0.00012533	0.00014307	0.00012077	0.00012500
10 g	10.00010232	0.00011101	0.00010625	0.00011162	0.00010282	0.00011724
5 g	5.00003440	0.00003872	0.00002884	0.00003168	0.00003560	0.00003177
3 g	3.00003972	0.00003842	0.00003708	0.00003304	0.00003504	0.00003400
2 g	1.99996428	0.99996261	0.99995928	0.99995932	0.99996079	0.99996016
1 g	1.00004155	0.00003925	0.00003784	0.00004023	0.00004216	0.00003938
500 mg	0.50001068	0.50001078	0.50000984	0.50000900	0.50000810	0.50000761
300 mg	0.30002103	0.30002025	0.30002117	0.30001979	0.30001940	0.30002005
200 mg	0.20001977	0.20001934	0.20001902	0.20001880	0.20001942	0.20001942
100 mg	0.10000850	0.10000912	0.10000933	0.10000870	0.10000855	0.10000792
50 mg	0.05001219	0.05001088	0.05001241	0.05001278	0.05001246	0.05001213
30 mg	0.02998942	0.02999127	0.02999016	0.02998985	0.02998999	0.02999007
20 mg	0.01996228	0.01996226	0.01996181	0.01996181	0.01996193	0.01996178
10 mg	0.00997522	0.00997565	0.00997527	0.00997534	0.00997546	0.00997491
5 mg	0.00500106	0.00500146	0.00500124	0.00500155	0.00500146	0.00500148
3 mg	0.00300325	0.00300336	0.00300338	0.00300323	0.00300346	0.00300362
2 mg	0.00200291	0.00200261	0.00200313	0.00200313	0.00200289	0.00200298
1 mg$_1$	0.00100149	0.00100175	0.00100123	0.00100176	0.00100182	0.00100172
1 mg$_2$	0.00099390	0.00099365	0.00099386	0.00099380	0.00099400	0.00099391

The awareness of the all-around excellence to be found in the California laboratory guaranteed the results would be comparable to a routine calibration performed by NIST.

However, Table 27.2 presents the more noteworthy statistics that are routinely generated by the fine statistical program written by Varner and Raybold.[7] The program is the result of a very large effort undertaken jointly by the Mass Group and the statistical laboratory at NBS during most of the 1960s. These studies[8] are the underpinning of the mass calibration program still in use at this writing.

The software used here is merely a menu-driven version of that so-called mass code written specifically for these measurements and for operators with limited computer knowledge. There are many calculations made by the analysis program, but a key few calculations are sufficient to judge adequacy of the assigned mass values. They are the statistical F-test and t-test, the results for which are given in Table 27.2 for the entire series of measurements.

A brief description of data reduction and analysis is given here for the reader who is not familiar with the process. If one reviews the two measurement designs we described earlier, a few cardinal features will be observed:

First, all mass values arise from the two starting standards, kilograms T_1 and T_2.

Second, all possible weight combinations are observed on the balance for the five 1's series whereas this is not true for the 5,3,2,1,1,1 series. However, both series have the common property of being overdetermined. That is, there is more information available than is really necessary for a simple mass assignment. This last feature allows us to apply the statistical test that we discuss here.

Finally, the calibration is a chain of measurements where each series is linked together by the value of mass assigned to summation in the decade above.

Table 27.2 *F*-Test and *t*-Test Values for All Six Calibrations (all values are within the prescribed limits)

	Run 1		Run 2		Run 3	
	F	*t*	*F*	*t*	*F*	*t*
			NIST			
1	1.531	0.65	1.309	0.41	0.763	1.59
2	0.520	−2.54	2.046	−0.41	2.661	−1.33
3	0.396	−0.46	0.778	−0.61	1.069	−1.38
4	2.807	0.17	1.894	−0.39	0.780	1.09
5	1.286	1.92	0.449	−2.33	0.848	0.90
6	1.080	−1.11	2.177	−2.44	1.701	−1.49
7	0.760	−1.57	0.404	−0.66	0.347	−1.49
			California			
1	0.606	0.91	0.568	0.57	1.224	0.75
2	1.033	1.04	0.187	1.76	1.423	0.48
3	0.717	−1.88	0.530	1.68	2.785	1.12
4	2.475	−1.35	2.581	0.52	2.727	0.33
5	0.897	1.22	1.422	−1.28	1.958	1.41
6	0.365	0.54	1.796	0.48	0.321	2.18
7	2.199	−1.94	0.866	−2.78	0.566	−2.80

In conclusion, the series is a sequence of force differences observed on a balance that is proportional to the gravitational forces imposed by the mass and its displacement volume when weighed in air.

When corrected for buoyancy, the system of equations can be solved for the masses involved and utilize the method known as "least squares" to take advantage of the additional information, as there are more observations than are necessary for a solution.

This weighing method consumes considerably more time than the simple "one-on-one" measurement but provides the statistics needed to ensure that the calibration is free of major errors, and provides the information needed to assign realistic estimates of uncertainty to the assigned mass values.

27.7 Analysis

The software (mass code) serves two primary functions in that it reduces the raw data to mass values and computes an associated uncertainty based on input parameters.

A third function is quality control. This is accomplished by comparing present balance peformance and check standard values to their historical values (*F*-test and *t*-test). These values are used by the operators as "go" and "no-go" criteria; i.e., values within the prescribed limits are indicative of a process with predictable behavior.

The limits used here are that the *F*-test value must be less than 2.81 and the absolute value of *t* less than 3. The selection of the limits and the details of the test can be found in the literature.[8]

Table 27.2 gives the values computed for the NIST measurements and those of California; all are within the limits. The reader should note that California used different balances and check standards than NIST except for the 1-kg level.

Table 27.3 presents the starting kilogram check ($T_1 - T_2$) data in a more interesting format. That is, the accepted value and the observed value are given for each measurement. Except for the starting check ($T_1 - T_2$), the check standards provided to California are not the same as those used at NIST and were assigned mass values on the basis of six mass determinations each, as opposed to hundreds at NIST.

More work in assigning mass values to the check standards would not be cost-effective. Beyond the *t*-test values already discussed, there is nothing to be gained in further analysis of the data.

Table 27.3 The Starting Kilogram Check Values Determined by the Measurements at Both Locations (Accepted Value $T_1 - T_2 = 1.149$ mg)

Cal1	Cal2	Cal3	NIST1	NIST2	NIST3
1.177	1.167	1.173	1.167	1.160	1.192
Average = 1.172 mg			Average = 1.173 mg		

Table 27.4 Differences between the Mean Values Assigned at NIST and California (in mg)

Nominal Mass	NIST Mean Correction, mg	CA Mean Correction, mg	Type A Uncertainty[a] mg	Difference NIST – CA
1 kg$_1$	8.219	8.286	0.025	−0.067
1 kg$_2$	7.212	7.212	0.025	0
500 g	2.288	2.289	0.016	−0.001
300 g	1.521	1.555	0.014	−0.034
200 g	0.881	0.908	0.012	−0.027
100 g	−0.227	−0.229	0.015	−0.002
50 g	1.035	1.041	0.008	−0.006
30 g	1.173	1.174	0.006	−0.001
20 g	0.123	0.130	0.005	−0.007
10 g	0.106	0.110	0.006	−0.004
5 g	0.034	0.033	0.003	0.001
3 g	0.038	0.034	0.002	0.004
2 g	−0.038	−0.040	0.001	−0.002
1 g	0.040	0.041	0.001	−0.001
500 mg	0.0104	0.0082	0.0004	0.0022
300 mg	0.0208	0.0197	0.0003	0.0011
200 mg	0.0194	0.0192	0.0003	0.0002
100 mg	0.0090	0.0084	0.0002	0.0006
50 mg	0.0118	0.0125	0.0002	−0.0007
30 mg	−0.0097	−0.0100	0.0001	−0.0003
20 mg	−0.0379	−0.0382	0.0001	−0.0003
10 mg	−0.0246	−0.0248	0.0001	−0.0002
5 mg	0.0012	0.0015	0.0001	−0.0003
3 mg	0.0033	0.0034	0.0001	−0.0001
2 mg	0.0029	0.0030	0.0001	−0.0001
1 mg$_1$	0.0015	0.0018	0.0001	−0.0003
1 mg$_2$	−0.0062	−0.0061	0.0001	0.0001

[a]The Type A uncertainty is identical for both laboratories but does not include a between-time component.

It is instructive to look at the mean mass values for each weight in the set by location. Table 27.4 presents the mean value for each location, the difference in the mean values, and the corresponding standard deviation (Type A uncertainty).

To judge the adequacy of the assigned mass values, one must know the uncertainties and then judge in terms of the expected application. The second part of this postulate is answered on the basis of need. One must be able to compare the uncertainties with those required by specific application.

From experience there are always users of the measurement system who want to improve on the basis of ego and not real need. That is, these users desire to be on par with NIST regardless of the economic cost or usefulness.

There are real limits imposed by characteristics of the equipment ancillary to each application. For example, the mass of piston-gauge weights need not be known to much better than 10 ppm. Many gravimetric applications are in the range between 100 and 1 ppm, such as volumetric calibration and the density of solids and liquids. In fact, there are few requirements below 1 ppm other than scientific applications.

Most mass measurements are directed toward compliance with weight adjustment tolerances such as those set forth by American Society for Testing and Materials (ASTM), International Bureau of Legal Metrology (OIML), or balance calibration requirements.

Except for the very uncommon weight classes that are only achievable by a state-of-practice calibration, almost all of these tolerances range from 5 to 500 ppm. Probably the largest industrial application is the certification of balances, and here again the limits are similar to those of weights.

Of course, as balances improve, the calibration requirements will be more demanding and will require mass standards with very small uncertainties.

If surrogate laboratories are able to meet the uncertainty requirement, and the demand develops, then the work decribed here would be useful.

The uncertainty of the mass of a standard is determined from the random and systematic uncertainties associated with the measurement process, Type A and Type B uncertainties, respectively.

Some of these errors are readily determined and others are not. For example, the correction of the thermometer used to measure air temperature is obtainable through calibration. However, the application of the thermometer to the measurement process may be incorrect or not understood, leading to a temperature error.

The latter situation may be unknown to those directly involved and the true uncertainty may be unknown. Through careful analysis and long-term effort, the performance of a measurement process may become fairly well known. Then, one appeals to an accepted statistical method for calculating the measurement uncertainty.

This uncertainty in the case of a routine NIST mass determination means that the artifact behavior is assumed to be similar to that of the NIST check standards that are constantly being recalibrated; therefore, one can assign a reasonable estimate of uncertainty based on a single calibration of a client's artifact.

In a sense, these artifacts are fragile and it is very difficult, without the proper facilities and techniques, to ascertain the ongoing validity of an assigned mass value. The statement of uncertainty does not reveal any information regarding the use of the artifact in another measurement process other than that it becomes a source of Type B error when used as the standard.

The uncertainty, in grams, assigned to each of the California state weights based on a single calibration is given in Table 27.5. The uncertainty is based on the International Standards Organization (ISO) recommendation along with the uncertainty as calculated theretofore (see Chapter 7). In essence, the newer method adds the Type A and Type B errors in quadrature and multiplies the result by two for a 95% confidence level, and the old method is $3A + B$ for a 99% confidence level.

The column titles refer to the nominal value of the weight, the 1967 value of the mass adjusted to 1990 (g), the average 1992 value of the mass (g), column 2 − column 3 (g), the old value of the uncertainty of the 1993 value of the mass (3 × the standard deviation + the estimate of the systematic uncertainty)(g), and the ISO uncertainty of the post-1993 value of the mass (g).

Type A uncertainties arise from the measurement process, random uncertainties, and Type B from systematic uncertainties. Since weighing designs are used in the calibration process, the detail may be obscured. However, in terms of the starting kilograms, T_1 and T_2, the uncertainty of the work performed in California is comparable to the NIST calibrations.

If the integrity of the starting standards is ensured, then all is well. We note that the average check values of these standards for both locations indicate a shift of about 20 μg. This offset between the expected and observed values is not alarming but does merit recalibration.

Table 27.5 Uncertainty Assigned to Each of the California Weights Based on a Single Calibration

Nominal Value	Mass 1967, Adjusted to 1990, g	Mass 1992, Average, g	Difference, g	Uncertainty 3SD + SU, 1993, g	Uncertainty ISO, Post 1993, g
1 kg$_1$	1000.008249	1000.008253	−0.000004	0.000087	0.000043
1 kg$_2$	1000.007185	1000.007211	−0.000026	0.000087	0.000043
500 g	500.002749	500.002288	0.000461	0.000047	0.000024
300 g	300.001629	300.001538	0.000091	0.000034	0.000018
200 g	200.000959	200.000895	0.000064	0.000026	0.000014
100 g	99.999950	99.999772	0.000178	0.000025	0.000015
50 g	50.001065	50.001038	0.000027	0.000014	0.000008
30 g	30.001181	30.001173	0.000008	0.000010	0.000006
20 g	20.000170	20.000126	0.000044	0.000008	0.000005
10 g	10.000117	10.000108	0.000009	0.000008	0.000005
5 g	5.000049	5.000034	0.000015	0.000004	0.000003
3 g	3.000041	3.000036	0.000005	0.000003	0.000002
2 g	1.999966	1.999961	0.000005	0.000002	0.000002
1 g	1.000043	1.000040	0.000003	0.000003	0.000002

27.8 Conclusions

It was concluded from this exercise that the California State Laboratory was on a par with a routine NIST calibration when using the standards and methods described here. However, the method does illustrate the critical dependence on the integrity of the starting kilogram standards. To achieve the uncertainties stated here requires a state-of-the-art calibration of the starting kilogram standards, and probably a rigorous monitoring program thereafter. Otherwise, through wear (see Chapter 8), these fragile artifacts will introduce systematic errors in the calibration process. Some significant errors can remain undetected for some time.

The reader can obtain some idea of the system stabilty by comparing prior calibrations of the California weights to the present ones. The former calibrations of the California weights were adjusted to compensate for the 1990 mass scale shift and compared the values to the average of all six measurments of this work in Table 27.5.

Considering the 26-year time span and an even longer period of inattention given to the NIST working standards, the system is quite stable. One can only conclude that the State of California has treated the kilograms with care.

The implementation of a surrogate laboratory program would entail an expensive effort to maintain and track the starting kilogram pairs. In addition, tracking of the entire measurement process to assure that the uncertainty statements have validity would be a requirement.

The method of the circulating package, if properly designed and managed, would fill an obvious training gap that exists at near state-of-the-art mass measurement in the country today. Furthermore, the examination of laboratories for certification at this high level of accuracy would require the kind of laboratory examination techniques developed as part of this program.

References

1. Schoonover, R. M., Advanced Mass Calibration in State Laboratories, NBSIR 83-2752, 1983.
2. Schoonover, R. M. and Taylor, J. E., Some recent developments at NBS in mass measurements, presented at IEEE Instrumentation and Measurement Technology Conference, Boulder, CO, 1986.

3. Schoonover, R. M. and Taylor, J. E., An Investigation of a User-Operated Mass Calibration Package, NISTIR 88-3876, 1988.
4. Schoonover, R. M., Davis, R. S., Driver, R. G., and Bower, V. E., A practical test of the air density equation in standards laboratories at differing altitude, *J. Res. Natl. Bur. Stand.* (U.S.), 85, 27, 1980.
5. Schoonover, R. M. and Keller, J., A surface-dependent thermal effect in mass calibration and what you can do about it, in *Proc. 68th National Conference on Weights and Measures,* Sacramento, CA, 1983.
6. Glaser, M., Response of apparent mass to thermal gradients, *Metrologia*, 27, 95, 1990.
7. Varner, R. N. and Raybold, R. C., National Bureau of Standards Mass Calibration Computer Software, NBS Technical Note 1127, 1980.
8. Pontius, P. E., Notes on the Fundamentals of Measurement and Measurement as a Production Process, NBSIR 74-545, 1974.

28

Highly Accurate Direct Mass Measurements without the Use of External Standards*

28.1 Introduction

The usual method of determining the mass of an object is by comparing the nominally equal forces exerted on a balance pan by the object and by a mass standard. The small difference in mass between the unknown object and the mass standard is then expressed as the solution of two force equations, which include terms for displacement volumes of the objects, and the air density.

In this chapter, the concept[1] that the mass of an object can be adequately determined (for most applications) by direct weighing on an electronic balance without the use of external standards is examined. The only requirements are that the mass and density of the built-in weight of the balance be known adequately with respect to the SI units,[2] and that the balance be linear or corrected for nonlinearity.

28.2 The Force Detector

The electronic balance can be considered to be a highly linear and precise force detector. An overview of the electronic balance is given in Refs. 3 and 4. A short summary of the principles of operation is given here. Detailed knowledge of the electronic circuits is unnecessary.

When a downward force is applied to the balance pan (loaded with an object) it is opposed by a magnetic force generated by the interaction of two magnetic fields. One field is generated by a permanent magnet and the other by a controllable electromagnet. Usually, the magnetic force is applied through a multiplying lever and not by direct levitation. Sufficient magnetic force is generated to restore the mechanism (pan) to its unloaded position or null point relative to the balance structure. Obviously, the device is electromechanical, and one should expect errors (both random and systematic) associated with both electrical and mechanical sources to arise in the use of these instruments.

It is desirable in common weighing applications to tie the magnetic force to the unit of mass through calibration of the electronic circuit. The circuit is adjusted such that the algebraic sum of the gravitational and buoyant forces produces a balance indication approximately equal to the nominal value of the applied mass.

---

*Chapter is based on Ref. 1.

It is common practice for high-precision balances to be supplied with a built-in weight of density about 8 g/cm³ and with its mass adjusted to a nominal value. This practice provides for a uniform response among balances to the given load at a given location.

When the built-in calibration weight is tied to the SI mass unit by a calibration and its density is accurately known, the electronic balance provides a convenient way to multiply and divide the mass unit within the capacity of the instrument. This built-in weight and the high degree of precision and linearity of the electronic force balance *eliminates* the need for a calibrated set of mass standards (i.e., a weight set).

In calibrating the balance, the manufacturer forces the no-load indication to be zero and, when the built-in calibration weight is engaged, has adjusted its electronic circuit to indicate the nominal value of the built-in weight.

The ideal balance response is, of course, a straight line connecting the no-load indication to the built-in weight indication. For some balances, linearity is preserved with extrapolation beyond these bounds. Usually, the response of balances is only approximately ideal, and therefore, for some applications, observations not at calibrated points may require correction for nonlinearity.

In the following discussion, it is assumed that the correction for nonlinearity has been applied to the balance observations for the unknown object. The subject of balance linearity is discussed in detail in Appendix C. However, we note that high-quality electronic balances are available for which nonlinearity errors are less than 1 ppm of the capacity of the balance.

28.3   Discussion of the Method

The calibration function of the analytical-quality electronic balance is totally or partially controlled by its microprocessor. In the calibration process, the pan-empty balance indication is set equal to zero. Similarly, the balance indication is adjusted to the nominal mass of the built-in weight when it is loaded on the balance. Customarily, the balance manufacturer adjusts the balance response to indicate the conventional value of weighing in air (see Chapter 15) of the built-in weight or the mass of the built-in weight if the density of the weight is 8.0 g/cm³. This adjustment procedure ignores the opposing buoyant force on the built-in weight. In the latter case, the manufacturer adjusts the mass to be within one display count of the nominal mass.

During the calibration using the built-in weight, the balance response (indication) is proportional to the force imposed on the balance mechanism by the built-in weight. This force is expressed by the following equation:

$$\mathbf{F}_s = S\left[1-\left(\rho_a/\rho_s\right)\right]g = kO_c. \tag{28.1}$$

The force is $\mathbf{F}_s$, S is the mass of the *built-in weight* and ρ_s is its density, ρ_a is the density of air, g is the local acceleration due to gravity, k is a constant of proportionality, and O_c is the balance response in balance units.

Similarly, the force imposed on the balance pan by an object of unknown mass is expressed as

$$\mathbf{F}_X = M_X\left[1-\left(\rho_a/\rho_X\right)\right]g = k\left(O_L-O_E\right). \tag{28.2}$$

The force is $\mathbf{F}_X$; M_X is the unknown mass (to be determined) of the object and ρ_X is its density; k is unchanged; O_L is the balance response under load; and O_E is the empty-pan balance indication. Usually, but not necessarily, O_E is adjusted to zero at the beginning of the weighing process. It is imperative that the proportionality constant, k, remain unchanged for both the calibration cycle and the weighing cycle. The modern electronic balance maintains its calibration, provided that the balance is left undisturbed and that the environmental conditions are stable. Solution of Eqs. (28.1) and (28.2) for M_X yields

$$M_X = S\left[1-\left(\rho_a/\rho_s\right)\right]\Big/\left\{\left[O_C\big/\left(O_L-O_E\right)\right]\left[1-\left(\rho_a/\rho_X\right)\right]\right\}. \tag{28.3}$$

The quantity, $[O_C/(O_L - O_E)]$, is the ratio of the force imposed on the balance by S to the force imposed by M_X. Normally, ρ_s and ρ_X are known at some reference temperature, t_r. To obtain densities at the test temperature, t, corrections must be made for the thermal expansion or contraction of the built-in weight and object using the following equation:

$$\rho_t = \rho_r\Big/\left[1+3\alpha\left(t-t_r\right)\right], \tag{28.4}$$

where α is the coefficient of linear thermal expansion of the material of which the built-in weight or the object is constructed, and t_r is the reference temperature.

ρ_t has been used to represent either ρ_s or ρ_X at the test temperature. Similarly, ρ_r represents either density at the reference temperature. Either ρ_s or ρ_X can be determined using the balance by performing a hydrostatic weighing (see Chapter 13).[5]

In all cases, the linearity corrections (if significant) are subsequently added to or subtracted from the mass calculated from the balance indication. For those balances that extrapolate beyond the calibration point, the same technique can be applied to the extrapolated region. One need only load the balance pan with a weight of known mass nominally equal to S and then perform a linearity test.

There are hybrid balances that use a series of tare weights (built-in weights) in conjunction with the electronic force balance to increase the capacity. These weights require calibration by the method described in Ref. 6 to assure linearity.

28.4   Uncertainties

The measurement uncertainties are propagated by the method described by Ku.[7] The general propagation equation for the uncertainty of M_X is

$$\left(\text{SD}_{MX}\right)=\left\{\Sigma_i\left[\left(\partial M_X/\partial y_i\right)^2\left(\text{SD}_i\right)^2\right]\right\}^{1/2}, \tag{28.5}$$

where $\partial M_X/\partial y_i$ is the partial derivative of the equation for the unknown mass, M_X, with respect to the parameter, y_i. SD_i is the estimate of standard deviation for each parameter, y_i. The parameters are S, ρ_S, ρ_X, ρ_a, O_C, O_L, and O_E. Second-order effects such as the thermal expansion or contraction of the weights, and the covariance terms in the Ku equation are considered negligible. Referring now to Eq. (28.3), the partial derivatives are

$$\partial M_X/\partial S =\left[1-\left(\rho_a/\rho_S\right)\right]\Big/\left\{\left[O_C\big/\left(O_L-O_E\right)\right]\left[1-\left(\rho_a/\rho_X\right)\right]\right\}, \tag{28.6}$$

$$\partial M_X/\partial O_C =-S\left[1-\left(\rho_a/\rho_S\right)\right]\left[\left(O_L-O_E\right)\big/O_{C2}\right]\Big/\left[1-\left(\rho_a/\rho_X\right)\right], \tag{28.7}$$

$$\partial M_X/\partial\left(O_L-O_E\right)=S\left[1-\left(\rho_a/\rho_S\right)\right]\Big/O_C\left[1-\left(\rho_a/\rho_X\right)\right], \tag{28.8}$$

$$\partial M_X/\partial\rho_S =\left(S\rho_a/\rho_S^2\right)\Big/\left\{\left[O_C\big/\left(O_L-O_E\right)\right]\left[1-\left(\rho_a/\rho_X\right)\right]\right\}, \tag{28.9}$$

Table 28.1   Error Propagation

	Value	SD	$(\partial M_X/\partial y_i)$	$[SD(y_i)^2 \cdot (\partial M_X/\partial y_i)^2]^{1/2}$
S	100 g	5.0×10^{-5} g	2.00073	1.0×10^{-4} g
ρ_S	8 g/cm³	3.2×10^{-3} g/cm³	3.75×10^{-3} cm³	1.2×10^{-6} g
ρ_X	2.329 g/cm³	4.0×10^{-6} g/cm³	4.428×10^{-2} cm³	1.8×10^{-7} g
O_C	100 g	4.9×10^{-5} g	2.00073	9.8×10^{-5} g
$(O_L - O_E)$	200 g	$1.4 \times 10^{-4}/\sqrt{6}$	1.000366	5.6×10^{-5} g
ρ_a	0.0012 g/cm³	8.6×10^{-7} g/cm³	-60.87659 cm³	5.2×10^{-5} g
			RSS = 0.00016 g (0.8 ppm)	

$$\partial M_X/\rho_a = -S\left\{\left(1/\rho_S\right)\left[1-\left(\rho_a/\rho_X\right)\right]+\left(1/\rho_X\right)\left[1-\left(\rho_a/\rho_S\right)\right]\right\}$$
$$\times\left\{\left[O_C/\left(O_L-O_E\right)\right]\left[1-\left(\rho_a/\rho_X\right)\right]^2\right\},\tag{28.10}$$

$$\partial M_X/\partial_x = \left\{S\left(\rho_a/\rho_X^2\right)\left[1-\left(\rho_a/\rho_S\right)\right]\right\}/\left\{\left[1-\left(\rho_a/\rho_X\right)\right]\left[O_C/\left(O_L-O_E\right)\right]\right\}.\tag{28.11}$$

For a typical weighing, the weighing of a 200-g silicon crystal, Table 28.1 lists the values for the estimates of standard deviation (SD), the partial derivatives, and their products.

The root-sum-square (RSS) of the products in Table 28.1 is the estimate of the uncertainty in the determination of M_X with a coverage factor of 1.

The SD of S is the calibration uncertainty (provided by a standards laboratory, for example) with a coverage factor of 1.

The SDs of ρ_S and ρ_S are similarly provided or determined, again with a coverage factor of 1.

The value of $(O_L - O_E)$ in the table is the mean of repeated determinations performed at one sitting.

The SD of the mean is the SD determined from the repeated determinations divided by the square root of the number, n, of repeated determinations, six in this case.

The SD of O_C is the SD of the balance with a 100-g load; this cannot be reduced by repeating the *automatic* balance calibration process. The SD of a single determination of $(O_L - O_E)$ is the same as the SD of O_C, that is, the same as the SD of the balance.

In the example above, the SD of the lower half of the balance range is constant. In the better balances, this is in general not true; the SD is smaller at the lower end of the range of the balance and the user might wish to take advantage of this fact.

The balance SD and linearity correction for the 200-g load were determined from a set of six weighings. The mean SD of the balance was combined with the mean SD of the linearity correction by the RSS method for an estimate of the effective SD.

28.5   Balance Selection

The parameters in Table 28.1 that contribute the dominant uncertainties for a mass determination within the capacity of the balance are the SD of the balance, (O_C), and of S. The uncertainty of S can be reduced significantly by a rigorous calibration.

Depending on the desired accuracy for mass determinations, a balance with a lesser or greater SD might be chosen. After the SD on S, and on O_C, the limiting parameter becomes $(O_L - O_E)$.

In selecting a balance, the error propagation table, Table 28.1, is useful in determining the desired SD of a balance. Having determined the desired SD, one then depends on the specified SDs provided by manufacturers to select an appropriate balance.

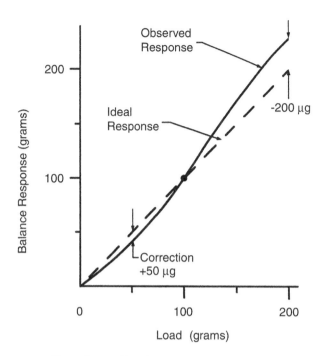

FIGURE 28.1 Balance response illustrating nonlinearity.

28.5.1 Determining the Estimate of Standard Deviation of the Balance

After an appropriate balance is acquired, one then should determine, and use, the SD of the balance in calculating weighing uncertainties. The SD can be determined by multiple weighings of the object to be weighed, or of any *stable weight* within the range of the balance capacity. It is preferred that the mass of the stable weight be near that of the object(s) to be weighed. The SD (defined in Chapter 6) is determined in the usual manner from the values from the multiple weighings.

Returning now to Table 28.1, the SD to be assigned to O_C and to $(O_L - O_E)$ are those determined above. The SD assigned to O_C cannot be reduced by n repeated calibrations; that is, it cannot be divided by the square root of n. However, the SD assigned to $(O_L - O_E)$ can be so reduced.

28.5.2 Linearity Test and Correction

Ideally, the response of the balance would be linear; that is, observations would fall on the straight line between zero and the calibration point. For example, if an object of the same density as that of the calibration weight and ½ its mass were placed on the pan of the balance, the balance would indicate ½ the mass of the calibration weight. Failure to do this would indicate that the balance response is nonlinear. A linearity test should be performed to determine whether the nonlinearity requires correction.

Figure 28.1 is a plot of balance response illustrating nonlinearity.

A linearity test is described in detail in Appendix C.

28.5.3 Data

Five independent determinations of the mass of a 200-g silicon crystal were made using an electronic force balance and the method described here. The five determinations of the mass of the crystal, in grams, are listed below:

199.4266, 199.4264, 199.4267, 199.4266, 199.4273

The mean of these values is 199.42672 g, and the estimate of SD is 0.00034 g. The relative SD is 1.7 ppm. The SD is about two times as large as the predicted uncertainty (see Table 28.1). The lack of thermal equilibrium of the object and the surrounding air is considered to be the major cause of this difference.

The quantitative success of the above measurements can be misleading in the conventional use of the electronic balance because in the example the densities of the objects are very well known. If the densities of S and M_X were only known to 1 part per thousand (a crude density determination), a table similar to Table 28.1 would yield a predicted uncertainty of 1 ppm rather than 0.8 ppm.

28.6 Discussion

It has been shown that very accurate measurements of mass can be achieved by the proper use of an appropriate electronic balance. In this treatment, the need for calibrated laboratory weight sets is *eliminated*. Traceability to NIST is attained through calibration of the *built-in* calibration weight of the electronic balance by NIST or by other standards laboratories.

Both the mass and density of the built-in weight should be provided by the standards laboratory. In most cases, a balance manufacturer can supply limits on the nonlinearity and the estimate of SD of the balance, and the density and the mass and uncertainty of the calibration built-in weight.

If one accepts the values of the nonlinearity and SD provided by the manufacturer, the uncertainties in these two quantities must be combined by RSS to calculate the estimated effective balance SD.

It has been shown how to make determinations of nonlinearity and of the SD of the balance. In the latter case, either an object to be weighed or another object of stable mass is used; a standard of mass is *not required*.

It has been shown in detail how to estimate and propagate uncertainties. The calculation and propagation of uncertainties have been demonstrated using data from weighings of an object of density 2.329 g/cm^3.

28.7 Direction of Future Developments in Electronic Balances and Their Uses

One of the shortcomings of electronics balances, at this writing, is the fact that repeated calibrations of the balance using the built-in weight cannot be averaged. As currently configured, O_c is the result of only one calibration. Therefore, the uncertainty in O_c cannot be reduced by the square root of the number of repeated calibrations.

Balance manufacturers at present have the opportunity to make a great step forward by incorporating a hollow (or low-density) weight[8] in the balance so that air density might be determined automatically by weighings of the built-in weight and of the hollow (or low-density) weight. This would eliminate the inconvenient and expensive practice of measuring pressure, temperature, and relative humidity to calculate air density.[9,10]

The algorithm for determining air density using the two weights and incorporating the air density to make an automatic buoyancy correction could be incorporated in the microprocessor of the balance. The balance would then query the user for the density of the object, the mass of which is being determined. And thus the balance would indicate mass directly, rather than an approximate mass uncorrected for air buoyancy in case the densities of the weighed object and built-in weight are not equal.

References

1. Schoonover, R. M. and Jones, F. E., The use of the electronic balance for highly accurate direct mass measurements without the use of external mass standards, NISTIR 5423, National Institute of Standards and Technology, May 1994.
2. Schoonover, R. M. and Jones, F. E., Examination of parameters that can cause errors in mass determinations, NISTIR 5376, National Institute of Standards and Technology, February 1994.
3. Schoonover, R. M., A look at the electronic analytical balance, *Anal. Chem.*, 52, 973A, 1992.

4. Schoonover, R. M. et al., The electronic balance and some gravimetric applications (the density of solids and liquids, pycnometry and mass), in *ISA 1993 Proceedings of the 39th International Instrumentation Symposium*, 1993, 299.

5. Schoonover, R. M., Hwang, M.-S., and Nater, R., The determination of density of mass standards; requirement and method, NISTIR 5378, February 1994.

6. Jaeger, K. B. and Davis, R. S., A Primer for Mass Metrology, NBS Special Publication 700-1, Industrial Measurement Series, 1984.

7. Ku, H. S., Statistical concepts in metrology, in *Handbook of Industrial Metrology*, American Society of Tool and Manufacturing Engineers, Prentice-Hall, Englewood Cliffs, NJ, 1967, chap. 3.

8. Schoonover, R. M., A simple gravimetric method to determine barometric corrections, *J. Res. Natl. Bur. Stand.* (U.S.), 85, 341, 1980.

9. Davis, R. S., Equation for the determination of the density of moist air (1981/1991), *Metrologia*, 29, 67, 1992.

10. Jones, F. E., The air density equation and the transfer of the mass unit, *J. Res. Natl. Bur. Stand.* (U.S.), 83, 419, 1978.

The Piggyback Balance Experiment: An Illustration of Archimedes' Principle and Newton's Third Law[1]

29.1 Introduction

Originating with Archimedes in the third century B.C., Archimedes' principle has been in the literature and is well known. It is often succinctly expressed by saying "that solids will be lighter in fluid by the weight of the fluid displaced." What has been described is the principle of buoyancy. The *Harper Encyclopedia of Science*[2] has a concise explanation of buoyancy:

> The principle of buoyancy has its origin in the law of fluid pressure, which says that pressure varies directly with depth. Thus the upward pressure on the bottom of a submerged solid (assumed rectangular for the sake of simplicity) is greater than the downward pressure on the upper face. The net upward, or buoyant, force is equal to the difference in weight between two fluid columns whose bases are the upper and lower faces of the solid. Hence the buoyant force is equal to the weight of the portion of fluid displaced by the solid. For a floating body, the buoyant force also equals the weight of the floating body itself. If a body is denser than the fluid in which it is submerged, buoyancy proves insufficient to support the body which thereupon sinks to the bottom.

From this description and the simple relationship,

$$\text{Density} = \text{Mass/Volume},$$

it can be shown that the buoyant force is the product of the fluid density, the volume of the object displacing the fluid, and the acceleration due to Earth's gravity.

In the practice of classical mass metrology, one must account for the buoyant force on a mass that is denser than the fluid in which it is immersed. The following thought problem was constructed to help those still struggling with Archimedes' buoyancy 23 centuries after publication of his work.

29.2 The Piggyback Thought Balance Experiment

Consider a test object O of mass M_O suspended by a massless fiber as shown in Figure 29.1. From Newton's second law, we know that the downward gravitational force, $\mathbf{F}_N$ exerted on O and transmitted through the fiber, is given by

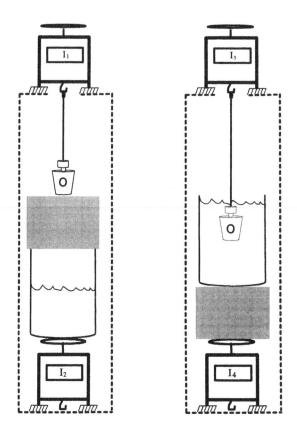

FIGURE 29.1 Experimental setup for piggyback balance.

$$\mathbf{F}_N = M_o g, \tag{29.1}$$

where M_O is the mass of the object and g is the local acceleration due to gravity. From Archimedes' principle just described, we know that the object also is operated on by an upward buoyant force the magnitude of which is given by

$$\mathbf{F}_B = \rho V_o g, \tag{29.2}$$

where ρ is the density of the fluid surrounding the object and V_O is the volume of the object.

The tension in the fiber is equal to the difference of the magnitudes of the two opposing forces. Taking the downward force to be positive, the tension, T, may be written as

$$T = \mathbf{F}_N - \mathbf{F}_B. \tag{29.3}$$

In the International System of Units (SI), mass is expressed in kilograms (kg), length in meters (m), and time in seconds (s). In SI units, volume is expressed in cubic meters (m³), density in kilograms per cubic meter (kg/m³), and the unit of force derived from Newton's second law is equal to one kilogram-meter per second per second (kg-m/s²) and is called the newton.

The experimental setup for a piggyback balance is shown in Figure 29.1. Now, consider the left side of Figure 29.1. A fiber connects test object O to an electronic balance above. Immediately below O is a second electronic balance the pan of which holds a beaker of water and a supporting block atop the beaker. For this experiment, possibilities such as evaporation of water and chipping of the block will be ignored, and it will be assumed that the tare force, $\mathbf{F}_T$, due to the beaker, block, pan, and water remains constant for the duration of the experiment.

With the apparatus set up in a laboratory, the force, $\mathbf{T}_1$, experienced by the upper balance will be given by

$$\mathbf{T}_1 = M_o g - \rho_a V_o g \tag{29.4}$$

$$= \left(M_o - \rho_a V_o\right)g, \tag{29.5}$$

where ρ_a is the density of the surrounding air.

The second stage of the experiment is shown on the right side of Figure 29.1. The beaker has been placed atop the block such that test object O is submerged in water (with no air bubbles clinging to it), and does not touch the beaker. Object O is now buoyed up by the water instead of air. One must now compute a new value for the force, $\mathbf{T}_2$, experienced by the upper balance:

$$\mathbf{T}_2 = \left(M_o - \rho_w V_o\right)g, \tag{29.6}$$

where ρ_w is the density of water. Since the density of water is approximately 800 times the density of air, $\mathbf{T}_2$ will be less than $\mathbf{T}_1$, and the difference:

$$\mathbf{T}_2 - \mathbf{T}_1 = \left(M_o - \rho_w V_o\right)g - \left(M_o - \rho_a V_o\right)g \tag{29.7}$$

$$= \left(\rho_a - \rho_w\right)V_o g \tag{29.8}$$

will be a negative number. It has been assumed that the mass of the test object is unaffected by its immersion. According to Newton's third law, the change in force seen by the upper balance must be matched by an equal and opposite change in force seen on the lower balance. This means that the new force, $\mathbf{F}_T$, on the lower balance is given by

$$\mathbf{F}_T = -\left(\rho_a - \rho_w\right)V_o g. \tag{29.9}$$

29.3 The Laboratory Experiment

Let I_{U1} be the number indicated by the upper balance prior to submersion of the test object.
Let I_{L1} be the number indicated by the lower balance prior to submersion of the test object.
Let I_{U2} be the number indicated by the upper balance with the test object submerged.
Let I_{L2} be the number indicated by the lower balance with the test object submerged.

The various values of I are proportional to the respective values of force exerted on the balances. The value of k, a proportionality constant, is the same for both balances by virtue of the calibration described below. For this experiment, one can write:

$$kI_{U1} = \left(M - \rho_a V_o\right)g \tag{29.10}$$

$$kI_{L1} = \mathbf{F}_T \tag{29.11}$$

$$kI_{U2} = \left(M - \rho_w V_o\right)g \tag{29.12}$$

$$kI_{L2} = \mathbf{F_T} + \left(\rho_w - \rho_a\right)V_o g \tag{29.13}$$

from the results of the previous discussion. With simple algebra:

$$I_{U1} = I_{U2} = -\left(I_{L1} - I_{L2}\right) \tag{29.14}$$

This expression is useful because it allows all the readings to be cross-checked. It was derived using only Archimedes' principle and Newton's second and third laws.

Prior to beginning the experiment, the two electronic balances are placed side by side and calibrated with a standard weight of mass S. After calibration, both balances give the same indication when the standard weight is placed on their respective pans and the air density is unchanged. One can predict the change in both balance indications in the experiment, in kilograms, if the terms are expressed in SI units.

$$\text{Upper balance prediction} = \left(\rho_w - \rho_a\right)V_o \tag{29.15}$$

$$\text{Lower balance prediction} = -\left(\rho_w - \rho_a\right)V_o \tag{29.16}$$

The density of water[3] (see Chapter 14) can be calculated from the water temperature measurement and the air density[4] (see Chapter 12) can be calculated from measurements of air temperature, barometric pressure, and relative humidity.

29.4   Experimental Results

The experiment was conducted using an 85-g silicon crystal of known volume.[1] (One could use a precision sphere and measure its diameter and calculate its volume.) The silicon crystal volume was 37.01596 cm³. The experiment was performed immediately after the balances were calibrated *in situ* with the standard weight.

First, the water temperature was measured and then the balances were adjusted to indicate zero just prior to loading. The crystal was attached to a hook on the upper balance for weighing below the pan and the beaker of water and the blocks were placed on the lower balance pan. All the remaining instrument indications were then recorded.

The blocks and the beaker of water were then arranged to submerge the crystal and after reaching stabilty the balance indications were again recorded.

We can now calculate for each balance the difference between indications and compare them to the observed difference. The calculated and observed differences are tabulated in Table 29.1.

Within experimental error, the balances indicate the equal and opposite responses in kilograms.

Table 29.1   Balance Responses

Calculated kg		Observed kg	
Upper	Lower	Upper	Lower
0.0369082	−0.0369082	0.0369079	−0.0369070
0.0369083	−0.0369083	0.0369105	−0.0369084
0.0368886	−0.0368886	0.0368881	−0.0368881
0.0368886	−0.0368886	0.0368887	−0.0368884

An ancient and useful method of volume determination was used to obtain the crystal volume from the observations made on each balance.

Electronic balances are usually calibrated[5] (see Chapter 10) by adjusting the balance to indicate zero when the pan is empty and indicate the nominal value, I_C, of the calibration weight of mass S when it is loaded on the mechanism. I_C and S are close to each other in value and the calibration weight has a density ρ_S of approximately 8.0 g/cm^3.

One can express the force imposed on the balance by an object of unknown mass, X, and the corresponding balance indication, I_X, as follows:

$$\left[S\left(1-\rho_a/\rho_s\right)I_X/I_C\right]g = \left(X-\rho_a V_X\right)g. \tag{29.17}$$

Eq. (29.17) is rearranged to obtain the volume V_X. M_O and V_O are substituted for X and V_X, respectively, in Eq. (29.17). The expression for the crystal volume (could be any object) is

$$V_o = \left[S\left(1-\rho_a/\rho_s\right)\left(I_1 - I_3\right)\right]/I_C\left(\rho_w - \rho_a\right). \tag{29.18}$$

The upper balance data were used to calculate the silicon crystal volume. The crystal volume determined from the four experiments is 37.0239 cm^3 and the standard deviation is 0.003 cm^3. The difference between the measured volume and the known volume is –0.008 cm^3 and is statistically significant. The difference was most likely caused by gas bubbles adhering to the submerged crystal.

However, the uncertainty of the measured volume is adequate for use in ordinary weighing.

Mass is not in the expression for volume above. Furthermore, when Eq. (29.17) is solved for the mass X, g is not present; the same is true of Eq. (29.18).

Although sufficient precision was not achieved in the experiment to observe the effect from the vertical separation of the balances, one would expect to see 0.0000003 kg/m/kg due to the gradient in the Earth's gravitational field.

29.5 Conclusion

The piggyback balance experiment is easy to perform and useful in teaching students about the opposition of the gravitational and buoyant forces. This can be especially useful for anyone engaged in high-accuracy gravimetric measurements. There is the additional advantage of teaching the importance of the ancillary measurements to achieve accurate results. This is especially true of the volume determination.

References

1. Schoonover, R. M., The piggyback balance experiment; an illustration of Archimedes' principle and Newton's third law, in *Proceedings of the Measurement Science Conference*, Pasadena, CA, 1994.
2. Newman, J. R., *The Harper Encyclopedia of Science*, Harper and Row Evanston and Sigma, New York, 1967, 223.
3. Kell, G. S., Density, thermal expansivity, and compressibilty of liquid water from 0 to 150°C: correlations and tables for atmospheric pressure and saturation reviewed and expressed on 1968 Temperature Scale, *J. Chem. Eng. Data*, 20, 97, 1975.
4. Davis, R. S., Equation for the determination of density of moist air (1981/91), *Metrologia*, 29, 67, 1992.
5. Schoonover, R. M., A look at the analytical balance, *Anal. Chem.*, 52, 973A, 1982.

30

The Application of the Electronic Balance in High-Precision Pycnometry[1]

30.1 Introduction

Pycnometers are essentially flasks whose internal capacity has been determined by weighing the vessel empty and again when filled with water. The pycnometer can then be filled with a liquid of unknown density and, from a similar set of weighings, the unknown density of the liquid is determined by dividing the mass of the liquid by the internal capacity of the pycnometer. Pycnometers are usually constructed from glass and designed to minimize the filling errors associated with setting the liquid level and with trapped gas.

Unlike the stable mass of a laboratory weight, the contained mass of water is likely to vary significantly from filling to filling. In addition, the density of water is about 1 g/cm³, one eighth that of the typical stainless steel mass standard used in balance calibration. The resulting difference in volume between the contained water and an equal mass of standards necessitates a large buoyancy correction.

Other pycnometer characteristics that make weighing the pycnometer more difficult than the weighing of laboratory weights are its propensity to become electrically charged and its hygroscopic surface.

30.2 Pycnometer Calibration

The pycnometer calibration consists of two parts:

1. Weighing the pycnometer when empty
2. Weighing the pycnometer again when filled with water

The first observation is of the force, $\mathbf{F}_{PE}$, imposed upon an electronic balance by the empty pycnometer less its buoyant force:

$$\mathbf{F}_{PE} = M_P\left[1-\left(\rho_a/\rho_p\right)\right]g = KO_E, \tag{30.1}$$

where
M_P = mass of empty pycnometer
ρ_a = air density
ρ_P = density of glass pycnometer body, handbook value
K = constant of proportionality
O_E = balance observation for empty pycnometer
g = local acceleration due to gravity

The second force, $\mathbf{F}_{PF}$, is similar to the above except that the pycnometer is now full of water:

$$\mathbf{F}_{PF} = \left\{ M_P\left[1 - \left(\rho_a/\rho_P\right)\right] + M_w\left[1 - \left(\rho_a/\rho_w\right)\right]\right\} g = KO_F, \qquad (30.2)$$

where
   M_w = mass of water
   ρ_w = water density
   O_F = balance observation for full pycnometer

Additional information is required to determine the mass of water contained in the pycnometer. The mass of an object (in this case water) must be tied to the International System of Units (SI) by calibration of the balance using a weight of known mass and density (calibration weight). Therefore, the pycnometer calibration process would begin with a balance calibration immediately prior to the weighing process.

The force, $\mathbf{F}_C$, imposed on the balance by the calibration weight, of mass S, during the calibration procedure can be expressed as follows:

$$\mathbf{F}_C = S\left[1 - \left(\rho_a/\rho_S\right)\right] g = KO_C, \qquad (30.3)$$

where
   ρ_S = density of calibration weight
   O_C = balance reading when calibration weight is engaged

Eqs. (30.1) through (30.3) are solved for the mass of the contained water, M_w.

The balance is set during the calibration process such that the indication is zero with the pan empty.

With knowledge of the water temperature, t, the water density can be calculated from the formula discussed later. The result is the pycnometer volume, V_t, given by the following equation:

$$V_t = S\left[1 - \left(\rho_a/\rho_s\right)\right]\left[\left(O_F - O_E\right)/O_C\right]/\left(\rho_w - \rho_a\right). \qquad (30.4)$$

The above solution for the pycnometer volume assumes that the air density has remained constant during the weighings and the balance calibration.

The air, the pycnometer, and its contents are assumed to be in thermal equilibrium with each other throughout the series of weighings. For this reason, the pycnometer displacement volume does not manifest itself and the pycnometer density term is not present.

In Eq. (30.4) it also has been assumed that the balance has been zeroed before the empty and full weighings and the capacity of the pycnometer is at the temperature of the water.

The nominal value of the built-in calibration weight, S, is usually adjusted by the balance manufacturer to be accurate within the least significant digit displayed by the balance and may not need additional calibration for this application.

Furthermore, it has been assumed that the weight is made from a material with a density near 8.0 g/cm³. If necessary, the weight can be calibrated *in situ* or removed from the balance for determining both its mass and density.[2,5] These topics are also covered in Chapters 28 and 13, respectively.

In practice, the pycnometer is filled in a constant-temperature water bath and the bath temperature, i.e., the temperature of the pycnometer and its contents, may be different from the temperature in the balance case at the time of weighing. The above weighing equations require modification to account for variations in temperature.

The empty weighing can again be expressed as a force equation as is the associated balance calibration:

$$\mathbf{F}_{PE} = M_P\left[1 - \left(\rho_{aE}/\rho_P\right)X\right] g = KO_E \qquad (30.5)$$

and the balance calibration:

$$\mathbf{F}_{CE} = S\left[1 - \left(\rho_{aE}/\rho_s Z_1\right)\right]g = KO_C, \tag{30.6}$$

where

ρ_{aE} = air density during empty weighing
X = $[1 + 3\beta(t_{aE} - t_{ref})]$, thermal expansion of glass pycnometer
α = linear thermal expansion coefficient of S
t_{aE} = air temperature during empty weighing
t_{ref} = reference temperature
Z_1 = $1/[1 + 3\alpha(t_{aE} - t_{ref})]$, thermal expansion of S
β = linear thermal expansion coefficient of glass pycnometer

Two similar force equations can be written for the weighing of the pycnometer when filled with water:

$$\mathbf{F}_{PF} = \left\{M_P\left[1 - \left(\rho_{aF}/\rho_P\right)Y\right]g + M_w\left[1 - \left(\rho_{aF}/\rho_{wF}\right)\right]\right\}g = KO_F, \tag{30.7}$$

$$\mathbf{F}_{CF} = S\left[1 - \left(\rho_{aF}/\rho_s Z_2\right)\right]g = KO_C, \tag{30.8}$$

where

ρ_{aF} = air density during full weighing
Y = $[1 + 3\beta(t_W - t_{ref})]$, thermal expansion of glass pycnometer
t_W = water temperature during full weighing
ρ_{wF} = water density during full weighing
Z_2 = $1/[1 + 3\alpha(t_{aF} - t_{ref})]$, thermal expansion of S
t_{aF} = air temperature during full weighing
ρ_S = density of calibration weight at t_{ref}
ρ_P = handbook density of pycnometer body at t_{ref}
ρ_{wf} = water density

One assumes that, with proper attention to thermal soaking, the water temperature is the same as the air temperature.

The volume of the contained water is the mass of the water divided by the density of the water. We note that the density of the water is calculated using the *temperature of the bath at the time of the filling*. The pycnometer capacity, V_{bt}, at the bath temperature is

$$V_{bt} = \left(S/O_C\right)\left\{O_F\left[1 - \left(\rho_{aF}/\rho_S Z_2\right)\right] - O_E\left[1 - \left(\rho_{aE}/\rho_S Z_1\right)\right]\left[1 - \left(\rho_{aF}/\rho_P\right)Y\right]\right/$$
$$\left[1 - \left(\rho_{aE}/\rho_P\right)X\right]\right\}\left/\left[1 - \left(\rho_{aF}/\rho_{wF}\right)\right]\rho_{wbt} \tag{30.9}$$

The balance observations are assumed to have been corrected for any nonlinear balance response if required. A detailed discussion of the balance linearity test and corrections is given in Appendix C.

The pycnometer volume is standardized to a reference temperature of 25°C. The following relationship is used:

$$V_{25} = V_{bt}\left[1 + 3\beta\left(25°C - bt\right)\right] \tag{30.10}$$

where

V_{25} = volume of pycnometer at 25°C
V_{bt} = volume of pycnometer at bt, bath temperature

30.3 Experimental Pycnometer Calibration

30.3.1 Apparatus

30.3.1.1 The Electronic Balance

A short summary of the principles of operation of an electronic balance is given here. A more thorough overview of these instruments is given in Chapter 28. Detailed knowledge of the electronic circuits is unnecessary.

Figures illustrating the basic principles of an electronic balance and a representative mechanical structure are shown in Chapter 13. When a downward force is applied to the balance pan (loaded with an object), it is opposed by a magnetic force generated by the interaction of two magnetic fields.

One field is generated by a permanent magnet and the other by a controllable electromagnet. Usually, the magnetic force is applied through a multiplying lever and not by direct levitation. Sufficient magnetic force is generated to restore the mechanism (pan) to its unloaded position (null point) relative to the balance structure.

It is desirable in common weighing applications to tie the magnetic force to the unit of mass through calibration of the electronic circuit. The circuit is adjusted such that the algebraic sum of the gravitational and buoyant forces produces a balance indication approximately equal to the nominal value of the applied mass.

In mass calibration work the applied mass, i.e., the calibration weight, usually has a density of about 8 g/cm^3.

In the pycnometer application here, a balance with a capacity of at least 150 g was required. A balance with a capacity of 200 g was selected.

The manufacturer's specifications were a standard uncertainty (estimated standard deviation) of 0.0001 g with a maximum nonlinearity of 0.0002 g. The level of repeatability was found to be better than the manufacturer's claim.

30.3.1.2 Pycnometer

The pycnometer is fabricated from borosilicate glass with a linear coefficient of thermal expansion of 0.0000033/°C. The configuration of the pycnometer is shown in Figure 30.1.

Each of the capillaries has an internal diameter of approximately 0.65 mm. The wire plugs shown do not fit perfectly (see Discussion). The exterior surface of the pycnometer was coated with a nearly invisible layer of tin oxide to prevent the buildup of static charge.

30.3.1.3 Constant-Temperature Water Bath

The constant-temperature water bath was set to a temperature very close to 25°C, the reference temperature for the pycnometer. The variation of the temperature in the bath was controlled to ±0.003°C/h.

The pycnometer was contained in a brass sleeve that supported the pycnometer in such a manner that the sleeve and the pycnometer were completely surrounded by the bathwater.

The sleeve and the pycnometer were raised to the water surface for insertion of the plugs and drying of the interior of the bowl. In this way the bowl of the device was briefly exposed for setting the water level.

30.3.1.4 Water Bath Temperature

A two-probe thermistor thermometer with a standard uncertainty of 0.003°C was used to measure the water temperature in the bath. The thermometer probes were inserted into the space between the wall of the brass sleeve and the pycnometer body. An average of the two readings was used in the calculations.

30.3.2 Air Density and Water Density

30.3.2.1 Air Density

There are three parameter measurements required to calculate the air density: air temperature, barometric presssure, and relative humidity. The air temperature measurements were made in the balance case using

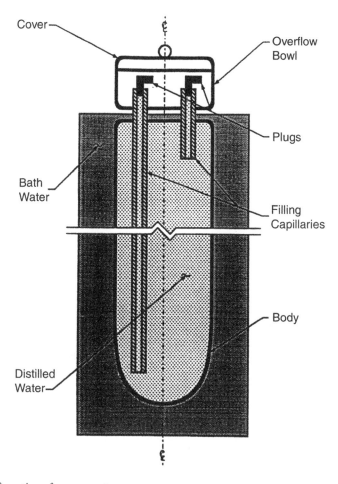

FIGURE 30.1 Configuration of a pycnometer.

a mercury-in-glass thermometer with an uncertainty of 0.05°C. Barometric pressure was measured using an aneroid barometer with a standard uncertainty of 13.3 Pa (0.1 mmHg), and relative humidity was measured with a capacitance-type humidity probe with a standard uncertainty of 5% relative humidity. The latter two instruments were located in the laboratory, close to the balance. A value of 0.043% was used for the content of atmospheric carbon dioxide present.

The air density equation used in this work is the CIPM 1981/91 recommendation.[6] This formula ties its predecessor, CIPM-81, to the International Temperature Scale of 1990 (ITS-90) and utilizes better estimates for the values of one of the constants and one of the other parameters.

Uncertainties in the values of the temperature, pressure, and relative humidity dominate the uncertainty of the calculated air density. A standard uncertainty estimate for the density of air is presented in Table 30.1.

Chapter 12 presents an alternative air density equation that is easier to implement.[1]

30.3.2.2 Water Density

The work of Kell[7] is generally regarded as the comprehensive treatment of water density at this writing. Water density is discussed in detail in Chapter 14. The Kell formula provides a value for the density of air-free water at 101.3250 kPa (1 atmosphere) of pressure with an estimated standard uncertainty of about 1.7 ppm.

The formula assumes the use of the IPTS-1968 (t_{68}) temperature scale and temperatures measured in terms of the ITS-1990 must be converted to IPTS-1968. This is readily accomplished in the range between 20 and 30°C from the following approximate relationship[8]:

$$t_{90} - t_{68} = -0.006°C$$

The water temperature measurements here are estimated to have a standard uncertainty of 0.003°C, with a negligible effect on calculated water density.

30.4   Analysis

The method described by Ku[9] has been used to propagate standard uncertainties in the functional relationship, $f(Y_1, Y_2, ..., Y_n)$, of the uncorrelated variables $Y_1, Y_2, ..., Y_n$.

Table 30.1 presents for each variable its value, the estimated standard uncertainty, u_i,[10] and an evaluation of the partial derivatives.

At the bottom of the table is the estimated combined standard uncertainty for the function as given by the following relationship, where V_t is the volume of the pycnometer:

$$\left(u_{cvt}\right) = \left[\Sigma_i \left(\partial V_t / \partial Y_i\right)^2 \left(u_i\right)^2\right]^{1/2} \tag{30.11}$$

The *effect* of each variable in Table 30.1 is calculated from the product of the standard uncertainty and the partial derivative for that particular variable. For convenience to the reader we list the partial derivatives below.

It has been chosen to propagate the standard uncertainties through the simple form of Eq. (30.4) rather than the more complex form of Eq. (30.9).

The resulting uncertainty analysis from the following equations is nearly the same and is easier to perform. Readers can use these derivatives to evaluate their own standard uncertainty using the applicable standard uncertainties for their equipment. One important parameter in the uncertainty analysis is the balance reproducibility, as measured by the standard uncertainty.

$$\left(\partial V / \partial S\right) = \left[1 - \left(\rho_a / \rho_S\right)\right] \left[\left(O_F - O_E\right) / O_C\right] / \left(\rho_w - \rho_a\right). \tag{30.12}$$

$$\left(\partial V / \partial \rho_s\right) = S\left[\left(O_F - O_E\right) / O_C\right] \left(\rho_a / \rho_S^2\right) / \left(\rho_w - \rho_a\right). \tag{30.13}$$

$$\left(\partial V / \partial \rho_a\right) = S\left[\left(O_F - O_E\right) / O_C\right] \left[1 - \left(\rho_w / \rho_S\right)\right] / \left(\rho_w - \rho_a\right)^2. \tag{30.14}$$

$$\left(\partial V / \partial \rho_w\right) = -S\left[1 - \left(\rho_a / \rho_S\right)\right] \left[\left(O_F - O_E\right) / O_C\right] / \left(\rho_w - \rho_a\right)^2. \tag{30.15}$$

$$\left(\partial V / \partial O_E\right) = -S\left[1 - \left(\rho_a / \rho_S\right)\right] / \left[O_C \left(\rho_w - \rho_a\right)\right]. \tag{30.16}$$

$$\left(\partial V / \partial O_F\right) = S\left[1 - \left(\rho_a / \rho_S\right)\right] / \left[O_C \left(\rho_w - \rho_a\right)\right]. \tag{30.17}$$

$$\left(\partial V / \partial O_C\right) = -S\left[1 - \left(\rho_a / \rho_S\right)\right] \left(O_F - O_E\right) / \left[O_C^2 \left(\rho_w - \rho_a\right)\right]. \tag{30.18}$$

The balance used here, like many electronic balances, performs better when lightly loaded, as in the case of the pycnometer weighings. Its standard uncertainty was found to be 42 µg from 0 to 130 g and

Table 30.1 Standard Uncertainty Budget for the Pycnometer Calibration

	Value, Y_i	SD_i	$\partial V_{bt}/\partial Y_i$	$SD_i \times \partial V_{bt}$, cm^3
S	100 g	0.00005 g	70.07359 cm^3/g	0.000035
ρ_s	8.0 g/cm^3	0.00032 g/cm^3	0.00131 cm^6/g	0.00000042
ρ_a	0.0012 g/cm^3	0.0000003 g/cm^3	61.39726 cm^6/g	0.000018
ρ_w	1.0 g/cm^3	0.0000017 g/cm^3	−70.15778 cm^6/g	−0.000119
O_E	64 g	0.000042/$\sqrt{6}$ g	−1.001051 cm^3/g	−0.000017
O_F	134 g	0.000042/$\sqrt{6}$ g	1.001051 cm^3/g	0.000017
O_C	100 g	0.000049 g	−0.7007359 cm^3/g	−0.000034
				SD = RSS 0.003506 cm^3

118 µg at 200 g. These standard uncertainties are combined in quadrature with the standard uncertainty of the linearity correction.[10]

With hindsight (see Discussion), the application of a linearity correction to the balance observations was not justified with respect to the reproducibility of the pycnometer volume achieved here.

Six balance weighings were made during each weighing cycle; this is reflected in Table 30.1.

It is noteworthy that the balance calibration reproducibilty (42 µg) is not improved by repeated calibration cycles and therefore the balance calibration is performed only once.

This standard uncertainty cannot be obtained explicitly but, from the nature of digital circuits, it is known to be less than ½ count, i.e., 50 µg. The standard uncertainty component would then be [(1/2)/$\sqrt{3}$] because all values between −½ count and +½ count are equally probable. The value used here (42 µg) was determined by repeated weighings at the 100-g level.

The pycnometer calibration results did not justify this much rigor. The root-sum-square (RSS) given in Table 30.1 provides a satisfactory standard uncertainty estimate for the volume of the pycnometer. However, we note again and discuss later that the pycnometer volume is not constant but varies, with filling errors that can only be determined experimentally.

Based upon the analysis given in Table 30.1, one would expect to determine the pycnometer volume with a standard uncertainty of about 2 ppm.

30.5 Data

Five independent determinations of the volume of the pycnometer were made. That is, after each set of empty and full weighings the pycnometer was emptied, cleaned, and refilled.

Each filling was accomplished by siphoning hot distilled water through the pycnometer body by way of the capillaries until the water overflowed into the overflow bowl to a level such that it would cover the ends of the capillaries.

The bowl was then capped and the pycnometer and its contents were soaked in a constant-temperature water bath until thermal equilibrium was achieved. Then, the water temperature was recorded and the overflow bowl was partially emptied.

The plugs were then promptly inserted into the capillary openings and the rest of the excess moisture was "wicked" away with small pieces of filter paper. The overflow bowl was then covered with a sealed lid and the exterior of the pycnometer was carefully dried. The pycnometer was then stored in the balance case for 24 h to achieve thermal and hygroscopic equilibrium with the air in the balance case, before the weighing commenced.

For the empty weighings, the pycnometer was inverted and emptied until no visible water was present. The interior was dried by passing dry nitrogen gas through the capillaries. The same attention must be paid to the thermal soaking of the empty pycnometer in the balance case.

This procedure was repeated five times and the results are reported in Table 30.2. The combined standard uncertainty, u_c, was found to be 40 ppm with 4 degrees of freedom (DF). Note that if the first measurement were to be omitted the standard uncertainty would be lowered to 24 ppm.

Table 30.2   Calibration Results for
the High-Precision Pycnometer

Run	Capacity at 25°C, cm³
1	70.86937
2	70.87351
3	70.87429
4	70.87736
5	70.87438
Mean	70.873782
u_c (4 DF)	0.0029
	(40 ppm)

Although there are insufficient data to justify discarding the first data point, we feel that we were still learning in that portion of the exercise and that repeated measurements would validate this assertion.

Based upon an analysis of Table 30.1, a standard uncertainty of about 1.9 ppm was projected; an experimental standard uncertainty of 40 ppm was observed.

The model did not account for filling errors. Possible improvements to reduce these standard uncertainties are discussed later.

30.6   Discussion

There is a 38 ppm difference between the standard uncertainty projection of Table 30.1 and the experimental standard uncertainty found in Table 30.2. It is believed that the major source of the discrepancy results from filling errors. It is also believed that with further experience an experimental standard uncertainty of 20 ppm could easily be attained without any changes in the present procedure. This would lower the difference between the projected and the experimental standard uncertainty to about 18 ppm. To obtain further improvement, the filling errors must be reduced.

The plugs used in this experiment did not provide a fixed pycnometer volume. This condition could be improved by replacing the straight plugs with tapered glass plugs made from the same borosilicate glass. This improvement, along with tapered seats in the ends of the capillaries would be expected to yield an experimental standard uncertainty of 10 ppm or better.

When the pycnometer is made to work reliably at the 10 ppm level it would be competitive with high-accuracy hydrostatic weighing methods used to determine liquid densities, and would have fewer attendant difficulties.

It is not unusual for a high-precision electronic balance to have a nearly ideal linear response. The lack of linearity, at its worst, is usually not more than 2 ppm of full scale and somewhat better on the higher-quality balances. Therefore, it is unnecessary to apply linearity corrections without significant improvement in the pycnometer performance.

The needs of many users of electronic balances would be well served by accepting the mass and density values reported by the balance manufacturer for the built-in calibration weight. Users requiring higher accuracy can determine the mass and density of the built-in weight themselves or by the use of an appropriate laboratory.

References

1. Schoonover, R. M., Hwang, M.-S., and Crupe, W. E., The application of the elctronic balance in high precision pycnometry, NISTIR 5422, May 1994.
2. Schoonover, R. M. and Jones, F. E., The use of the electronic balance for highly accurate direct mass measurements without the use of external mass standards, presented at National Council of Standards Laboratories 1994 Workshop and Symposium, Chicago, IL, 1994.
3. Schoonover, R. M., A look at the electronic analytical balance, *Anal. Chem.*, 52, 973A, 1982.

4. Schoonover, R. M., Taylor, J., Hwang, M.-S., and Smith, C., The electronic balance and some gravimetric applications, presented at Instrument Society of America 1993 Proceedings of the 39th International Instrumentation Symposium, 1993, 299.

5. Schoonover, R. M., Hwang, M.-S., and Nater, R., The Determination of Density of Mass Standards: Requirements and Method, NISTIR 5378, February 1994.

6. Davis, R. S., Equation for the determination of the density of moist air (1981/91), *Metrologia*, 29, 67, 1992.

7. Kell, G. S., Density, thermal expansivity, and compressibility of liquid water from 0 to 150°C: correlations and tables for atmospheric pressure and saturation reviewed and expressed on 1968 temperature scale, *J. Chem. Eng. Data*, 20, 97, 1975.

8. Mangum, B. W. and Furukawa, G. T., Guidelines for Realizing the International Temperature Scale of 1990 (ITS-90), NIST Technical Note 1265, 1990.

9. Ku, H. S., Statistical concepts in metrology, in *Handbook of Industrial Metrology*, American Society of Tool and Manufacturing Engineers, Prentice-Hall, Englewood Cliffs, NJ, 1967, chap. 3.

10. Taylor, B. N. and Kuyatt, C. E., Guidelines for Evaluating the Uncertainty of NIST Measurement Results, NIST Technical Note 1297, 1994.

APPENDIX A

Buoyancy Corrections in Weighing Course*

I. Objectives

Terminal Objective:

Upon completion of the course, the trainee will be able to demonstrate an understanding of buoyancy corrections, how to make them, and how to determine when it is necessary to make them.

Intermediate Objectives:

1. Identify the origin of buoyancy effects, Archimedes' principle.
2. Determine the form of buoyancy correction factors and buoyancy corrections.
3. Determine how much of the mass of an object is supported by the air in which the weighing is made.
4. Apply buoyancy to weighing on a single-pan balance.
5. Define "apparent mass" (conventioanl value of weighing in air) and calculate it.
6. Apply buoyancy to weighing on an electronic balance.
7. Apply buoyancy correction factors in usual cases in which the air density is not the reference value.
8. Determine the extremes of values of buoyancy corrections.
9. Determine the extremes and variability of air density.
10. Identify equations to be used to calculate air density.
11. Identify the environmental variables that determine the value of air density.
12. Understand the use of constant values of compressibility factor (Z), apparent molecular weight of air (M_a), and enhancement factor (f).
13. Recapitulate air density equations.
14. Identify recommended values and practices in calculating air density.
15. Make sample calculation of air density.
16. Apply buoyancy correction to calibration of volumetric flask.

1. The Origin of Buoyancy Effects

In this section the source of buoyancy forces will be identified and equations will be developed that include the effects.

Forces on an Object on a Pan of a Balance

In general, an object being weighed on a balance in a fluid, air, for example, experiences primarily two *forces*:

---

*Title of a course prepared by Frank E. Jones.[1]

1. The gravitational force on the object
2. The buoyant force exerted on the object by the air in which the weighing is made

The gravitational force, $\mathbf{F}_g$, is

$$\mathbf{F}_g = Mg_L, \tag{A.1}$$

where M is the mass of the object and g_L is the local acceleration due to gravity.

Other forces could be present, such as electrostatic, magnetic, and drag force from convection; however, efforts should be made to exclude or minimize these other forces.

The Buoyant Force

The source of the buoyant force will now be investigated. Consider an object immersed in air.

The *mass of air*, M_{a1}, above an area on a plane at the *base* of the object is equal to

$$M_{a1} = \rho_a V_1, \tag{A.2}$$

where ρ_a is the density of the air and V_1 is the volume of air above the area on the plane.

The *mass of air*, M_{a2}, above an area on a plane at the *top* of the object is

$$M_{a2} = \rho_a V_2, \tag{A.3}$$

where V_2 is the volume of air above the area on the plane.

The forces on the two areas are

$$\mathbf{F}_1 = M_{a1} g_L \tag{A.4}$$

and

$$\mathbf{F}_2 = M_{a2} g_L. \tag{A.5}$$

The pressures on the two areas on the object are

$$P_1 = \mathbf{F}_1 / A_1 = M_{a1} g_L / A_1 = \rho_a V_1 g_L / A_1 \tag{A.6}$$

and

$$P_2 = \mathbf{F}_2 / A_2 = M_{a2} g_L / A_2 = \rho_a V_2 g_L / A_2, \tag{A.7}$$

where A_1 is the area at the base of the object and A_2 is the area at the top of the object.

For simplicity's sake, with no lack of generality, consider the object to be a cylinder with a base of A_1, an upper face of area A_2 (equal to $A_1 = A$), and a height of h. Then,

$$\left(P_1 - P_2\right) = \rho_a g_L \left(V_1 - V_2\right) / A = \rho_a g_L h, \tag{A.8}$$

since $(V_1 - V_2) = V_o$ is the volume of the object.

The force corresponding to $(P_1 - P_2)$ is

$$\mathbf{F}_b = \left(P_1 - P_2\right)A = \rho_a V_o g_L. \tag{A.9}$$

$\rho_a V_o$ is the mass, M_{air}, of the air displaced by the object. Therefore,

$$\mathbf{F}_b = M_{air} g_L. \tag{A.10}$$

Because $M_{air} = \rho_a V_o$ and $M = \rho_m V_o$, the mass of air displaced is also equal to

$$M_{air} = M\rho_a / \rho_m, \tag{A.11}$$

where M is the mass of the object and ρ_m is the density of the object.
Thus,

$$\mathbf{F}_b = M\left(\rho_a / \rho_m\right)g_L \tag{A.12}$$

is the force with which the air *supports* (opposes Mg_L) the object, and we call it a *buoyant force*.

2. The Form of Buoyancy Correction Factors and Buoyancy Corrections

The difference,

$$\left(\mathbf{F}_g - \mathbf{F}_b\right) = Mg_L - M\left(\rho_a / \rho_m\right)g_L = M\left(1 - \rho_a / \rho_m\right)g_L, \tag{A.13}$$

is the net force that the object exerts on the pan of the balance. The balance would interpret this force as the force exerted by the mass M, and a balance indication in response to $(\mathbf{F}_g - \mathbf{F}_b)$ would be in error — it would be too low.
Dividing Eq. (A.13) by g_L,

$$M\left(1 - \rho_a / \rho_m\right)$$

is that part of the mass of the object that is supported by the balance pan.

$$M\left(\rho_a / \rho_m\right)$$

is that part of the mass of the object that is *supported by the air*. That is, the buoyant force supports $M(\rho_a / \rho_m)$.
Note that M/ρ_m is the volume of the object, V_m, and thus the volume of air displaced by it, V_a. Therefore, $M(\rho_a / \rho_m)$ = mass of air displaced, M_a. And $M(1 - \rho_a / \rho_m) = (M - M_a)$ is the mass of the *object* minus the mass of *air* displaced.
The quantity $(1 - \rho_a / \rho_m)$ is a *simple buoyancy correction factor* and $M(\rho_a / \rho_m)$ is a *simple buoyancy correction* and can be written as $\rho_a V_m$.
The mass supported by the air is seen to depend on the value of ρ_a and on ρ_m, the density of the object.

3. Determination of How Much of the Mass of an Object Is Supported by the Air in Which the Weighing Is Made

Illustrations

Let us illustrate with simple examples. The mass of the object is taken to be 100 g; the density of the air, ρ_a, is taken to be 0.00118 g/cm³; and the density of the object, ρ_m, is taken to be 8.0 g/cm³.

$$M\left(1-\rho_a/\rho_m\right)=100\left[1-\left(0.00118/8.0\right)\right]=100\left(1-0.0001475\right) \qquad (A.14)$$

The mass supported by the pan of the balance is 99.98525 g; the mass supported by the air is 0.01475 g, which is 0.01475% of the mass of the object.

However, if the density of the object were 1.0 g/cm³ (the approximate density of water),

$$100\left[\left(1-\left(0.00118/1.0\right)\right)\right]=100\left(1-0.00118\right) \qquad (A.15)$$

and the mass supported by the air is 0.118 g, which is 0.118% of the mass of the object.

Calculation of Mass of Air Displaced

For the above illustrations the mass of air displaced by the object is now calculated. The volume of air displaced, also the volume of the object, is, for $\rho_m = 8.0$ g/cm³,

$$V_a = V_m = M/\rho_m = 100/8.0 = 12.5 \text{ cm}^3. \qquad (A.16)$$

The mass of the volume of air displaced is

$$M_{\text{air}} = \rho_a V_a = 0.00118 \times 12.5 = 0.01475 \text{ g}, \qquad (A.17)$$

which is equal to that part of the mass of the object that is supported by the air.

Archimedes' Principle

Above, Archimedes' principle is derived and illustrated. The principle, formulated in the third century B.C., is translated:[2]

> Solids heavier than the fluid, when thrown into the fluid, will be driven downward as far as they can sink, and will be lighter [when weighed] in fluid [than their weight in air] by the weight of the portion of fluid having the same volume as the solid.

In the case of weighing in the simple case, the object being weighed is denser than the air in which the weighing is made, and a mass of the object equal to the mass of the air displaced by the object will be supported by the air. The balance will indicate that the mass of the object is less by this amount than the true mass (see Chapter 15) of the object.

Buoyancy Correction

It has been shown here that for the weighing of an object of density 8.0 g/cm³ (approximately that of stainless steel) in air of density 0.00118 g/cm³, the buoyancy correction is 0.01475% of the mass of the object. If, in this simple case, the buoyancy correction were not made, the measured mass (the balance indication) would be in *error* by 0.01475%. The balance indication would be *too low*; therefore, it would be necessary to *add* the buoyancy correction to the balance indication to determine the mass of the object.

Table A.1 Buoyancy Correction Factors and Ratios
$[\rho_a = 0.0012 \text{ g/cm}^3, A = [1 - (\rho_a/\rho_x)], B = (\rho_a/\rho_x)]$

ρ_x, g/cm^3	A	B		
		%	ppm	mg/100 g
0.7	0.998286	0.1714	1714	171.4
1.0	0.9988	0.12	1200	120.0
1.5	0.9992	0.08	800	80.0
2.0	0.9994	0.06	600	60.0
3.0	0.9996	0.04	400	40.0
4.0	0.9997	0.03	300	30.0
5.0	0.99976	0.024	240	24.0
6.0	0.9998	0.020	200	20.0
7.0	0.999829	0.0171	171	17.1
8.0	0.99985	0.015	150	15.0
9.0	0.999867	0.0133	133	13.3
10.0	0.99988	0.012	120	12.0
11.0	0.999891	0.0109	109	10.9
12.0	0.9999	0.01	100	10.0
13.0	0.999908	0.0092	92	9.2
14.0	0.999914	0.0086	86	8.6
15.0	0.99992	0.0080	80	8.0
16.0	0.999925	0.0075	75	7.5
16.5	0.999927	0.0073	73	7.3
17.0	0.999929	0.0071	71	7.1
18.0	0.999933	0.0067	67	6.7
19.0	0.999937	0.0063	63	6.3
20.0	0.999940	0.0060	60	6.0
21.0	0.999943	0.0057	57	5.7
22.0	0.999945	0.0055	55	5.5

For the weighing of an object of density 1.0 g/cm^3 (approximately that of water) in air of density 0.00118 g/cm^3, the buoyancy correction is 0.118% of the mass of the object. The balance indication would be too low, and an *error* of 0.118% would be incurred if the buoyancy correction were *not* made.

In Table A.1, for a reference air density of 0.0012 g/cm^3 and substance densities of 0.7 to 22.0 g/cm^3, values of $A = (1 - \rho_a/\rho_x)$ and $B = \rho_a/\rho_x$ are tabulated. Values of B are expressed as %, ppm (parts per million), and mg/100 g. We note that the values of A range from 0.998286 to 0.999945 and that the values of B range from 171.4 mg/100 g to 5.5 mg/100 g.

The significance of the error that would result if buoyancy corrections were not made depends on the desired accuracy for the particular substance in the particular situation and on the precision of the balance, among other things.

4. Application of the Simple Buoyancy Factor to Weighing on a Single-Pan Balance

In a simple case, using a single-pan balance to make a determination of the unknown mass of an object, M_x, the balance indication is equaled by the balance indication, I_s, for an assemblage of objects (standards) of known mass, M_s.

The force, $\mathbf{F}_x$, exerted on the balance pan by the object of mass M_x is

$$\mathbf{F}_x = M_x\left(1 - \rho_a/\rho_x\right)g_L, \tag{A.18}$$

where ρ_a is the density of the air in which the weighing is made and ρ_x is the density of the object.

The known force, $\mathbf{F}_s$, exerted on the balance pan by an assemblage of weights of total mass M_s is

$$\mathbf{F}_s = M_s\left(1 - \rho_a/\rho_s\right)g_L. \tag{A.19}$$

When the forces $\mathbf{F}_X$ and $\mathbf{F}_S$ are equal,

$$\mathbf{F}_x = \mathbf{F}_s, \tag{A.20}$$

$$M_x\left(1 - \rho_a/\rho_x\right) = M_s\left(1 - \rho_a/\rho_s\right), \tag{A.21}$$

$$M_x = M_s\left(1 - \rho_a/\rho_s\right)\big/\left(1 - \rho_a/\rho_x\right). \tag{A.22}$$

Now calculations are made of M_x for several values of ρ_x for the following fixed quantities:

1. $\rho_a = 0.00118$ g/cm^3.
2. $\rho_s = 8.0$ g/cm^3.
3. $M_s = 100$ g.

For $\rho_x = 1.0$ g/cm^3, the approximate density of water,

$$M_x = 100\left(1 - 0.00118/8.0\right)\big/\left[1 - \left(0.00118/1.0\right)\right] \tag{A.23}$$

$$M_x = 100.103372 \text{ g}.$$

That is, under these conditions, 100.103372 g of object X of density 1.0 g/cm^3 would balance 100 g of S weights of density 8.0 g/cm^3. The difference between the mass of X and the S weights, 0.10337 g, is due to the buoyant forces on the object X and the weights S.

This is, of course, due to the difference in density of the object and the standard weights, and consequently (a la Archimedes) to the difference in volume of air displaced by them.

Similarly,

For $\rho_x = 0.7$ g/cm^3, $M_x = 100.156265$ g.
For $\rho_x = 2.7$ g/cm^3, $M_x = 100.028966$ g.
For $\rho_x = 8.0$ g/cm^3, $M_x = 100$ g.
For $\rho_x = 16.6$ g/cm^3, $M_x = 99.992358$ g.
For $\rho_x = 19.0$ g/cm^3, $M_x = 99.991460$ g.
For $\rho_x = 22.0$ g/cm^3, $M_x = 99.906131$ g.

Note that for these last three cases, M_x is less than 100 g. This is, of course, because ρ_x is greater than 8.0 g/cm^3 in these three cases and the buoyancy correction changes sign.

5. Identification of Apparent Mass and Its Calculation

Apparent Mass

We will not stress apparent mass, but we will briefly discuss it to dispel some misconceptions (see Chapter 15).

For

1. $\rho_a = \rho_o = 0.0012$ g/cm^3
2. $t = t_o = 20°C$
3. $\rho_r = 8.0$ g/cm^3

we have the conditions that *define* apparent mass.

The "apparent mass" M_x of an object X is equal to the "true mass" of just enough reference material to produce a balance reading equal to that produced by X if the measurements are done at temperature t_o [20°C] in air of density ρ_o [0.0012 g/cm³].[3]

The density of the reference material, ρ_r, is equal to 8.0 g/cm³. The reference temperature, 20°C, has been specified because the volume of a weight depends on temperature.

In the form of an equation, the definition of apparent mass is

$$_AM_x = {}_TM_r = {}_TM_x\left(1-\rho_o/\rho_x\right)\big/\left(1-\rho_o/\rho_r\right), \tag{A.24}$$

where the subscripts A and T indicate apparent mass and true mass, respectively.

The apparent mass of an object can thus be obtained by multiplying the true mass of the object by the ratio of two buoyancy correction factors. Conversely, the true mass can be obtained by multiplying the apparent mass by the ratio of two buoyancy correction factors (the inverse of the above ratio).

The denominator of Eq. (A.24) is the constant,

$$\left(1-0.0012\big/8.0\right)=0.99985.$$

Apparent Mass and the Single-Pan Two-Knife Direct Reading Analytical Balance

In this section, we shall quote from the paper by Schoonover and Jones ("Air Buoyancy Correction in High-Accuracy Weighing on Analytical Balances")[4]:

In this type of balance [analytical] the load on the pan is balanced by weights which are built into the balance and which are manipulated through external controls and by a functional characteristic of the balance which has been adjusted to approximately indicate in mass units the remaining weight beyond the least of the built-in weights. The nominal values of the built-in weights are indicated as dial readings or other direct readout, and remaining weight is indicated on a ground glass screen or displayed electronically.

The built-in weights are usually accurately adjusted by the manufacturer to one of the "apparent mass" [see Chapter 15] scales.

It is necessary to convert from the apparent mass to the approximate true mass of the built-in weights by using the equation

$$B = I\left[\left(\left(1-\left(0.0012\big/\rho_I\right)\right)\big/\left(\left(1-0.0012\big/\rho_B\right)\right)\right)\right]=IQ, \tag{A.25}$$

where B is the approximate true mass of the built-in weight, I is the mass of the hypothetical reference material (i.e., the dial reading), ρ_B is the density of the built-in weight, and ρ_I is the hypothetical density. The values of ρ_B and ρ_I are supplied by the manufacturer, $\rho_I = 8.0$ or 8.4 g/cm³.

The relationship between the mass, A, of the material being weighed on the balance and the mass, B, of the corresponding built-in weights, taking into account the buoyant forces, is

$$A = B\left[1-\left(\rho_a/\rho_B\right)\right]\big/\left[1-\left(\rho_a/\rho_A\right)\right], \tag{A.26}$$

where ρ_A and ρ_B are the densities of the material being weighed and the built-in weights, respectively.

In summary, according to Schoonover,[5] emphasis added, "The balance reading is *not* the mass of the sample being weighed and is therefore not the desired result. The balance manufacturer has built the balance to indicate the *apparent* mass of the material being weighed."

Apparent Mass and the Electronic Analytical Balance

There are two prevalent types of electronic balances in use at this writing:

1. The hybrid balance
2. The electronic force balance

Mechanically and electronically generated forces are used in the hybrid balance. The electronic force balance uses only electronically generated forces. In this section, we concentrate on the electronic force balance.

In an electronic force balance:

1. An electromagnetic *force* is generated to oppose the net gravitational and buoyant force imposed by the object being weighed.
2. The readout of the balance is proportional to the *current* in a servomotor coil.
3. In calibration of the balance, a calibrating weight is used and the electronic circuitry is adjusted so that the readout indicates the *apparent mass* of the *calibrating* weight.

6. Application of Buoyancy to Weighing on an Electronic Balance Electronic Balance Calibration and Use

We now investigate the calibration and performance of an electronic balance with a built-in calibrating weight.

The *true mass* of the calibrating weight is assumed to be *100 g*.

Throughout this discussion, M with no pre-M subscript refers to true mass. The density of the calibrating weight, ρ_b, is assumed to be *8.0 g/cm³*, ρ_r. The temperature at which the balance was calibrated at the factory is assumed to be *20°C*, and the air density at the factory is assumed to be *0.0012 g/cm³*.

We return now to the defining equation for apparent mass:

$$_A M_x = {_T}M_r = {_T}M_x\left(1-\rho_o/\rho_x\right)\big/\left(1-\rho_o/\rho_r\right). \qquad (A.27)$$

Under the above conditions,

$$_T M_x = 100 \text{ g} = {_T}M_b$$

$$\rho_o = 0.0012 \text{ g}/\text{cm}^3$$

$$\rho_r = 8.0 \text{ g}/\text{cm}^3 = \rho_b$$

$$_T M_r = 100 \text{ g}$$

$$\rho_x = 8.0 \text{ g}/\text{cm}^3$$

Thus,

$$_A M_b = 100\left(1-0.0012/8.0\right)\big/\left(1-0.0012/8.0\right)= {_T}M_b = 100 \text{ g}. \qquad (A.28)$$

Therefore, at the factory under the above conditions, the apparent mass of the calibrating weight is equal to the true mass of the calibrating weight. At the factory, the balance is calibrated using the built-in calibrating weight to indicate *apparent mass*.

In the balance, the electromotive force, **F**, is generated (sometimes through a fixed-ratio level system) to equal and oppose the net force impressed on the balance pan by the gravitational force minus the buoyant force. The electromotive force, $\mathbf{F}$, is generated by the current, $\mathbf{I}$, passing through the coil of an electromotive force cell. $\mathbf{F}$ is proportional to $\mathbf{I}$. The indication of the balance, U, is proportional to $\mathbf{I}$ at equilibrium. Thus,

$$\mathbf{F} = k\mathbf{I}, \tag{A.29}$$

$$U = c\mathbf{I}, \tag{A.30}$$

where k and c are constants of proportionality.

$$\mathbf{I} = \mathbf{F}/k = U/c, \tag{A.31}$$

$$U = \left(c/k\right)\mathbf{F} = K\mathbf{F}, \tag{A.32}$$

where $(c/k) = K$.

Again, at the factory,

$$\rho_a = \rho_o = 0.0012 \ \mathrm{g/cm^3} \ \left(\text{assumed}\right)$$

$$g = g_f = \text{the acceleration due to gravity at the}$$
$$\text{balance location in the factory}$$

$$\rho_x = \rho_r = 8.0 \ \mathrm{g/cm^3} \ \left(\text{assumed}\right)$$

$$t = t_o = 20°\mathrm{C} \ \left(\text{assumed}\right)$$

$$1/K = K_o$$

The force, $\mathbf{F}_b$, exerted on the balance by the built-in calibrating weight of mass M_b and density 8.0 g/cm³ is

$$\mathbf{F}_b = M_b\left(1 - \rho_o/\rho_b\right)g_f = UK_o. \tag{A.33}$$

At the factory, the electronics are adjusted in such a way that the indication of the balance, U, is equal to the apparent mass of the built-in weight (100 g, for example) with the built-in weight introduced to the balance. We shall refer to this operation as the *adjusting* of the balance rather than the *calibration* of the balance.

Thus, under the above conditions,

$$U = {}_A M_b = M_b. \tag{A.34}$$

That is, the indication of the balance is the apparent mass of the built-in calibrating weight, which is also its true mass.

Then $\mathbf{F}_b$ is given by

$$\mathbf{F}_b = U\left(1 - \rho_o/\rho_b\right)g_f = UK_o, \tag{A.35}$$

and, thus,

$$K_o = \left(1 - \rho_o/\rho_b\right)g_f. \tag{A.36}$$

Having thus determined the value of K_o, for a weight of unknown mass M_x on the balance, the force exerted on the balance is

$$\mathbf{F}_x = M_x\left(1-\rho_o/\rho_x\right)g_f = U_x K_o = U_x\left(1-\rho_o/\rho_b\right)g_f \tag{A.37}$$

and

$$M_x = U_x\left(1-\rho_o/\rho_b\right)\Big/\left(1-\rho_o/\rho_x\right). \tag{A.38}$$

Application of Buoyancy Corrections in Usual Cases in Which the Air Density Is Not the Reference Value

We now consider the more usual case for which the air density, ρ_a, is not equal to the reference value, $\rho_o = 0.0012$ g/cm^3.

In the *laboratory*, for adjusting the balance using the built-in weight and the conditions:

$$\rho_a = \rho_a$$

$$t = \text{room temperature}$$

$$g = g_L$$

$$\rho_b = \text{approximately } 8.0 \text{ g}/\text{cm}^3$$

the force exerted on the balance by the introduction of the built-in weight of mass M_b is

$$\mathbf{F}_b = M_b\left(1-\rho_a/\rho_b\right)g_L = U_b K_L, \tag{A.39}$$

where K_L is K under laboratory conditions.

The electronics of the balance are adjusted so that the scale indication, U_b, is equal to the approximate apparent mass of the built-in weight. Then,

$$K_L = \left(1-\rho_a/\rho_b\right)g_L. \tag{A.40}$$

For a *standard* weight of true mass M_s on the pan of the balance, the force exerted by the standard weight is

$$\mathbf{F}_s = M_s\left(1-\rho_a/\rho_s\right)g_L = U_s\left(1-\rho_a/\rho_b\right)g_L, \tag{A.41}$$

and

$$U_s = M_s\left(1-\rho_a/\rho_s\right)\Big/\left(1-\rho_a/\rho_b\right), \tag{A.42}$$

and

$$M_s = U_s\left(1-\rho_a/\rho_b\right)\Big/\left(1-\rho_a/\rho_s\right). \tag{A.43}$$

Therefore, if the balance were operating perfectly, with the standard weight on the pan, the balance indication would be equal to the right side of Eq. (A.42). Deviation of the balance indication from this value would represent a *weighing error* or a *random deviation*.

For a weight of unknown mass, M_x, on the balance, the force exerted on the balance is

$$\mathbf{F}_x = M_x\left(1-\rho_a/\rho_x\right)g_L = U_x\left(1-\rho_a/\rho_b\right)g_L, \tag{A.44}$$

and

$$U_x = M_x\left(1-\rho_a/\rho_x\right)\big/\left(1-\rho_a/\rho_b\right), \tag{A.45}$$

and

$$M_x = U_x\left(1-\rho_a/\rho_b\right)\big/\left(1-\rho_a/\rho_x\right). \tag{A.46}$$

Therefore, if the balance were operating perfectly, with the unknown weight on the pan, the balance indication would be equal to the right side of Eq. (A.45), and the true mass of the unknown would be calculated using Eq. (A.46).

Note: It is the *true mass* of the unknown that is the required mass quantity for most metrological purposes, *not* the *indication* of the balance, and *not* the *apparent mass* of the unknown. Even if the apparent mass were measured perfectly, a calculation must be made to determine the true mass.

In Table A.2, errors due to the use of the buoyancy factor corresponding to that in the definition of apparent mass rather than that appropriate for ρ_a not equal to ρ_o are tabulated for an object of density 1.0 g/cm³. That is, for example, for an ambient air density of 0.00110 g/cm³ the use of the apparent mass buoyancy factor would result in an error of 0.0088%, or 88 ppm, or 8.8 mg/100 g.

Tables A.3, A.4, and A.5 give similar results for object densities of 0.7, 3.0, and 16.5 g/cm³, respectively.

In Table A.6, uncertainties in the weighing of a 100-g object due to uncertainties of 1 and 5% in ρ_x are tabulated using the buoyancy correction factor defined in the table. For example, an uncertainty in ρ_x of −5% at $\rho_x = 0.7$ g/cm³ results in an uncertainty of (−)9.0 mg for a 100-g object.

Uses of the Standard Weight

If the balance were adjusted using the built-in weight, the standard weight could be used to assess the accuracy of the balance, at the mass of the standard weight. That is, the equation

$$M_s = U_s\left(1-\rho_a/\rho_b\right)\big/\left(1-\rho_a/\rho_s\right) \tag{A.47}$$

could be used to calculate the measured value of M_s from the indication of the balance with the standard weight on the pan.

Rearranging Eq. (A.47),

$$U_s = M_s\left(1-\rho_a/\rho_s\right)\big/\left(1-\rho_a/\rho_b\right). \tag{A.48}$$

The standard weight could be used to adjust the balance by placing the standard weight on the pan and adjusting the electronics until the balance indication was equal to the right side of Eq. (A.48), where the parameters are known (except for ρ_a, which is calculated from environmental measurements, see Chapter 12).

Then, with an object of unknown mass, M_x, on the pan, the balance indication would be

$$U_x = M_x\left(1-\rho_a/\rho_x\right)\big/\left(1-\rho_a/\rho_b\right), \tag{A.49}$$

and the indicated true mass of the object of unknown mass would be

$$M_x = U_x\left(1-\rho_a/\rho_b\right)\big/\left(1-\rho_a/\rho_x\right). \tag{A.50}$$

Table A.2 Error Due to Differences between $A = [1 - (0.0012/\rho_x)]/[1 - (0.0012/8.0)]$ and $B = [1 - (\rho_a/\rho_x)]/[1 - (\rho_a/8.0)]$ for Various Values of ρ_a, for $\rho_x = 1.0$ g/cm^3

ρ_a, g/cm^3	B	[1 − (B/A)]		
		%	ppm	mg/100 g
0.0012	0.998950	0.0000	0	0
0.00119	0.998959	0.0009	9	0.9
0.00118	0.998967	0.0017	17	1.7
0.00117	0.998976	0.0026	26	2.6
0.00116	0.998985	0.0035	35	3.5
0.00115	0.998994	0.0044	44	4.4
0.00114	0.999002	0.0052	52	5.2
0.00113	0.999011	0.0061	61	6.1
0.00112	0.999020	0.0070	70	7.0
0.00111	0.999029	0.0079	79	7.9
0.00110	0.999037	0.0088	88	8.8
0.00109	0.999046	0.0096	96	9.6
0.00108	0.999055	0.0105	105	10.5
0.00107	0.999064	0.0114	114	11.4
0.00106	0.999072	0.0122	122	12.2
0.00105	0.999081	0.0131	131	13.1
0.00104	0.999090	0.0140	140	14.0
0.00103	0.999099	0.0149	149	14.9
0.00102	0.999107	0.0157	157	15.7
0.00101	0.999116	0.0166	166	16.6
0.00100	0.999125	0.0175	175	17.5
0.00099	0.999134	0.0184	184	18.4
0.00098	0.999142	0.0192	192	19.2
0.00097	0.999151	0.0201	201	20.1
0.00096	0.999160	0.0210	210	21.0
0.00095	0.999169	0.0219	219	21.9
0.00094	0.999177	0.0227	227	22.7
0.00093	0.999186	0.0236	236	23.6
0.00092	0.999195	0.0245	245	24.5
0.00091	0.999204	0.0254	254	25.4

Subsequent to adjusting the balance using a standard weight, the standard weight could be used to monitor the performance of the balance, at that value of mass (or a series of standard weights could be used to monitor the performance of the balance over a range of mass). With the standard weight on the pan, the true mass of the standard weight would be calculated using Eq. (A.47). That is, the measured value of M_s would be equal to the right side of Eq. (A.47).

If the air density changed significantly between the time the balance was adjusted and the time the performance of the balance was monitored, the values of ρ_a in the two sets of parentheses in Eq. (A.47) would be different.

Again, the calculation of the true mass of an object from the balance indication involves the application of buoyancy corrections. Modern electronic balances may not have a provision to adjust the balance readout as described above.

Extremes of the Values of Buoyancy Corrections

We now examine extremes of the values of buoyancy correction factors. For a group of selected cities in the United States, the average air densities range from 0.00092 (Denver, CO) to 0.00119 (near sea level) g/cm^3.[6] *Actual* values of air density may differ from the *average* value by as much as 3%.[6]

We now substitute these extreme values in equations to calculate buoyancy correction factors. We use 8.0 g/cm^3 for ρ_b and ρ_s. We use 0.7 and 22.0 g/cm^3 as the extremes of ρ_x.

Table A.3 Error Due to Difference between $A = [1 - (0.0012/\rho_x)]/$ $[1 - (0.0012/8.0)]$ and $B = [1 - (\rho_a/\rho_x)]/[1 - (\rho_a/8.0)]$ for Various Values of ρ_a, for $\rho_x = 0.7$ g/cm^3

ρ_a, g/cm^3	B	[1 − (B/A)]		
		%	ppm	mg/100 g
0.0012	0.998435	0.0000	0	0
0.00119	0.998449	0.0014	14	1.4
0.00118	0.998463	0.0028	28	2.8
0.00117	0.998475	0.0040	40	4.0
0.00116	0.998488	0.0052	52	5.2
0.00115	0.998501	0.0066	66	6.6
0.00114	0.998514	0.0079	79	7.9
0.00113	0.998527	0.0092	92	9.2
0.00112	0.998540	0.0105	105	10.5
0.00111	0.998553	0.0118	118	11.8
0.00110	0.998566	0.0131	131	13.1
0.00109	0.998579	0.0144	144	14.4
0.00108	0.998592	0.0157	157	15.7
0.00107	0.998605	0.0170	170	17.0
0.00106	0.998618	0.0183	183	18.3
0.00105	0.998631	0.0196	196	19.6
0.00104	0.998644	0.0209	209	20.9
0.00103	0.998657	0.0222	222	22.2
0.00102	0.998670	0.0235	235	23.5
0.00101	0.998683	0.0248	248	24.8
0.00100	0.998696	0.0261	261	26.1
0.00099	0.998709	0.0274	274	27.4
0.00098	0.998722	0.0287	287	28.7
0.00097	0.998735	0.0300	300	30.0
0.00096	0.998748	0.0313	313	31.3
0.00095	0.998761	0.0326	326	32.6
0.00094	0.998774	0.0339	339	33.9
0.00093	0.998788	0.0353	353	35.3
0.00092	0.998801	0.0366	366	36.6
0.00091	0.998814	0.0379	379	37.9

The values of $(1 - \rho_a/\rho_b)$ and $(1 - \rho_a/\rho_s)$ using the above values of ρ_a, ρ_b, and ρ_s are

$$\left[1 - \left(0.00092/8.0\right)\right] = 0.999885$$

and

$$\left[1 - \left(0.00119/8.0\right)\right] = 0.999851.$$

The values of $(1 - \rho_a/\rho_x)$ using the above values of ρ_a, ρ_b, ρ_s, and ρ_x are

$$\left[1 - \left(0.00092/0.7\right)\right] = 0.998686,$$

$$\left[1 - \left(0.00119/0.7\right)\right] = 0.998300,$$

$$\left[1 - \left(0.00092/22.0\right)\right] = 0.999958,$$

and

Table A.4   Error Due to Difference between $A = [1 - (0.0012/\rho_x)]/$
$[1 - (0.0012/8.0)]$ and $B = [1 - (\rho_a/\rho_x)]/[1 - (\rho_a/8.0)]$ for Various
Values of ρ_a, for $\rho_x = 3.0$ g/cm^3

ρ_a, g/cm^3	B	[1 − (B/A)]		
		%	ppm	mg/100 g
0.0012	0.999750	0.0000	0	0
0.00119	0.999752	0.0002	2	0.2
0.00118	0.999755	0.0005	5	0.5
0.00117	0.999756	0.0006	6	0.6
0.00116	0.999758	0.0008	8	0.8
0.00115	0.999760	0.0010	10	1.0
0.00114	0.999762	0.0012	12	1.2
0.00113	0.999765	0.0015	15	1.5
0.00112	0.999767	0.0017	17	1.7
0.00111	0.999769	0.0019	19	1.9
0.00110	0.999771	0.0021	21	2.1
0.00109	0.999773	0.0023	23	2.3
0.00108	0.999775	0.0025	25	2.5
0.00107	0.999777	0.0027	27	2.7
0.00106	0.999779	0.0029	29	2.9
0.00105	0.999781	0.0031	31	3.1
0.00104	0.999783	0.0033	33	3.3
0.00103	0.999785	0.0035	35	3.5
0.00102	0.999787	0.0037	37	3.7
0.00101	0.999790	0.0040	40	4.0
0.00100	0.999792	0.0042	42	4.2
0.00099	0.999794	0.0044	44	4.4
0.00098	0.999796	0.0046	46	4.6
0.00097	0.999798	0.0048	48	4.8
0.00096	0.999800	0.0050	50	5.0
0.00095	0.999802	0.0052	52	5.2
0.00094	0.999804	0.0054	54	5.4
0.00093	0.999806	0.0056	56	5.6
0.00092	0.999808	0.0058	58	5.8
0.00091	0.999810	0.0060	60	6.0

$$\left[1 - \left(0.00119/22.0\right)\right] = 0.999946.$$

The values of $[(1 - \rho_a/\rho_b)/(1 - \rho_a/\rho_x)]$ and $[(1 - \rho_a/\rho_s)/(1 - \rho_a/\rho_x)]$ are then

$$1.001201$$

and

$$0.999927$$

for $\rho_a = 0.00094$ g/cm^3, and

$$1.001554$$

and

$$0.999905$$

for $\rho_a = 0.00119$ g/cm^3.

The extremes of these last four values are

$$0.999905$$

and

$$1.001554.$$

Table A.5 Error Due to Differences between $A = [1 - (0.0012/\rho_x)]/[1 - (0.0012/8.0)]$ and $B = [1 - (\rho_a/\rho_x)]/[1 - (\rho_a/8.0)]$ for Various Values of ρ_a, for $\rho_x = 16.5$ g/cm^3

ρ_a, g/cm^3	B	\multicolumn{3}{c}{$[1 - (B/A)]$}
		%	ppm	mg/100 g
0.0012	1.000077	0	0	0
0.00119	1.000077	0	0	0
0.00118	1.000076	−0.0001	−1	−0.1
0.00117	1.000075	−0.0002	−2	−0.2
0.00116	1.000075	−0.0002	−3	−0.3
0.00115	1.000074	−0.0003	−4	−0.4
0.00114	1.000073	−0.0004	−4	−0.4
0.00113	1.000073	−0.0004	−5	−0.5
0.00112	1.000072	−0.0005	−6	−0.6
0.00111	1.000071	−0.0006	−6	−0.6
0.00110	1.000071	−0.0006	−7	−0.7
0.00109	1.000070	−0.0007	−7	−0.7
0.00108	1.000070	−0.0007	−8	−0.8
0.00107	1.000069	−0.0008	−9	−0.9
0.00106	1.000068	−0.0009	−9	−0.9
0.00105	1.000068	−0.0009	−10	−1.0
0.00104	1.000067	−0.0010	−11	−1.1
0.00103	1.000066	−0.0011	−11	−1.1
0.00102	1.000066	−0.0011	−12	−1.2
0.00101	1.000065	−0.0012	−13	−1.3
0.00100	1.000064	−0.0013	−13	−1.3
0.00099	1.000064	−0.0013	−14	−1.4
0.00098	1.000063	−0.0014	−15	−1.5
0.00097	1.000062	−0.0015	−15	−1.5
0.00096	1.000062	−0.0015	−16	−1.6
0.00095	1.000061	−0.0016	−16	−1.6
0.00094	1.000061	−0.0016	−17	−1.7
0.00093	1.000060	−0.0017	−17	−1.7
0.00092	1.000059	−0.0018	−18	−1.8
0.00091	1.000059	−0.0018	−18	−1.8

For a mass M_x of 100 g, the indication of the balance would vary from 99.844841 to 100.009501 g (if the balance displayed a sufficient number of digits) using these extreme values.

Therefore, the errors (on this account) in determinations of M_x would range from +9.5 to −155.2 mg in 100 g. Expressed in %, these errors are +0.0095 to −0.1552%; in ppm, they are +95 to −1552 ppm.

Therefore, failing to make appropriate buoyancy corrections (that is, failing to use the appropriate values of ρ_a, ρ_x, ρ_s, and ρ_b), weighings in various locations in the United States could be in error (on this account) by as much as 0.1552%, using the values above.

Effects of Variations about the Value of Air Density

We now investigate the effects of a variation of ±3% about the average value of ρ_a. In these examples, we use an average value of ρ_a of 0.00117 g/cm^3, varying from 0.00113 to 0.00121 g/cm^3 (a variation of ±3% about the mean value). We use the values of ρ_b, ρ_s, and ρ_x given above.

The values of $(1 - \rho_a/\rho_b)$ and $(1 - \rho_a/\rho_s)$ using values of ρ_a of 0.00113 and 0.00121 g/cm^3 are 0.999859 and 0.999849.

The values of $(1 - \rho_a/\rho_x)$ for the extreme values of ρ_x of 0.7 and 22.0 g/cm^3 are 0.998386, 0.998271, 0.999949, and 0.999945.

Table A.6 Uncertainties (Unc) in the Weighing of a 100-g Object Due to an Uncertainty of 1 or 5% in ρ_x, Using the Buoyancy Correction Ratio (BCR), $[1 - (0.0012/\rho_x)]/[1 - (0.0012/8.0)]$

ρ_x, g/cm^3	BCR	BCR, $(\rho_x - 1\%\rho_x)$	Unc, 1%, mg	Unc, 5%, mg
0.7	0.998435	0.998418	−1.7	−9.0
1	0.998950	0.998938	−1.2	−6.3
2	0.999550	0.999544	−0.6	−3.2
3	0.999750	0.999746	−0.4	−2.1
4	0.999850	0.999847	−0.3	−1.6
5	0.999910	0.999908	−0.2	−1.3
6	0.999950	0.999948	−0.2	−1.0
7	0.999979	0.999977	−0.2	−0.9
8	1.0	0.999985	−0.1	−0.8
9	1.000017	1.000015	−0.1	−0.7
10	1.000030	1.000029	−0.1	−0.6
11	1.000041	1.000040	−0.1	−0.6
12	1.000050	1.000049	−0.1	−0.5
13	1.000058	1.000057	−0.1	−0.5
14	1.000064	1.000063	−0.1	−0.5
15	1.000070	1.000069	−0.1	−0.4
16	1.000075	1.000074	−0.1	−0.4
16.5	1.000077	1.000077	0	−0.4
17	1.000079	1.000079	0	−0.4
18	1.000083	1.000083	0	−0.3
19	1.000087	1.000086	−0.1	−0.3
20	1.000090	1.000089	−0.1	−0.3
21	1.000093	1.000092	−0.1	−0.3
22	1.000096	1.000095	−0.1	−0.3

For $\rho_x = 0.7$ g/cm^3, the values of $[(1 - \rho_a/\rho_b)/(1 - \rho_a/\rho_x)]$ and $[(1 - \rho_a/\rho_s)/(1 - \rho_a/\rho_x)]$ are 1.001475 for $\rho_a = 0.00113$ g/cm^3, and 1.001580 for $\rho_a = 0.00121$ g/cm^3. For M_x of 100 g, the difference in ratios corresponds to 0.0105 g or 0.0105%.

For $\rho_x = 22.0$ g/cm^3, the values of the two ratios are 0.999910 for $\rho_a = 0.00113$ g/cm^3 and 0.999904 for $\rho_a = 0.0012$ g/cm^3. For M_x of 100 g, the difference in ratios corresponds to 0.000626 g or 0.000626%.

For an average value of ρ_a of 0.00117 g/cm^3, the values of $(1 - \rho_a/\rho_x)$ for ρ_x of 8.0, 0.7, and 22.0 g/cm^3 are 0.999854, 0.998329, and 0.999947, respectively.

For $\rho_x = 0.7$ g/cm^3, the value of the ratio $[(1 - \rho_a/\rho_b)/(1 - \rho_a/\rho_x)]$ or $[(1 - \rho_a/\rho_s)/(1 - \rho_a/\rho_x)]$ is 1.001528; for $\rho_x = 22.0$ g/cm^3, the ratio is 0.999907.

For $\rho_x = 0.7$ g/cm^3, the difference between the ratios between the average value of ρ_a of 0.00117 and 0.00113 g/cm^3 for M_x of 100 g corresponds to 0.005232 g or 0.005232%. For $\rho_x = 22.0$ g/cm^3, the difference between the ratios corresponds to 0.000318 g or 0.000318%.

Therefore, for variations of ±3% about the average value of ρ_a of 0.00117 g/cm^3, the variation in the determination of M_x of 100 g is 0.005232 g or 0.005232% for $\rho_x = 0.7$ g/cm^3 and 0.000318 g or 0.000318% for $\rho_x = 22.0$ g/cm^3.

Thus, if the average air density at a location were 0.00117 g/cm^3 and if the air density varied within ±3% of the average, the consequence of using the average value of air density for calculating buoyancy correction factors would be errors between 0.000318 and 0.005232% in mass determinations for objects of density ranging between 0.7 and 22.0 g/cm^3. If errors of such magnitude are not significant, then the effort of measuring ambient pressure, temperature, and relative humidity and calculating air density can be avoided.

Similarly, for variations of ±3% about the minimum value of air density, 0.00094 g/cm^3, the variation in the determination of M_x of 100 g is 0.003686 g or 0.003686% for $\rho_x = 0.7$ g/cm^3 or 0.000224 g or 0.000224% for $\rho_x = 22.0$ g/cm^3. Again, if errors of such magnitude are not significant, the mean value of air density can be used for calculating buoyancy correction factors.

Calculation of Air Density

The air density equation developed by Jones[7] is

$$\rho_a = (PM_a/RTZ)\left[1-(1-M_w/M_a)(U/100)(fe_s/P)\right],\tag{A.51}$$

where

ρ_a = density of air
P = ambient pressure in the weighing chamber
M_a = effective molecular weight of dry air
R = universal gas constant
T = temperature in the weighing chamber in kelvins (temperature t in °C + 273.15)
Z = compressibility of air
M_w = molecular weight of water vapor
M_a = molecular weight of dry air
U = relative humidity in percent
f = enhancement factor
e_s = saturation vapor pressure of water

Substituting values for R and M_w, Eq. (A.51) becomes

$$\rho_a = 0.000120272(PM_a/TZ)\left[1-(1-18.0152/M_a)(U/100)(fe_s/P)\right],\tag{A.52}$$

where

$$M_a = 28.963+12.011(x_{CO2}-0.00033),\tag{A.53}$$

and x_{CO_2} is the concentration of carbon dioxide in the air expressed as mole fraction.

For T = 293.15 K (20°C), P = 101325 Pa (760 mmHg, 14.69595 PSI), 50% relative humidity ($U = 50$), and M_a = 28.963 g/mol, the air density calculated using Eq. (A.52) is 1.1992 kg/m³ = 0.0011992 g/cm³ = 1.1992 mg/cm³.

Use of Constant Values of *F, Z,* and *M$_a$* in the Air Density Equation

By considering the expected variations in pressure, temperature, and relative humidity in the laboratory, it might be possible to use constant values of f, Z, and M_a.

For example, in the Mass Laboratory of the National Institutes of Standards and Technology (NIST), constant values of f (1.0042), Z (0.99966), and M_a (28.964) are considered to be adequate. With these values of f, Z, and M_a, the resulting equation for calculating air density is, for P and e_s in pascals, and absolute temperature T = 273.15 + t (°C),

$$\rho_a = (0.0034847/T)(P-0.00379Ue_s)Ue_s).\tag{A.54}$$

For P and e_s in PSI and t in °C,

$$\rho_a = \left[24.026/(t+273.15)\right](P-0.0037960Ue_s).\tag{A.55}$$

For P and e_s in mmHg and t in °C,

$$\rho_a = \left[0.46459/(t+273.15)\right](P-0.0037960Ue_s).\tag{A.56}$$

For $P = 101325$ Pa $= 14.69595$ PSI $= 760$ mmHg, $T = 293.15$ (t $= 20°C$), $U = 50$, and $e_s = 2337.82$ Pa $= 0.339072$ PSI $= 17.5309$ mmHg, $\rho_a = 1.1992$ kg/m^{-3} $= 0.0011992$ g/cm^{-3} $= 1.1992$ mg/cm^{-3} for Eqs. (A.55) and (A.56).

Mole Fraction of Carbon Dioxide, x_{CO_2}

If the mole fraction of carbon dioxide (CO_2) departs from the reference level of 0.00033, the adjusted M_a becomes

$$M_a = M_a(033) + 12.011\left[x(CO_2) - 0.00033\right], \tag{A.57}$$

where $M_a(033)$ is the apparent molecular weight of dry air with a CO_2 mole fraction of 0.00033 and $x(CO_2)$ is the mole fraction of CO_2. For example, the mean value of CO_2 mole fraction in three samples of air taken from a glove box in the Mass Laboratory at NIST was 0.00043. The mean value of the adjusted M_a was then *28.964* g/mol.

Saturation Water Vapor Pressure, e_s

e_s is calculated using the following equations:

$$e_s\left(\text{pascals}\right) = 1.7526 \times 10^{11} \times e^{\left(-5315.56/T\right)}, \tag{A.58}$$

where $e = 2.71828 \ldots$ is the base of Naperian logarithms, and

$$e_s\left(\text{in mmHg}\right) = 1.31456 \times 10^9 \times 10^{\left(-2308.52/T\right)}. \tag{A.59}$$

Values of e_s in millimeters of mercury calculated using Eq. (A.58) and converting from pascals to millimeters of mercury are tabulated in Table A.7.

Enhancement Factor, f

The enhancement factor, f, is a function of temperature and pressure. It can be calculated using the equation:

$$f = 1.00070 + 3.113 \times 10^{-8} P + 5.4 \times 10^{-7} t^2, \tag{A.60}$$

where P is pressure in pascals and t is temperature in °C.
For pressure in millimeters of mercury,

$$f = 1.00070 + 4.150 \times 10^{-6} P + 5.4 \times 10^{-7} t^2. \tag{A.61}$$

Over the temperature range 19.0 to 26.0°C and the pressure range 525.0 to 825.1 mmHg, f ranges from 1.0031 to 1.0045. The maximum variation of f from a nominal value of 1.0042 is equal to 0.11% of the nominal value. The corresponding relative variation of air density is equal to 0.00040%, which is negilgible. Therefore, a constant value of f of *1.0042* can be used in the calculation of air density.

Compressibility Factor, Z

The value of Z depends on temperature, pressure, and relative humidity (RH). For mixtures containing reasonable amounts of carbon dioxide, values of Z can be taken from Table A.8.

Table A.7 Satuation Vapor Pressure of Water Calculated Using Eq. (A.59)

t, °C	e_s, mmHg	t, °C	e_s, mmHg
19.0	16.480	23.0	21.071
19.1	16.583	23.1	21.199
19.2	16.686	23.2	21.327
19.3	16.790	23.3	21.457
19.4	16.895	23.4	21.587
19.5	17.000	23.5	21.718
19.6	17.106	23.6	21.849
19.7	17.212	23.7	21.982
19.8	17.319	23.8	22.114
19.9	17.427	23.9	22.248
20.0	17.535	24.0	22.383
20.1	17.644	24.1	22.518
20.2	17.753	24.2	22.653
20.3	17.863	24.3	22.790
20.4	17.974	24.4	22.927
20.5	18.085	24.5	23.065
20.6	18.197	24.6	23.204
20.7	18.309	24.7	23.344
20.8	18.422	24.8	23.484
20.9	18.536	24.9	23.625
21.0	18.650	25.0	23.767
21.1	18.765	25.1	23.909
21.2	18.880	25.2	24.052
21.3	18.996	25.3	24.196
21.4	19.113	25.4	24.341
21.5	19.231	25.5	24.487
21.6	19.349	25.6	24.633
21.7	19.467	25.7	24.780
21.8	19.587	25.8	24.928
21.9	19.707	25.9	25.077
22.0	19.827	26.0	25.226
22.1	19.949		
22.2	20.071		
22.3	20.193		
22.4	20.317		
22.5	20.441		
22.6	20.565		
22.7	20.691		
22.8	20.817		
22.9	20.943		

Alternatively, Z can be calculated using the following equations:

For P in pascals and t in °C,

$$Z = 1.00001 - 5.8057 \times 10^{-9}\,P + 2.6402 \times 10^{-16}\,P^2 - 3.3297 \times 10^{-7}\,t$$

$$+ 1.2420 \times 10^{-10}\,Pt - 2.0158 \times 10^{-18}\,P^2 t + 2.4925 \times 10^{-9}\,t^2 - 6.2873 \times 10^{-13}\,Pt^2 \quad (A.62)$$

$$+ 5.4174 \times 10^{-21}\,P^2 t^2 - 3.5 \times 10^{-7}\,(RH) - 5.0 \times 10^{-9}\,(RH)^2;$$

Table A.8 Compressibility Factor, Z, Calculated Using Eq. (A.63)

t, °C	P, mmHg	RH,% 0	25	50	75	100
19.0	712.6	0.99964	0.99963	0.99961	0.99960	0.99957
	750.1	0.99962	0.99961	0.99959	0.99958	0.99956
	760.0	0.99962	0.99960	0.99959	0.99957	0.99956
	787.6	0.99960	0.99959	0.99958	0.99956	0.99952
	825.1	0.99959	0.99957	0.99956	0.99954	0.99952
20.0	712.6	0.99965	0.99964	0.99962	0.99960	0.99958
	750.1	0.99963	0.99962	0.99960	0.99958	0.99956
	760.0	0.99963	0.99962	0.99960	0.99958	0.99956
	787.6	0.99961	0.99960	0.99958	0.99957	0.99954
	825.1	0.99959	0.99958	0.99957	0.99955	0.99953
21.0	712.6	0.99966	0.99965	0.99963	0.99961	0.99958
	750.1	0.99964	0.99963	0.99961	0.99959	0.99956
	760.0	0.99964	0.99962	0.99961	0.99959	0.99956
	787.6	0.99962	0.99961	0.99959	0.99957	0.99955
	825.1	0.99960	0.99959	0.99958	0.99956	0.99953
22.0	712.6	0.99967	0.99965	0.99963	0.99961	0.99958
	750.1	0.99965	0.99964	0.99962	0.99960	0.99957
	760.0	0.99965	0.99963	0.99961	0.99959	0.99956
	787.6	0.99963	0.99962	0.99960	0.99958	0.99955
	825.1	0.99962	0.99960	0.99958	0.99956	0.99954
23.0	712.6	0.99968	0.99966	0.99964	0.99962	0.99959
	750.1	0.99966	0.99964	0.99962	0.99960	0.99957
	760.0	0.99965	0.99964	0.99962	0.99960	0.99957
	787.6	0.99964	0.99963	0.99961	0.99958	0.99956
	825.1	0.99963	0.99961	0.99959	0.99957	0.99954
24.0	712.6	0.99968	0.99967	0.99965	0.99962	0.99959
	750.1	0.99967	0.99965	0.99963	0.99961	0.99957
	760.1	0.99966	0.99965	0.99963	0.99960	0.99957
	787.6	0.99965	0.99964	0.99962	0.99959	0.99956
	825.1	0.99964	0.99962	0.99960	0.99957	0.99954
25.0	712.6	0.99969	0.99968	0.99965	0.99962	0.99959
	750.1	0.99968	0.99966	0.99964	0.99961	0.99958
	760.0	0.99967	0.99966	0.99963	0.99961	0.99957
	787.6	0.99966	0.99964	0.99962	0.99960	0.99956
	825.1	0.99965	0.99963	0.99961	0.99958	0.99955
26.0	712.6	0.99970	0.99968	0.99966	0.99963	0.99959
	750.1	0.99969	0.99967	0.99964	0.99961	0.99958
	760.0	0.99968	0.99966	0.99964	0.99961	0.99957
	787.6	0.99967	0.99965	0.99963	0.99960	0.99956
	825.1	0.99966	0.99964	0.99961	0.99959	0.99955

For P in millimeters of mercury and t in °C,

$$Z = 1.00001 - 7.7403 \times 10^{-7} P + 4.6929 \times 10^{-12} P^2 - 3.3297 \times 10^{-7} t$$

$$+ 1.6559 \times 10^{-8} Pt - 3.5831 \times 10^{-14} P^2 t + 2.4925 \times 10^{-9} t^2 - 8.3824 \times 10^{-11} Pt^2 \quad \text{(A.63)}$$

$$+ 9.6293 \times 10^{-17} P^2 t^2 - 3.5 \times 10^{-7} \left(\text{RH}\right) - 5.0 \times 10^{-9} \left(\text{RH}\right)^2.$$

Recapitulation of Air Density Equations

The general equation for calculating air density is Eq. (A.52). For pressure in millimeters of mercury and inserting the value of 8.314471 J/K/mol for the universal gas constant, Eq. (A.52) becomes explicitly

$$\rho_a = \left[0.0160350\left(M_a/TZ\right)\right]\left[P - \left(1 - 18.0152/M_a\right)\left(Ufe_s/100\right)\right]. \tag{A.64}$$

Table A.8 is adequate for determining the compressibility factor, Z, and should be used.

A constant value of 1.0042 can be used for the enhancement factor, f.

Table A.7 should be used for determining the saturation water vapor pressure, e_s, in millimeters of mercury.

If the carbon dioxide mole fraction in the air in the vicinity of the balance is not known, the value of the apparent molecular weight of dry air, M_a, appropriate for the Mass Laboratory of NIST, 28.964, should be used. If the carbon dioxide mole fraction is known, Eq. (A.57) should be used to make the adjustment.

Sample Calculation of Air Density

For $t = 25°C$ ($T = 298.15$ K), P = 760 mmHg, U(RH) = 50(%), Z from Table A.8 is 0.99963, f calculated using Eq. (A.61) is 1.0042, and e_s from Table A.7 is 23.767 mmHg. Inserting these values and $M_a = 28.964$ into Eq. (A.64), $\rho_a = 1.1773$ kg/m^3 = 1.1773 mg/cm^3 = 0.0011773 g/cm^3.

Recommended Values and Practices in Calculating Air Density

1. If the carbon dioxide mole fraction is not known, use *28.964* for M_a; otherwise, use Eq. (A.57) to calculate the adjustment.
2. Use *1.0042* for f.
3. Use Table A.7 to determine e_s.
4. Use Table A.8 to determine Z.
5. Use the resulting equation to calculate the density of air, ρ_a:

$$\rho_a = \left[0.46444/\left(t + 273.15\right)Z\right]\left(P - 0.0037960Ue_s\right). \tag{A.65}$$

For each location, determination of the ranges of temperature, pressure, and relative humidity at the location and use of Table A.8 could provide a constant value of Z that could used in the calculation of ρ_a.

Inserting the values of the parameters used in the sample calculation into Eq. (A.65), the resulting value of ρ_a is 0.0011773 g/cm^3.

Application of Buoyancy Correction to Calibration of Volumetric Flask

As in other measurements, the *true mass* not apparent mass is required to calculate flask volume from the mass of water contained.

In the conventional calibration of a volumetric flask, the flask is weighed empty (balance indication UE), a mass of water is introduced into the flask and the filled flask is weighed (balance indication UF), and the difference (UF – UE) is divided by the density of the water to determine the volume.

However, the true mass of the water contained by the flask is determined by applying a bouyancy correction to (UF – UE) using the equation:

$$M_w = \left(UF - UE\right)\left(1 - \rho_a/8.0\right)/\left(1 - \rho_a/\rho_w\right), \tag{A.66}$$

where M_w is the true mass of the water and ρ_w is the density of water at the ambient temperature.

For a laboratory with a temperature of 72°F (22.2°C), ambient pressure of 752 mmHg, and a relative humidity of 50%, the value of ρ_a is *0.0011770* g/cm³ and ρ_w is *0.997734* g/cm³.

Inserting these values in Eq. (A.66),

$$M_w = 1.00103\left(UF - UE\right).\qquad\qquad(A.67)$$

That is, the value of the buoyancy correction for this case is +0.00103(UF − UE) or +0.103% of (UF − UE).

Consequently, if the buoyancy correction were not made, the mass determination for the water would be in error by 0.103% — too low.

In the calibration of a volumetric flask, the mass of water determined is divided by the density of the water to calculate the volume of the water contained and the volume of the flask (*V*) at the level of the water:

$$V = M_w/\rho_w.\qquad\qquad(A.68)$$

Thus, the percentage error in *V* is the same as the percentage error in the mass determination if the buoyancy correction were not made.

For this example, the volume of the flask would be too low by 0.103%. For an uncorrected volume of 100 ml, the true volume would be 100.103 ml.

CIPM-81 Air Density Equation

The equation developed by Jones, "with minor changes,"[8] was endorsed for use in mass metrology by CIPM (Comite International des Poids et Mesures) in 1981.[9] The equation given in Ref. 8 is now referred to as the CIPM-81 equation-of-state for moist air and is used for mass metrology by most national laboratories. Use of CIPM-81 instead of its predecessor[7] makes a negligible change in routine mass calibrations (see Chapter 12).[10]

The CIPM-81 equation[9] is

$$\rho = pM_a\left[1 - x_v\left(1 - M_v/M_a\right)\right]\big/ZRT,\qquad\qquad(A.69)$$

where *p* is pressure, x_v is the mole fraction of water vapor in moist air, M_v is the molar mass of water vapor in moist air (M_w in the Jones development), and M_a is molar mass of dry air (M_a in the Jones development). The mole fraction of water vapor, x_v, is equal to $(U/100)(fe_s/P)$ and is determined from the relative humidity, *U*, or the dew-point temperature.

CIPM 1981/1991 Equation[8]

The CIPM 1981/1991 equation is the same as the CIPM-81 equation, Eq. (A.69). Davis[8] tabulated amended constant parameters appropriate to the CIPM 1981/1991 equation. Davis stated that air densities calculated from the 1991 parameters are smaller by about 3 parts in 10^5 relative to calculations using the 1981 parameters, and that the overall uncertainty for air density calculated using the 1981/1991 equation is essentially the same as if the 1981 equation were used. Davis[8] noted that ITS-90 (International Temperature Scale, 1990) should be used.

The difference between the air density calculated using the CIPM 1981/191 and that calculated using the CIPM-81 are well within the practical uncertainty. If one prefers, the CIPM 1981/1991 equation can be used.

References

1. Course prepared by Frank E. Jones.
2. *Dictionary of Scientific Biography,* Vol. 1., C.S. Gillespie, Ed., Charles Scribner's Sons, New York, 1970.
3. Jaeger, K. B. and Davis, R. S., A Primer for Mass Metrology, NBS Special Publication 700-1, November 1984.
4. Schoonover, R. M. and Jones, F. E., Air buoyancy correction in high-accuracy weighing on analytical balances, *Anal. Chem.,* 53, 900, 1981.
5. Schoonover, R. M., A look at the electronic analytical balance, *Anal. Chem.,* 54, 973A, 1982.
6. Pontius, P. E., Mass and Mass Values, NBS Monograph 133, January 1974.
7. Jones, F. E., The air density equation and the transfer of the mass unit, *J. Res. Natl. Bur. Stand.* (U.S.), 83, 419, 1978.
8. Davis, R. S., Equation for the determination of the density of moist air (1981/1991), *Metrologia,* 28, 67, 1992.
9. Giacomo, P., Equation for the determination of the density of moist air (1981), *Metrologia,* 18, 33, 1982.
10. Davis, R. S., New assignment of mass values and uncertainties to NIST working standards, *J. Res. Natl. Bur. Stand.* (U.S.), 95, 79, 1990.

Appendix A.1: Examination for "Buoyancy Corrections in Weighing" Course

1. What are the two *forces* that an object being weighed experiences?

    ans. _____

    _____

2. Express the larger of these two forces in terms of the mass of the object being weighed, M, and the local acceleration due to gravity, g_L.

    ans. _____

3. Express the larger of these two forces in terms of the mass of the object being weighed, M, the density of air, ρ_a, the density of the object, ρ_m, and the local acceleration due to gravity, g_L.

    ans. _____

4. Express the differences between the two forces above in terms of M, ρ_a, ρ_m, and g_L.

    ans. _____

5. Express the simple buoyancy correction factor in terms of ρ_a and ρ_m. *Hint:* The answer will be dimensionless.

    ans. _____

6. Express the simple buoyancy correction in terms of M, ρ_a, and ρ_m.

    ans. _____

7. For:

$$M = 100 \text{ g}$$

$$\rho_a = 0.0012 \text{ g}/\text{cm}^3$$

$$\rho_m = 1.0 \text{ g}/\text{cm}^3$$

    a. Calculate the simple buoyancy correction factor.

    ans. _____

    b. Calculate the simple buoyancy correction, in g.

    ans. _____

8. For:

$$M = 100 \text{ g}$$

$$\rho_a = 0.0012 \text{ g}/\text{cm}^3$$

$$\rho_x = 1.0 \text{ g}/\text{cm}^3$$

a. What is the volume of the object?

ans. _____

b. What is the mass of the same volume of air; that is, the mass of the volume of air displaced?

ans. _____

c. What is the ratio of the mass of air to the mass of the object, M_x, in %?

ans. _____

9. For:

$$M = 100 \text{ g}$$

$$\rho_a = 0.0012 \text{ g}/\text{cm}^3$$

$$\rho_m = 1.0 \text{ g}/\text{cm}^3$$

a. Calculate how much of the mass, in g, is supported by the balance pan.

ans. _____

b. Calculate how much of the mass, in g, is supported by the air.

ans. _____

10. What is the relationship between the mass of air displaced by the object and the part of the mass of the object that is supported by the air?

ans. _____

11. a. In the case of the weighing on a balance of an object that is denser than air, in the simple case is the balance indication higher or lower than the mass of the object?

ans. _____

b. Why?

ans. _____

12. For an electronic balance adjusted using a built-in weight of density $\rho_b = 8.0$ g/cm³, and

$$M_x = 100 \text{ g}$$

$$\rho_o = 0.0012 \text{ g}/\text{cm}^3$$

$$\rho_x = 1.0 \text{ g}/\text{cm}^3$$

a. Use the following equation to calculate the balance indication, U_x, with an object of mass, M_x, on the balance pan:

$$U_x = M_x\left[1-\left(\rho_o/\rho_x\right)\right]/\left[1-\left(\rho_o/\rho_b\right)\right].$$

ans. _____

b. For the same parameters except that M_x is unknown and $U_x = 100$ g, use the following equation to calculate M_x:

$$M_x = U_x\left[1-\left(\rho_o/\rho_b\right)\right]\Big/\left[1-\left(\rho_o/\rho_x\right)\right].$$

ans. _____

c. For these examples, what is the magnitude of the buoyancy correction?

ans. _____

d. Is the balance indication less than or greater than the mass of the object?

ans. _____

13. For

$$M_x = \text{unknown}$$

$$U_x = 100 \text{ g}$$

$$\rho_a = 0.00118$$

$$\rho_x = 1.0 \text{ g}/\text{cm}^3$$

$$\rho_b = 8.0 \text{ g}/\text{cm}^3$$

use the following equation to calculate M_x:

$$M_x = U_x\left[1-\left(\rho_a/\rho_b\right)\right]\Big/\left[1-\left(\rho_a/\rho_x\right)\right].$$

ans. _____

14. For

$$M_b = 100 \text{ g}$$

$$\rho_a = 0.00118 \text{ g}/\text{cm}^3$$

$$\rho_b = 8.0 \text{ g}/\text{cm}^3$$

$$\rho_x = 0.7 \text{ g}/\text{cm}^3$$

a. Calculate the mass, M_x, of the object that will balance 100 g of built-in weights, M_b, using the equation:

$$M_x = M_b\left[1-\left(\rho_a/\rho_b\right)\right]\Big/\left[1-\left(\rho_a/\rho_x\right)\right].$$

ans. _____

b. What is the magnitude and sign of the buoyancy correction?

ans. _____

15. Which three of the following are the major atmospheric variables in the calculation of air density?
a. Pressure, P
b. Temperature, T
c. Relative humidity, RH or U

  d. Local acceleration due to gravity, g_L
  e. Carbon dioxide mole fraction, x_{CO_2}
  Please use the abbreviations in the answer.

       ans. _____

16. From a group of selected cities in the United States, the average air densities range from
    a. 0.0012 to 8.0 g/cm^3
    b. 0.0008 to 0.0012 g/cm^3
    c. 0.00092 to 0.00119 g/cm^3
    Please use values in the answer.

       ans. _____

17. Actual values of air density may differ from average values by as much as
    a. 1%
    b. 7%
    c. 3%
    d. 0.01%
    Please use the value in the answer.

       ans. _____

Appendix A.2: Answers for Examination Questions for "Buoyancy Corrections in Weighing" Course

  1. Gravitational force
     Buoyant force
  2. $F_g = Mg_L$
  3. $F_b = M(\rho_a/\rho_m)g_L$
  4. $F_g - F_b = M[1 - (\rho_a/\rho_m)]g_L$
  5. $[1 - (\rho_a/\rho_m)]$
  6. $M(\rho_a/\rho_m)$
  7. a. 0.9988
     b. 0.12 g
  8. a. 100 cm^3
     b. 0.12 g
     c. 0.12%
  9. a. 99.88 g
     b. 0.12 g
 10. They are equal.
 11. a. Lower
     b. Part of the mass is supported by the air
 12. a. 99.8950 g
     b. 100.105 g
     c. 0.105 g
     d. Less
 13. 100.1033 g
 14. a. 100.154 g
     b. +0.154 g
 15. P, T, RH or U
 16. 0.00092 to 0.00119 g/cm^3
 17. 3%

APPENDIX B

Table B.1 Maximum Permissible Errors (MPE), in mg

Nominal Value	Class E_1	Class E_2	Class F_1	Class F_2	Class M_1	Class M_2	Class M_3
50 kg	25	75	250	750	2500	7500	25,000
20 kg	10	30	100	300	1000	3000	10,000
10 kg	5	15	50	150	500	1500	5000
5 kg	2.5	7.5	25	75	250	750	2500
2 kg	1.0	3.0	10	30	100	300	1000
1 kg	0.5	1.5	5	15	50	150	500
500 g	0.25	0.75	2.5	7.5	25	75	250
200 g	0.10	0.30	1.0	3.0	10	30	100
100 g	0.05	0.15	0.5	1.5	5	15	50
50 g	0.030	0.10	0.30	1.0	3.0	10	30
20 g	0.025	0.080	0.25	0.8	2.5	8	25
10 g	0.020	0.060	0.20	0.6	2	6	20
5 g	0.015	0.050	0.15	0.5	1.5	5	15
2 g	0.012	0.040	0.12	0.4	1.2	4	12
1 g	0.010	0.030	0.10	0.3	1.0	3	10
500 mg	0.008	0.025	0.08	0.25	0.8	2.5	
200 mg	0.006	0.020	0.06	0.20	0.6	2.0	
100 mg	0.005	0.015	0.05	0.15	0.5	1.5	
50 mg	0.004	0.012	0.04	0.12	0.4		
20 mg	0.003	0.010	0.03	0.10	0.3		
10 mg	0.002	0.008	0.025	0.08	0.25		
5 mg	0.002	0.006	0.020	0.06	0.20		
2 mg	0.002	0.006	0.020	0.06	0.20		
1 mg	0.002	0.006	0.020	0.06	0.20		

The values in this table are taken from Organisation International de Metrologie Legale International Recommendation OIML R111.

Table B.2 Minimum and Maximum Limits for Density of Weights (ρ_{min}, ρ_{max}) (kg/m³)

Nominal Value	Class E$_1$	Class E$_2$	Class F$_1$	Class F$_2$	Class M$_1$	Class M$_2$
≥100 g	7.934–8.067	7.81–8.21	7.39–8.73	6.4–10.7	≥4.4	≥2.3
50 g	7.92–8.08	7.74–8.28	7.27–8.89	6.0–12.0	≥4.0	
20 g	7.84–8.17	7.50–8.57	6.6–10.1	4.8–24.0	≥2.6	
10 g	7.74–8.28	7.27–8.89	6.0–12.0	≥4.0	≥2.0	
5 g	7.62–8.42	6.9–9.6	5.3–16.0	≥3.0		
2 g	7.27–8.89	6.0–12.0	≥4.0	≥2.0		
1 g	6.9–9.6	5.3–16.0	≥3.0			
500 mg	6.3–10.9	≥4.4	≥2.2			
200 mg	5.3–16.0	≥3.0				
100 mg	≥4.4	≥2.3				
50 mg	≥3.4					
20 mg	≥2.3					

The values in this table are taken from Organisation Internationale de Metrologie Legale International Recommendation OIML R111.

Table B.3 Density and Coefficient of Linear Expansion of Pure Metals, Commercial Metals, and Alloys

Substance	Density at 25°C, g/cm³	Coefficient of Linear Expansion, 10^{-6}/°C
Aluminum	2.70	23.1
Chromium	7.15	4.9
Cobalt	8.86	13.0
Copper	8.96	16.5
Gold	19.3	14.2
Iridium	22.5	6.4
Nickel	8.90	13.4
Palladium	12.0	11.8
Rhodium	12.4	8.2
Silver	10.5	18.9
Tantalum	16.4	6.3
Tin	7.26	22.0
Titanium	4.51	8.6
Tungsten	19.3	4.5
Uranium	19.1	13.9
Zinc	7.14	30.2
Zirconium	6.52	5.7
Plain carbon steel, AISI-SAE 1020	7.86	11.7
Stainless steel, type 304	7.9	17.3
Inconel	8.25	11.5
Aluminum alloy 3003, rolled	2.73	23.2
Aluminum alloy 2014, annealed	2.8	23.0
Aluminum alloy 360	2.64	21.0
Copper, electrolytic (ETP)	8.94	16.5
Yellow brass (high brass)	8.47	20.3
Red brass, 85%	8.75	18.7
Nickel (commercial)	8.89	13.3
Titanium (commercial)	4.5	8.5
Zinc (commercial)	7.14	32.5
Zirconium		

Values in this table are taken from *Handbook of Physics and Chemistry*, 80th ed., Edited by David R. Lide, CRC Press, Boca Raton, FL, 1999-2000.

APPENDIX C
Linearity Test

The linearity test involves dividing the mass range between zero and the calibration mass into four equal segments. This test requires four test weights, two at approximately 50% of the range and two at approximately 25% of the range. The test weights should be fabricated of the same material of which the built-in weight is fabricated to ensure nearly equal densities. This effectively eliminates buoyancy terms in the following test.

One now *assigns* mass values to the test weights relative to the mass of the built-in weight by sum and difference weighings. These weighings can be performed on the balance under test or on another balance. If the built-in weight cannot be directly manipulated by the operator, it must be removed from the balance for this test.

One begins by comparing the sum of the two 50% weights with the built-in weight, after the balance has been calibrated. The comparison is performed by the method of substitution weighing (see Chapter 5). Sum and difference weighings between the built-in weight, S, and the two 50% weights, designated D and E, result in two equations:

$$S - (D + E) = \Delta_1,$$

$$D - E = \Delta_2,$$

where Δ_1 and Δ_2 are mass differences derived from balance indications.

The solution of these two equations for D is

$$D = (S + \Delta_2 - \Delta_1)/2.$$

The quantity, Δ_1, is the difference between the balance indication with S on the balance pan and the balance indication with D and E on the balance pan. Similarly, Δ_2 is the difference between the balance indication with D on the pan and the balance indication with E on the pan. The differences are small; therefore, they are unaffected by reasonable balance nonlinearity. It is not necessary to solve for E.

This procedure is repeated for the two 25% weights, F and G, where D or E serves the function of S above. As above, the following two equations are solved for the 25% weight, F:

$$D - (F + G) = \Delta_3$$

$$F - G = \Delta_4.$$

Solving these two equations for F yields

$$F = (D + \Delta_4 - \Delta_3)/2.$$

The uncertainties in the determinations of D and F relative to the built-in weight, S, can be reduced to trivial amounts by repeating the sequence through O_9. For example, if the sequence is repeated 15 additional times, the uncertainty in D or F is reduced by dividing each by the square root of 16.

The linearity measurement sequence, where the weight on the pan is expressed as a percentage of the built-in weight, is

$$0 \quad 25\% \quad 50\% \quad 75\% \quad 100\% \quad 75\% \quad 50\% \quad 25\% \quad 0$$

The 100% weight corresponds to both 50% weights being on the pan.

The observations, in mass units, corresponding to the above sequence are

$$O_1 \quad O_2 \quad O_3 \quad O_4 \quad O_5 \quad O_6 \quad O_7 \quad O_8 \quad O_9$$

This measurement sequence minimizes the effects of drift and hysteresis, if any.

The linearity correction is derived relative to the sum of the 50% weights, that is, at the 100% calibration point.

The linearity correction at the 50% point, $LC_{50\%}$, is

$$LC_{50\%} = D - \left[\left(O_3 - O_1 \right) + \left(O_7 - O_9 \right) \right] / 2$$

The linearity correction at the 25% point is

$$LC_{25\%} = F - \left[\left(O_2 - O_1 \right) + \left(O_8 - O_9 \right) \right] / 2$$

The linearity correction at the 75% point is

$$LC_{75\%} = \left(D + F \right) - \left[\left(O_4 - O_1 \right) + \left(O_6 - O_9 \right) \right] / 2$$

With the linearity correction determined at five points (it is zero at the 0 and 100% points), the shape of the linearity correction–mass curve is revealed by plotting the five points against mass. Intermediate points can be determined graphically or by mathematically fitting a curve to the points.

In this example, the 100% point is also the balance calibration point. The ideal response is a straight line from 0 to the 100% point. For cases in which the balance extrapolates beyond the calibration point, additional linearity weights are required.

The calibration point by definition has no linearity correction, neither does the 0 point. All other points may have linear corrections (see Chapter 28).

Index

Printed and bound by CPI Group (UK) Ltd, Croydon, CR0 4YY

23/10/2024

01778249-0011